T0360659

Tropical Value Distribution Theory
and Ultra-Discrete Equations

Tropical Value Distribution Theory
and Ultra-Discrete Equations

Risto Korhonen
University of Eastern Finland, Finland

Ilpo Laine
University of Eastern Finland, Finland

Kazuya Tohge
Kanazawa University, Japan

NEW JERSEY • LONDON • SINGAPORE • BEIJING • SHANGHAI • HONG KONG • TAIPEI • CHENNAI

Published by

World Scientific Publishing Co. Pte. Ltd.

5 Toh Tuck Link, Singapore 596224

USA office: 27 Warren Street, Suite 401-402, Hackensack, NJ 07601

UK office: 57 Shelton Street, Covent Garden, London WC2H 9HE

Library of Congress Cataloging-in-Publication Data
Korhonen, Risto.
    Tropical value distribution theory and ultra-discrete equations / by Risto Korhonen (University of Eastern Finland, Finland), Ilpo Laine (University of Eastern Finland, Finland), Kazuya Tohge (Kanazawa University, Japan).
        pages cm
    Includes bibliographical references and index.
    ISBN 978-9814632799 (hardcover : alk. paper)
    1. Value distribution theory. 2. Nevanlinna theory. 3. Difference equations. I. Laine, Ilpo.
II. Tohge, Kazuya. III. Title.
    QA331.K7418 2015
    515'.9--dc23
                                        2015004327

British Library Cataloguing-in-Publication Data
A catalogue record for this book is available from the British Library.

Printed in Singapore

Preface

Nevanlinna theory on value distribution of meromorphic functions in the complex plane has been one of the great achievements in mathematical research in the last century. Its key idea, to balance the proximity function and the enumerative function, has been applied, later on, at least in several complex variables, potential theory and minimal surfaces. In their recent paper, Halburd and Southall observed that this idea similarly applies to piecewise linear, continuous functions with integer slopes on the real line. Recalling the less investigated algebroid functions whose poles are of rational multiplicity instead of being integers, a proposal by Aimo Hinkkanen gave us the idea of using real numbers as possible multiplicities of poles.

The present lecture notes are prepared as an invitation to the tropical counterparts of the classical Nevanlinna and Cartan theories in the standard complex framework, including some preliminary observations about possible applications to ultra-discrete equations. Although we are not going into the history, origins of the terminology and recent developments in tropical mathematics, we assume, however, that the reader is willing to apply basic arithmetic of tropical analysis, i.e. elementary analysis in max-plus semi-rings. For the convenience of the reader, two rather detailed appendices have been included to recall Nevanlinna theory and some complex analysis of difference operators. These appendices serve the reader in making comparisons between tropical results with the corresponding classical framework. We have made an effort to point out situations where direct transfer of the classical reasoning into the tropical setting fails. In particular, the reader should always keep in mind that the multiplicities of poles are real numbers in the tropical setting, instead of being natural numbers in the classical function theory. One more appendix has been included to introduce the reader to the notion of ultra-discrete Painlevé equations, to enable easier study of the corresponding tropical counterparts in the actual text.

In principle, the presentation in this monograph is self-contained. To this end, some of the reasoning and the proofs have been presented in more detail than previously published. Note also that some of the contents in the first chapter is new, while treating polynomials and rational functions in the tropical setting. We have included this part here to provide the reader with a leisurely introduction first, to tropical entire functions and secondly, to the elements of the tropical Nevanlinna theory. To offer a quick insight of the chapters, including appendices, each of them starts with a short passage of introduction.

In addition to the elementary presentation of polynomials and rational functions in Chapter 1, we also offer basic properties of tropical exponential functions. This part serves as some kind of preparation to Chapter 2, where a rather extensive exposition of tropical entire functions can be found. One-dimensional tropical Nevanlinna theory will be developed in Chapter 3 and Chapter 4. Chapter 3 contains the two main theorems as well as the lemma on tropical quotients. Chapter 4 then offers tropical versions of the Clunie and Mohon'ko lemmas. Recalling that their classical versions have been extremely useful tools in the fields of complex differential and difference equations, our hope is that these tropical versions might find important applications in ultra-discrete equations. Chapter 5 then presents a multi-dimensional version of the tropical Nevanlinna theory, as an exposition of tropical holomorphic curves. Chapter 6 and Chapter 7 are then directed towards applications of ultra-discrete equations. Chapter 6 is in fact of a preparatory nature, while essentially giving representation results for tropical periodic functions. These representation results appear to be important in proceeding to actual applications in next chapter. This chapter, Chapter 7, also includes an analysis of ultra-discrete versions of Painlevé equations. Indeed, discrete Painlevé equations have very much been in the background of the recent interest to complex difference equations, difference variant of the Nevanlinna theory and, ultimately, to the present topic of tropical value distribution theory.

From the many colleagues who owe our gratitude, Rod Halburd, Christopher Ormerod and Zhuan Ye are the first to be mentioned. They volunteered to make careful reading of parts of the manuscript, giving us important proposals and comments to revise and improve the presentation. Moreover, Janne Gröhn kindly helped us in drawing the figures. In addition, several friends and colleagues helped to complete this project by their encouragements.

The authors are happy to acknowledge the financial support provided for us by the Väisälä Fund in Finland and the JSPS KAKENHI (grant numbers 22540181 and 25400131) in Japan, as well as the Academy of Finland (grant number 268009). We also acknowledge the University of Eastern Finland and the Kanazawa University for providing us good research environments. World Scientific is to be thanked for their decision to accept this manuscript into their mathematical lecture notes series. We also World Scientific staff for their patience in waiting our manuscript to be completed.

Last but not least, our warm thanks are due to our families for their understanding and patience during this writing process.

Joensuu and Kanazawa, January 2015
Risto Korhonen, Ilpo Laine and Kazuya Tohge

Contents

Preface   v

1. Tropical polynomials, rationals and exponentials   1

    1.1 Basic notions and elementary results for tropical
         polynomials and rationals   1
         1.1.1 Basic properties for tropical polynomials and
                rational functions   2
         1.1.2 Equivalence classes and compact forms of tropical
                polynomials   7
         1.1.3 Tropical version of the fundamental theorem of
                algebra   8
         1.1.4 Tropical rational functions   13
    1.2 Definitions of Nevanlinna functions   14
         1.2.1 Basic definitions   14
         1.2.2 Nevanlinna functions for tropical polynomials . . .   15
         1.2.3 Order of tropical meromorphic functions   17
         1.2.4 Nevanlinna functions for tropical exponentials . .   19

2. Tropical entire functions   29

    2.1 Definitions and basic results   30
         2.1.1 Preliminaries   30
         2.1.2 Growth order and tropical series expansions . . .   34
         2.1.3 Main theorem for tropical series expansions   37
         2.1.4 Proof of Theorem 2.4, first part   40
         2.1.5 Proof of Theorem 2.4, second part   42
         2.1.6 Representations of tropical entire functions   45

2.2   Examples of tropical entire functions   53
    2.2.1   Tropical entire functions of arbitrary order   53
    2.2.2   A q-analogue of the exponential function   55
    2.2.3   A tropical entire function related to q-difference
        equations   57
    2.2.4   A concluding remark   62

3.   One-dimensional tropical Nevanlinna theory   65

3.1   Poisson–Jensen formula in the tropical setting   65
3.2   Basic properties of Nevanlinna functions   67
    3.2.1   First main theorem   69
    3.2.2   Tropical Cartan identity   71
3.3   Auxiliary results from real analysis   75
    3.3.1   Borel-type theorems   75
3.4   Variants of the lemma on tropical quotients   78
3.5   Second main theorem   83
    3.5.1   General form of the second main theorem   83
    3.5.2   Variants of the second main theorem and
        deficiencies   92

4.   Clunie and Mohon'ko type theorems   97

4.1   Valiron–Mohon'ko and Mohon'ko lemmas in tropical
    setting   97
4.2   Tropical Clunie lemma   104

5.   Tropical holomorphic curves   111

5.1   Tropical matrixes and determinants   111
5.2   Tropical Casoratian   112
5.3   Tropical linear independence   114
5.4   Tropical holomorphic curves   121
5.5   Second main theorem for tropical holomorphic curves . . .   131
5.6   Ramification   136
5.7   Second main theorem as an application of the
    one-dimensional case   140

6.   Representations of tropical periodic functions   145

6.1   Representations of tropical periodic functions   145
6.2   Ultra-discrete theta functions   153

7. Applications to ultra-discrete equations   157

   7.1   First-order ultra-discrete equations   157

   7.2   Second-order ultra-discrete equations   160

   7.3   What is the general solution to ultra-discrete equations? .   169

   7.4   Slow growth criterion as a detector of ultra-discrete Painlevé equations   172

   7.5   Tropical rational solutions to ultra-discrete Painlevé equations   175

   7.6   Ultra-discrete hypergeometric solutions to ultra-discrete Painlevé equations   177

   7.7   An ultra-discrete operator   181

   7.8   Ultra-discrete hypergeometric function $_2\Phi_1$ 184

Appendix A   Classical Nevanlinna and Cartan theories   189

   A.1   Classical Nevanlinna theory   189

   A.2   Difference variant of Nevanlinna theory   197

   A.3   Cartan's version of Nevanlinna theory   200

   A.4   Difference variant of Cartan theory   210

Appendix B   Introduction to ultra-discrete Painlevé equations   219

   B.1   Painlevé equations   219

   B.2   Integrability of Painlevé equations and integrability testing   222

   B.3   Discrete Painlevé equations   223

   B.4   Discrete Painlevé equations and integrability testing . . .   226

   B.5   Ultra-discrete Painlevé equations   232

   B.6   Ultra-discrete Painlevé equations and integrability testing   236

   B.7   Hypergeometric solutions to Painlevé equations   242

Appendix C   Some operators in complex analysis   247

   C.1   Logarithmic order and type   247

   C.2   Some operators and related series expansions in complex analysis   250

       C.2.1   The case of difference operator   252

       C.2.2   The case of q-difference operator   253

Bibliography   257

Index   265

Chapter 1

Tropical polynomials, rationals and exponentials

Tropical value distribution theory, introduced by Halburd and Southall [46] and extended by Laine and Tohge [69], describes value distribution of continuous piecewise linear functions of a real variable. In many aspects, the presentation has similarities to what is known as the meromorphic function theory as described in the classical complex analysis. In these lecture notes, we start by looking at the one-dimensional setting, i.e. functions $f : \mathbb{R} \to \mathbb{R}$. Later on in Chapter 5, we shall consider corresponding functions defined on a tropical projective space.

1.1 Basic notions and elementary results for tropical polynomials and rationals

To start with, we shall now consider a max-plus semi-ring on $\mathbb{R} \cup \{-\infty\}$ endowed with (tropical) addition

$$x \oplus y := \max\{x, y\}$$

and (tropical) multiplication

$$x \otimes y := x + y.$$

We also use the notations $x \oslash y := x - y$ and $x^{\otimes \alpha} := \alpha x$, for $\alpha \in \mathbb{R}$. The identity elements for the tropical operations are $0_\circ = -\infty$ for addition and $1_\circ = 0$ for multiplication. Observe that such a structure is not a ring, since not all elements have tropical additive inverses.

Starting to proceed under this arithmetic, we first define

Definition 1.1. A continuous piecewise linear function $f : \mathbb{R} \to \mathbb{R}$ is said to be tropical meromorphic.

By this definition, it is obvious that a tropical meromorphic function f is differentiable outside of a discrete set of derivative discontinuities. Outside of this discrete set, the derivative f' is locally constant. Moreover, one-sided derivatives are well-defined at every point in $\mathbb{R}$. Analyzing what happens at the points of derivative discontinuity appears to be the key idea towards a tropical version of the Nevanlinna theory.

1.1.1   *Basic properties for tropical polynomials and rational functions*

A tropical polynomial means a finite (tropical) sum of type

$$f(x) := \bigoplus_{j=0}^{p} a_j \otimes x^{\otimes s_j} = a_0 \otimes x^{\otimes s_0} \oplus a_1 \otimes x^{\otimes s_1} \oplus \cdots \oplus a_p \otimes x^{\otimes s_p}, \quad (1.1)$$

where the exponents $s_0 < s_1 < \cdots < s_p$ and the corresponding coefficients $a_0, \ldots, a_p$ are real numbers. In the classical setting, $f(x)$ takes the form

$$f(x) = \max\{a_0 + s_0 x, a_1 + s_1 x, \ldots, a_p + s_p x\}. \quad (1.2)$$

For a recent exposition about tropical polynomials, called max-linear polynomials there, see Chapter 5 in the monograph [13] by Butkovič. Our presentation here is slightly different, due to our goal of heading towards the tropical Nevanlinna theory.

Remark 1.2. Observe that the graph of a polynomial $f(x)$ has real slopes in our setting, contrary to the case of positive integer slopes investigated by Tsai, see [106] for an exposition in that setting. This implies immediate differences. As a simple example,

$$f(x) = (x \oplus 0)^{\otimes 2} \oslash x = 2 \max\{x, 0\} - x = \max\{x, -x\} = x \oplus x^{\otimes(-1)} = |x|$$

is not a tropical polynomial in the Tsai setting, while we shall consider f as a polynomial. Moreover, Tsai constructs the **maximal representation** for a tropical polynomial with slopes between $[r, n]$ by adding into the representation all missing terms with suitable coefficients so that all slopes $r, r + 1, \ldots, n$ now appear while the graph remains unchanged. Trying to make use of the definition of the maximal representation of a tropical polynomial in [106], its uniqueness immediately fails in our real slopes setting. As a trivial example, we may consider

$$f(x) := x^{\otimes(1/2)} \oplus x^{\otimes(-1/2)} = \max\left\{\frac{1}{2}x, -\frac{1}{2}x\right\}.$$

Then it is clear that both of

$$f_1(x) := x^{\otimes(1/2)} \oplus x^{\otimes(1/4)} \oplus x^{\otimes(-1/2)} = \max\left\{\frac{1}{2}x, \frac{1}{4}x, -\frac{1}{2}x\right\}$$

and

$$f_2(x) := x^{\otimes(1/2)} \oplus x^{\otimes(-1/5)} \oplus x^{\otimes(-1/2)} = \max\left\{\frac{1}{2}x, -\frac{1}{5}x, -\frac{1}{2}x\right\}$$

may be considered as maximal representations of f, in the sense that the graphs of f, f_1, f_2 agree and if any one of the coefficients $\frac{1}{2}, \frac{1}{4}, -\frac{1}{2}$, resp. $\frac{1}{2}, -\frac{1}{5}, -\frac{1}{2}$ will be increased, then the graph is not any more the same as the graph of f, see [106] and the attached figures Fig. 1.1 and Fig. 1.2.

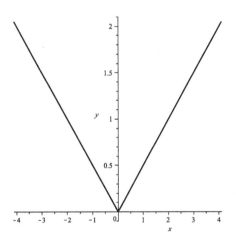

Fig. 1.1 Graph of $f(x)$, $f_1(x)$ and $f_2(x)$.

We say that two tropical polynomials

$$f(x) := \bigoplus_{j=0}^{p} a_j \otimes x^{\otimes s_j}, \qquad g(x) := \bigoplus_{j=0}^{p} b_j \otimes x^{\otimes s_j}$$

agree, denoted $f = g$, if the corresponding coefficients are equal, $a_j = b_j$.

For a simple example, look at

$$\tilde{f}(x) := x^{\otimes \pi} \oplus 1 \oplus x^{\otimes(-1/2)} = \max\{\pi x, 1, (-1/2)x\}$$

and

$$\tilde{g}(x) := x^{\otimes \pi} \oplus (1 - 1/\pi) \otimes x \oplus 1 \oplus x^{\otimes(-1/2)} = \max\{\pi x, x + 1 - 1/\pi, 1, (-1/2)x\},$$

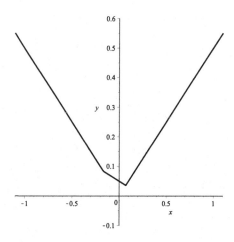

Fig. 1.2   Graph of $\max\{x/2, -x/5 + 1/20, -x/2\}$.

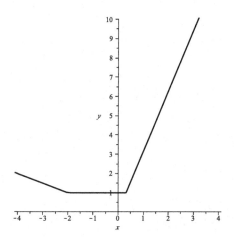

Fig. 1.3   Graph of $\tilde{f}(x)$ and $\tilde{g}(x)$.

see Fig. 1.3. Clearly, $\tilde{f} \neq \tilde{g}$, although their graphs agree. This gives rise to define (functional) equivalence of tropical polynomials:

Two tropical polynomials are called **equivalent**, provided they have the same graph. Obviously, this implies that the family of tropical polynomials will be divided into equivalence classes. The notation $f \sim g$ will be used for two tropical polynomials f, g that belong to the same equivalence class. This notion will be analyzed in some detail in the next subsection.

Tropical rational functions are just tropical quotients of tropical polynomials:

$$r(x) := f(x) \oslash g(x) = \left(\bigoplus_{j=0}^{p} a_j \otimes x^{\otimes s_j} \right) \oslash \left(\bigoplus_{j=0}^{q} b_j \otimes x^{\otimes t_j} \right). \quad (1.3)$$

In the classical setting, we have

$$r(x) = \max\{a_0 + s_0 x, a_1 + s_1 x, \dots, a_p + s_p x\}$$
$$- \max\{b_0 + t_0 x, b_1 + t_1 x, \dots, b_q + t_q x\}.$$

Observe that tropical polynomials, resp. tropical rational functions, are piecewise linear, continuous functions $\mathbb{R} \to \mathbb{R}$ such that their graphs consist of finitely many distinct linear segments. The corner-points of the graph determined by these segments are called either roots (resp. zeros) or poles according to the following definition:

Definition 1.3. Given a tropical polynomial, resp. a tropical rational function, $f(x)$, we define

$$\omega_f(x) := \lim_{\varepsilon \to 0+} \{f'(x + \varepsilon) - f'(x - \varepsilon)\}$$

for each $x \in \mathbb{R}$. If now $\omega_f(x) > 0$, then x is called a **root**, resp. a **zero**, of f with multiplicity $\omega_f(x)$. Similarly, if $\omega_f(x) < 0$, then x is called a **pole** of f with multiplicity $-\omega_f(x)$. In what follows, we denote by $\tau_f(x)$ the multiplicity of a pole, resp. a root, of f at x.

As an example, see Fig. 1.4, the tropical rational function

$$\tilde{r}(x) := \max \left\{ 3 - x, -1 + \frac{3}{2}x, 4 - \frac{1}{5}x \right\} - \max \left\{ -2 + \frac{1}{2}x, 4x, 1 - \frac{1}{3}x \right\}$$

has a pole of multiplicity $13/3$ at $x = 3/13$, and roots of multiplicity $4/5$, resp. $17/10$, at $x = -5/4$, resp. $x = 50/17$. Observe that multiplicities of roots, resp. of poles, are positive real numbers, not necessarily integers as it is the case in classical complex analysis.

Before proceeding, we should perhaps remark that tropical polynomials are convex functions:

Proposition 1.4. *Tropical polynomials are convex functions, and the graphs they determine in the plane are convex curves. Conversely, if f*

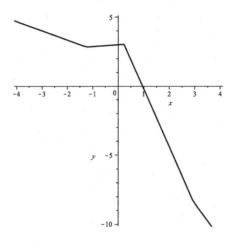

Fig. 1.4   Graph of $\tilde{r}(x)$.

is a tropical rational function, and if its graph is a convex curve, then f is a tropical polynomial.

Proof. Suppose first that $f(x)$ is a tropical polynomial, with the representation (1.2). From this representation, it immediately follows that the (continuous) graph of $f(x)$ consists of finitely many linear segments. To prove this part of the claim, it is sufficient to show that the slope of the graph is increasing along with increasing x. Since the slope remains constant within each of the segments, it suffices to see what happens at the corners. Clearly, there are finitely many corner points, say $x_1 < \cdots < x_q$, in the graph. By (1.2) again, $f(x) = a_p + s_p x$ for all x large enough, hence for all $x \geq x_q$. By the continuity, $f(x_q) = a_p + s_p x_q = a_{j_q} + s_{j_q} x_q$, hence $s_p - s_{j_q} > 0$ and the slope increases. Moreover, $s_{j_q} < s_j < s_p$ for all $j = j_q + 1, \ldots, p - 1$. Proceeding to the next corner x_{q-1}, we now have $f(x_{q-1}) = a_{j_q} + s_{j_q} x_{q-1} = a_{j_{q-1}} + s_{j_{q-1}} x_{q-1}$, where $s_{j_{q-1}} < s_{j_q}$, and we are done for this corner. Proceeding inductively, we finally arrive at the first corner x_1, and we then clearly have $s_{j_1} = s_0$, completing this part.

Next suppose that f has a convex graph. We may assume that the representation (1.2) of f consists of minimally few components in the sense that whenever one of these components removed, then the graph changes. By convexity, the slopes are increasing as the argument x increases at the corner points. Therefore, the tropical representation of f takes the form (1.1), and we are done. $\qquad\square$

1.1.2 *Equivalence classes and compact forms of tropical polynomials*

As already mentioned in the beginning of this section, tropical polynomials are divided into equivalence classes in the sense that such an equivalence class is characterized by the fact that all tropical polynomials in this class have the same graph. This fact gives rise in [106] to considering maximal representations of tropical polynomials. This representation is unique in the sense that all tropical polynomials in an equivalence class have the same maximal representation and conversely. As already mentioned in Remark 1.2, the idea of constructing the maximal representation for a tropical polynomial with slopes (positive integers!) between $[r, n]$ is to add into the representation all missing terms with suitable coefficients so that the slopes $r, r + 1, \ldots, n$ are all included and at the same time the graph remains unchanged. However, this process does not work in our case of real slopes. Therefore, we prefer to apply the idea of **compact forms**, see [106], p. 701–702, to characterize the equivalence classes of tropical polynomials.

Suppose we consider a tropical polynomial of the form

$$f(x) := \bigoplus_{j=0}^{p} a_j \otimes x^{\otimes s_j},$$

with strictly increasing sequence of slopes $s_0, \ldots, s_r$, written in the classical form

$$f(x) = \max\{a_0 + s_0 x, a_1 + s_1 x, \ldots, a_p + s_p x\}. \tag{1.4}$$

Its compact form may now be constructed by removing from its linear components $a_0 + s_0 x, a_1 + s_1 x, \ldots, a_p + s_p x$ all those that are not needed to obtain the linear segments in the graph of f. Obviously, this is a unique process to determine the graph of f. We are now ready to offer the real-slope variant of [106], Corollary 5.2:

Proposition 1.5. *Let*

$$f(x) := \bigoplus_{j=0}^{p} a_j \otimes x^{\otimes s_j}$$

be a tropical polynomial, presented in its compact form with $s_0 < \ldots < s_p$. Then the points

$$x_j := (a_{j+1} - a_j)/(s_j - s_{j+1}), \qquad j = 0, \ldots, p-1$$

are the roots of f, with the corresponding multiplicities being

$$\omega_f(x_j) = s_{j+1} - s_j, \qquad j = 0, \ldots, p-1.$$

Proof. The claim immediately follows from the equality

$$a_j + s_j x_j = a_{j+1} + s_{j+1} x_j,$$

valid at a corner point x_j, due to the continuity of the graph. $\square$

1.1.3   *Tropical version of the fundamental theorem of algebra*

We next study a possible tropical counterpart of the fundamental theorem of algebra by factorizing tropical polynomials with real slopes into a tropical finite product of linear factors. This is an extension of the case for tropical polynomials with integer slopes as in [106], Theorem 4.1.

We start by presenting first a special form of the tropical fundamental theorem of algebra as follows:

Lemma 1.6. *Assume that a tropical polynomial $g(x)$ of the form*

$$g(x) := \bigoplus_{j=0}^{p} b_j \otimes x^{\otimes t_j} = b_0 \oplus b_1 \otimes x^{\otimes t_1} \oplus \cdots \oplus b_{p-1} \otimes x^{\otimes t_{p-1}} \oplus 0 \otimes x^{\otimes t_p} \quad (1.5)$$

is monic in the sense that $b_p = 0$ and normalized so that $t_0 = 0$. We also assume that g is represented in its compact form with $0 < t_1 < \ldots < t_p$ and $b_0, \ldots, b_{p-1} \in \mathbb{R}$. Then we can factorize $g(x)$ into the following tropical product of exactly p factors:

$$g(x) = \bigotimes_{j=1}^{p} \left\{ x \oplus \left(-\frac{b_j - b_{j-1}}{t_j - t_{j-1}} \right) \right\}^{t_j - t_{j-1}}, \qquad t_0 = b_p = 0.$$

Proof. Drawing the graph of the function $g(x)$, we recall that it is formed with $p+1$ line segments, each of which on the lines given by the equations

$$y = b_0, \quad y = t_1 x + b_1, \quad \ldots, \quad y = t_{p-1} x + b_{p-1}, \quad y = t_p,$$

respectively. There are exactly p corner points of the convex curve on $\mathbb{R}$, each of which is the intersection of the pair of two lines given by $y =$

$t_j x + b_j$ and $y = t_{j+1} x + b_{j+1}$ for $j = 0, 1, \ldots, p-1$, respectively. Recalling Proposition 1.5, we know that the corner points of the graph of g are given by $x_j = (b_j - b_{j+1})/(t_{j+1} - t_j)$.

Now form a tropical product $h(x)$ as in the statement above:

$$h(x) = \bigotimes_{j=1}^{p} \left\{ x \oplus \left(-\frac{b_j - b_{j-1}}{t_j - t_{j-1}} \right) \right\}^{t_j - t_{j-1}}, \qquad t_0 = b_p = 0.$$

Here we evaluate $h(x)$ as

$$h(x) = \sum_{j=1}^{p} (t_j - t_{j-1}) \max \left\{ x, \left(-\frac{b_j - b_{j-1}}{t_j - t_{j-1}} \right) \right\}$$

$$= \max\{t_1 x, b_0 - b_1\} + \sum_{j=2}^{p-1} \max\{(t_j - t_{j-1})x, b_{j-1} - b_j\}$$

$$+ \max\{(t_p - t_{p-1})x, b_{p-1}\}.$$

Now we need to verify $g(x) = h(x)$ for each $x \in \mathbb{R}$. For this purpose, assume first that x is between

$$(b_{j-1} - b_j)/(t_j - t_{j-1}) \leq x \leq (b_j - b_{j+1})/(t_{j+1} - t_j)$$

for some $j \in \{1, 2, \ldots, p\}$. Then we have

$$\max \left\{ x, \left(-\frac{b_i - b_{i-1}}{t_i - t_{i-1}} \right) \right\} = \begin{cases} x & (1 \leq i \leq j) \\ -\frac{b_i - b_{i-1}}{t_i - t_{i-1}} & (j+1 \leq i \leq p), \end{cases}$$

so that, using $t_0 = b_p = 0$, we obtain

$$h(x) = \sum_{i=1}^{j} (t_i - t_{i-1})x + \sum_{i=j+1}^{p} (b_{i-1} - b_i) = t_j x + b_j = g(x). \qquad (1.6)$$

In addition, if $x \leq (b_0 - b_1)/t_1$, or $x \geq b_{p-1}/(t_p - t_{p-1})$, we see that

$$h(x) = \sum_{j=1}^{p} (-b_j + b_{j-1}) = -b_p + b_0 = b_0 = g(x)$$

and

$$h(x) = \sum_{j=1}^{p} (t_j - t_{j-1})x = (t_p - t_0)x = t_p x + b_p = g(x),$$

respectively, completing the proof. $\qquad \square$

Remark 1.7. The number of distinct roots of $g(x)$ above is exactly p and these roots are of multiplicity $t_i - t_{i-1}$ ($1 \le i \le p$). Therefore the total number of the roots of the $g(x)$ on $\mathbb{R}$ is

$$\sum_{i=1}^{p} (t_i - t_{i-1}) = t_p - t_0 = t_p,$$

when the multiplicities are taken into account.

On the other hand, by defining $s^+ := \max\{s, 0\}$ for $s \in \mathbb{R}$, the degree of the monic polynomial $g(x)$ in (1.5) may be given as

$$\deg g = (t_p)^+ + (-t_0)^+ = t_p,$$

similarly as in the classical theory. Note that, according to the definition in [106], Definition 2.6, this normalized polynomial $g(x)$ does not possess a root at the point $-\infty$, so that the equality is also true, even if we regard the function $g(x)$ as a tropical polynomial on the extended real line $\mathbb{T} = \mathbb{R} \cup \{-\infty\}$.

We now proceed to considering a general tropical polynomial given in compact form (1.1), that is,

$$f(x) = \bigoplus_{j=0}^{p} a_j \otimes x^{\otimes s_j} = a_0 \otimes x^{\otimes s_0} \oplus a_1 \otimes x^{\otimes s_1} \oplus \cdots \oplus a_p \otimes x^{\otimes s_p}$$

$$= \max\{a_0 + s_0 x, a_1 + s_1 x, \dots, a_p + s_p x\},$$

where the exponents $s_0 < s_1 < \cdots < s_p$ and $a_0, \dots, a_p \in \mathbb{R}$ and the number of the linear segments $a_j + s_j x$ that appear in the expression is minimal. We now rewrite $f(x)$ as

$$f(x) = a_p \otimes x^{\otimes s_0} \otimes g(x)$$

with

$$g(x) = (a_0 - a_p) \oplus (a_1 - a_p) \otimes x^{\otimes(s_1 - s_0)} \oplus \cdots$$
$$\cdots \oplus (a_{p-1} - a_p) \otimes x^{\otimes(s_{p-1} - s_0)} \oplus x^{\otimes(s_p - s_0)}.$$

We can write

$$f(x) = a_p + s_0 x +$$
$$+ \max\Big\{ (s_p - s_0)x, (a_{p-1} - a_p) + (s_{p-1} - s_0)x, \dots$$
$$\cdots, (a_1 - a_p) + (s_1 - s_0)x, a_0 - a_p \Big\}$$

in the classical form. Put $t_i := s_i - s_0$ and $b_i := a_i - a_p$ $(0 \le i \le p)$. Since $b_i \in \mathbb{R}$ and $0 = t_0 < t_1 < \cdots < t_{p-1} < t_p$, the application of the previous Lemma 1.6 implies the expression

$$f(x) = a_p \otimes x^{\otimes s_0} \bigotimes_{j=1}^{p} \left\{ x \oplus \left(-\frac{b_j - b_{j-1}}{t_j - t_{j-1}} \right) \right\}^{t_j - t_{j-1}}, \quad t_0 = b_p = 0.$$

Here $b_{i-1} - b_i = (a_{i-1} - a_p) - (a_i - a_p) = a_{i-1} - a_i$ and $t_i - t_{i-1} = (s_i - s_0) - (s_{i-1} - s_0) = s_i - s_{i-1}$ $(1 \le i \le p)$, and therefore

$$f(x) = a_p \otimes x^{\otimes s_0} \bigotimes_{j=1}^{p} \left\{ x \oplus \left(-\frac{a_j - a_{j-1}}{s_j - s_{j-1}} \right) \right\}^{s_j - s_{j-1}}.$$

Remark 1.8. In the case of integer slopes, this coincides with Tsai's Tropical Fundamental Theorem of Algebra [106, Theorem 4.1]. For the convenience of the reader, we recall this result from [106]:

Theorem 1.9. *Let*

$$f(x) = a_n \otimes x^{\otimes n} \oplus a_{n-1} \otimes x^{\otimes (n-1)} \oplus \cdots \oplus a_r \otimes x^{\otimes r}$$

be a maximally represented polynomial. Then $f(x)$ can be factored into

$$f(x) = a_n \otimes x^{\otimes r} \otimes (x \oplus d_{r+1}) \otimes (x \oplus d_{r+2}) \otimes \cdots \otimes (x \oplus d_n),$$

where $d_k = a_{k-1} - a_k$ for all $k = r+1, r+2, \ldots, n$.

In order to unify the expressions for $f(x)$ here and in Lemma 1.6, put $p = n - r$ and

$$s_j = r + j \quad (j = 1, 2, \ldots, p)$$

and rewrite a_j by a_{r+j} in (1.1). Then each of the exponents $s_j - s_{j-1}$ is 1 and $-\frac{a_j - a_{j-1}}{s_j - s_{j-1}}$ are rewritten as $a_{r+j-1} - a_{r+j}$. This coincides with d_{r+j} for $j = 1, 2, \ldots, n - r$ above. Hence the expressions are same in these both cases.

Observe, however, that we need here a different interpretation for the notion of maximal representation in [106, Definition 3.3] for integer slopes. Instead, we assume that the tropical polynomial of real slopes is in *compact form* as defined in [106, Definition 5.1].

As remarked after Lemma 1.6, the polynomial $f(x)$ has exactly p distinct roots in $\mathbb{R}$ and the total number of them with multiplicities is exactly $s_p - s_0$ which is positive. Let us now take the point $-\infty$ into account, recall here [106] again. If $s_0 \le 0$, then we can ignore this point, since then it is a pole of $f(x)$ when $s_0 < 0$ or its ordinary point when $s_0 = 0$, respectively. If $s_0 > 0$, $f(x)$ has a root at $-\infty$ and the multiplicity is $s_0 = (s_0)^+$. Thus, in total, $f(x)$ has exactly $(s_p - s_0) + (s_0)^+ = s_p + (-s_0)^+$ roots together with multiplicities on $\mathbb{T} \cup \{-\infty\}$. Further, if $s_p > 0$, that is, $s_p = (s_p)^+$, the total number of roots coincides with the degree. When $s_p \leq 0$, we need to take $+\infty$ into consideration similarly like $-\infty$ has been treated in [106]. Indeed, it is natural to define a pole, resp. a root, resp. an ordinary point of $f(x)$ at $+\infty$ when s_p is positive, resp. negative, resp. zero. Then the total number of the roots of f on $\mathbb{T} \cup \{+\infty\} = \mathbb{R} \cup \{\pm\infty\}$ equals to $s_p + (-s_0)^+ + (-s_p)^+ = (s_p)^+ + (-s_0)^+$. Actually, this is the natural way to define the degree of $f(x)$, see the next subsection.

Hence, we have now completed the tropical version of the fundamental theorem of algebra:

Theorem 1.10. *A tropical polynomial f has as much roots on $\mathbb{R} \cup \{\pm\infty\}$ as its degree $\deg f = (s_p)^+ + (-s_0)^+$.*

Remark 1.11. Note that a monomial $a + sx$ ($a \in \mathbb{T}, s \in \mathbb{R}$) is of degree $s^+ + (-s)^+ = |s|$, while it has no roots in $\mathbb{R}$. However, it has a root of multiplicity s^+ at $-\infty$ and of multiplicity $(-s)^+$ at $+\infty$.

To close this subsection, we add here an observation related to the factorization. At the same time, this observation also serves as a preparation to the tropical Poisson-Jensen formula below, see also [46] and [69] in relation to the factorization. Let us restrict our tropical polynomial $f(x)$ on a closed interval $[-R, R]$ where $f(x)$ has no finite roots outside of $(-R, R)$. By a simple computation, we obtain a concrete expression of $f(x)$ for any $x \in (-R, R)$ by means of the roots $-R < c_1 < c_2 < \ldots < c_p < R$ together with their multiplicities $m_j \in \mathbb{R}$ ($1 \leq j \leq p$) in the following form:

$$f(x) = \frac{f(R) + f(-R)}{2} + \frac{f(R) - f(-R)}{2R}x$$
$$- \sum_{j=1}^{p} \left(\frac{m_j R}{2} - \frac{m_j}{2} \max\{c_j - x, x - c_j\} - \frac{m_j c_j}{2R}x \right).$$

Further by using $\max\{c_j - x, x - c_j\} = 2\max\{x, c_j\} - (x + c_j)$, we see that three terms in the parentheses above become

$$-m_j \max\{x, c_j\} + \frac{m_j(R - c_j)}{2R}x + \frac{m_j(R + c_j)}{2}.$$

Then, as its tropical form, we arrive at the same expression as in the previous factorization formula:

$$f(x) = \left(a_p \otimes x^{\otimes s_0}\right) \bigotimes_{j=1}^{p} \left(x \oplus c_j\right)^{\otimes m_j}$$

with

$$a_p = \frac{f(R) + f(-R)}{2} - \frac{1}{2} \sum_{j=1}^{p} m_j(R + c_j),$$

and

$$s_0 = \frac{f(R) - f(-R)}{2R} + \frac{1}{2R} \sum_{j=1}^{p} m_j(-R + c_j).$$

An advantage of this expression is the description of the non-vanishing factor $a_p \otimes x^{\otimes s_p}$ by not only the roots c_j and their multiplicities m_j but also the values of $f(x)$ at $\pm R$. We could think it natural to examine the possibility of roots also at the boundary points $\pm\infty$.

1.1.4   *Tropical rational functions*

First recall the definition of tropical rational functions:

$$f(x) := P(x) \oslash Q(x) = \left(\bigoplus_{j=0}^{p} a_j \otimes x^{\otimes s_j}\right) \oslash \left(\bigoplus_{j=0}^{q} b_j \otimes x^{\otimes t_j}\right),$$

and the corresponding classical representation:

$$f(x) = \max\{a_0 + s_0 x, a_1 + s_1 x, \ldots, a_p + s_p x\}$$
$$- \max\{b_0 + t_0 x, b_1 + t_1 x, \ldots, b_q + t_q x\}.$$

From this representation, $f(x) = P(x) - Q(x)$ is nothing but the distance between the points $(x, P(x))$, $(x, Q(x))$ on the graphs of the numerator, resp. denominator, of f. Therefore, it is geometrically trivial to assume, if needed, that $P(x)$ and $Q(x)$ have been expressed in their compact forms. Of course, the graph of $f(x)$ also remains unchanged under this assumption.

We may now refer to [67], Lemma 3.1, to see that a tropical meromorphic function is tropical rational if and only if it has finitely many poles and roots only.

Proposition 1.12. *The following properties are equivalent for a tropical rational function f:*
(a) the tropical rational f is not a tropical polynomial,
(b) the graph of f is not convex,
(c) the function f has at least one pole.

Proof. If f is not a tropical polynomial, then by Proposition 1.4 its graph is not convex. Suppose then that the graph of f is not convex around a point x, say. Then it is geometrically obvious that f has a pole at x. The reversed argument is obvious. □

Corollary 1.13. *A tropical rational function is a tropical polynomial if and only if it has no poles.*

1.2 Definitions of Nevanlinna functions

1.2.1 *Basic definitions*

Recall first from the very beginning (Definition 1.1) that a continuous piecewise linear function $f : \mathbb{R} \to \mathbb{R}$ is called to be a tropical meromorphic function.

We now point out the simple observations that imply the tropical version of Nevanlinna theory. To this end, observe that the definitions for poles and roots, being local, see Definition 1.3, remain valid for tropical meromorphic functions as well: A point x of derivative discontinuity of a tropical meromorphic function f is a pole of f of multiplicity $\tau_f(x) := -\omega_f(x)$ whenever

$$\omega_f(x) := \lim_{\varepsilon \to 0+} \{f'(x + \varepsilon) - f'(x - \varepsilon)\} < 0,$$

and a root of multiplicity $\tau_f(x) := \omega_f(x)$, if $\omega_f(x) > 0$.

Basic notions of the tropical Nevanlinna functions are now easily set up as follows, see [46]:

The tropical **proximity function** for tropical meromorphic functions is defined as

$$m(r, f) := \frac{1}{2} \left\{ f^+(r) + f^+(-r) \right\}, \tag{1.7}$$

where $f^+(x) := \max\{f(x), 0\}$ for $x \in \mathbb{R}$. Denoting by $n(r, f)$ the number of poles of f in the interval $(-r, r)$, each pole counted by its multiplicity τ_f, the tropical **counting function** for the poles in $(-r, r)$ is defined as

$$N(r, f) := \frac{1}{2} \int_0^r n(t, f) dt = \frac{1}{2} \sum_{|b_\nu| < r} \tau_f(b_\nu)(r - |b_\nu|). \tag{1.8}$$

Defining then the tropical **characteristic function** $T(r, f)$ as usual,

$$T(r, f) := m(r, f) + N(r, f), \tag{1.9}$$

the tropical Poisson–Jensen formula, see [46], p. 5–6, to be proved below, readily implies the tropical Jensen formula

$$T(r, f) - T(r, -f) = f(0) \tag{1.10}$$

as a special case.

1.2.2 *Nevanlinna functions for tropical polynomials*

As an example of tropical Nevanlinna functions in a simple situation, we return back to considering tropical polynomials:

Consider first tropical polynomials

$$f(x) := \bigoplus_{j=0}^{p} a_j \otimes x^{\otimes s_j} = \max\{a_0 + s_0 x, a_1 + s_1 x, \ldots, a_p + s_p x\},$$

where $s_0 < s_1 < \ldots < s_p$ and $a_j \in \mathbb{R}$ $(j = 0, 1, \ldots, p)$. We may assume that f has been written in its compact form. In order to compute $T(r, f) = m(r, f) = \frac{1}{2}(f^+(r) + f^+(-r))$, it is clearly enough to assume that r is sufficiently large so that we only need to consider the first and the last term. Then we clearly have

$$f(r) = a_p + s_p r, \qquad f(-r) = a_0 - s_0 r. \tag{1.11}$$

By the assumption on the s_j, we need to observe the following five cases to be treated separately:

(All of the following five cases hold for all sufficiently large r so that (1.11) is valid.)

(1) If $s_0 > 0$ and thus $s_p > 0$, then $f(-r) < 0$ and $f(r) > 0$, hence

$$T(r, f) = m(r, f) = \frac{1}{2}(a_p + s_p r) = \frac{1}{2}(s_p^+ + (-s_0)^+)r + \frac{1}{2}a_p.$$

(2) If $s_0 < 0$ and $s_p > 0$, then $f(-r) > 0$ and $f(r) > 0$ and so

$$T(r, f) = m(r, f) = \frac{1}{2}(a_p + s_p r + a_0 - s_0 r) = \frac{1}{2}(s_p^+ + (-s_0)^+)r + \frac{1}{2}(a_p + a_0).$$

(3) If $s_p < 0$ and thus $s_0 < 0$, then $f(-r) > 0$ and $f(r) < 0$, and we now have

$$T(r, f) = m(r, f) = \frac{1}{2}(a_0 - s_0 r) = \frac{1}{2}(s_p^+ + (-s_0)^+)r + \frac{1}{2}a_0.$$

(4) If $s_0 = 0$ and thus $s_p > 0$, then $f(-r) = a_0^+$, while $f(r) > 0$. Therefore,
$$T(r, f) = m(r, f) = \frac{1}{2}(s_p^+ + (-s_0)^+)r + \frac{1}{2}(a_p + a_0^+).$$

(5) Finally, if $s_p = 0$ and thus $s_0 < 0$, then we get
$$T(r, f) = m(r, f) = \frac{1}{2}(s_p^+ + (-s_0)^+)r + \frac{1}{2}(a_0 + a_p^+).$$

To obtain an expression that covers all five cases simultaneously, we need to define
$$\delta(s, a) := \begin{cases} a & (s > 0) \\ a^+ & (s = 0) \\ 0 & (s < 0) \end{cases}$$

Then we may formally write
$$T(r, f) = m(r, f) = \frac{1}{2}\left(s_p^+ + (-s_0)^+\right)r + \frac{1}{2}\left(\delta(s_p, a_p) + \delta(-s_0, a_0)\right).$$

As mentioned in Subsection 1.1.3, the degree of a tropical polynomial $f(x)$ is defined as
$$\deg f := \frac{1}{2}(s_p^+ + (-s_0)^+).$$

Then we may shortly write
$$T(r, f) = m(r, f) = (\deg f)r + O(1).$$

To determine the counting function for the roots of a polynomial, we may apply the tropical Jensen formula, recalling that $f(x)$ has no poles:

$$N(r, -f) = T(r, -f) - m(r, -f) = m(r, f) - m(r, -f) - f(0).$$

To determine $m(r, -f)$, we interchange the role of s_0, s_p by $-s_p, -s_0$, as well as a_0, a_p by $-a_p, -a_0$, and repeat a similar computation as made for $m(r, f)$ above. Then this now results in

$$m(r, -f) = \frac{1}{2}(s_0^+ + (-s_p)^+)r + \frac{1}{2}\{\delta(s_0, -a_0) + \delta(-s_p, -a_p)\}.$$

We may also find this in such a way that $f(x)$ has $(-s_0)^+ + s_p^+$ poles at $\pm\infty$ together with multiplicities. Hence $-f(x)$ has $(-s_p)^+ + \left(-(-s_0)\right)^+ = (s_0)^+ + (-s_p)^+$ poles there, which implies the above.

Observing
$$\delta(s, a) - \delta(-s, -a) \equiv a$$

and substituting into the Jensen formula above, we get

$$N(r, -f) = \frac{1}{2}(s_p - s_0)r + \frac{1}{2}(a_p + a_0) - \max\{a_0, \ldots, a_p\}. \qquad (1.12)$$

Here we had to note that $s_p = s_p^+ + (-s_p)^+$, and similarly for s_0. Moreover, $f(0) = \max\{a_0, \ldots, a_p\}$.

1.2.3 *Order of tropical meromorphic functions*

Following the usual classical notion, a tropical meromorphic function f is said to be of finite order of growth if $T(r, f) \leq r^\sigma$ for some positive number σ, and for all r sufficiently large. Of course, this enables us to define the **order** $\rho(f)$ of a tropical meromorphic function f in the usual way as

$$\rho(f) := \limsup_{r \to \infty} \frac{\log T(r, f)}{\log r}. \tag{1.13}$$

In several instances below, we may profit of using the notion of **hyper-order** to treat tropical meromorphic functions of infinite order that have, nevertheless, a somewhat reasonably modest growth:

$$\rho_2(f) := \limsup_{r \to \infty} \frac{\log \log T(r, f)}{\log r}. \tag{1.14}$$

As an example of using the notion of order, we prove

Theorem 1.14. *A tropical meromorphic function f is tropical rational if and only if its characteristic function is of type $T(r, f) = O(r)$ and the set of the multiplicities of all the roots and poles of f is bounded below away from zero.*

Remark 1.15. Observe that assuming $T(r, f) = O(r)$ as $r \to \infty$ only is not a necessary and sufficient condition for $f(z)$ to be a tropical rational function. Therefore, a remark proposed in [69, Remark (4), p. 888] is false. In fact, Halburd [39] proposed the following counter-example:

Example 1.16. Consider the tropical meromorphic function f such that $f(x) = x - 1$ for all $x < 0$ and for all $N = 1, 2, 3, \ldots$ and all $x \in [N - 1, N)$,

$$f(x) = \frac{x - N - 1}{2^N}.$$

Clearly $f(x) < 0$ for all x, so $m(r, f) \equiv 0$. The poles of f are at the non-negative integers N, where each pole has multiplicity $2^{-(N+1)}$. So for any $r > 0$,

$$n(r, f) < \sum_{N=0}^{\infty} 2^{-(N+1)} = 1.$$

Therefore

$$T(r, f) = N(r, f) = \frac{1}{2} \int_0^r n(t, f) dt < \frac{r}{2} = O(r), \quad \text{as } r \to \infty.$$

But f is not rational, because it has infinitely many poles.

The additional assumption in Theorem 1.14 that the poles and roots of f should all be of multiplicity $\geq n_0$, say, implies that

$$\overline{n}(r, f) + \overline{n}(r, 1_\circ \oslash f) \leq \frac{1}{n_0}\{n(r, f) + n(r, 1_\circ \oslash f)\},$$

where $\overline{n}(r, g)$ denotes the number of poles of a tropical meromorphic function g in the interval $[-r, r]$, ignoring multiplicity. Indeed,

$$n(r, f) + n(r, 1_\circ \oslash f) = \sum_{|c_\nu| \leq r} |\omega_f(c_\nu)| \geq n_0 \sum_{|c_\nu| \leq r} 1 = n_0\{\overline{n}(r, f) + \overline{n}(0, 1_\circ \oslash f)\},$$

where the $\{c_\nu\}$ is the set of all the distinct poles and roots of f. Hence, if both of $n(r, f)$ and $n(r, 1_\circ \oslash f)$ are of order $O(1)$ as $r \to \infty$, then so is the sum $\overline{n}(r, f) + \overline{n}(r, 1_\circ \oslash f)$ and one concludes that the total number of the poles and roots of f should be finite.

Proof of Theorem 1.14. Suppose first that $T(r, f) \leq Kr$ for some $K > 0$ and for all r sufficiently large. We first observe that whenever $k > 1$, then

$$n(r, f) \leq \frac{2}{(k-1)r}N(kr, f).$$

In fact,

$$(k-1)rn(r, f) = n(r, f)\int_r^{kr} dt \leq \int_r^{kr} n(t, f)dt \leq \int_0^{kr} n(t, f)dt = 2N(kr, f).$$

Therefore, we conclude that

$$n(r, f) \leq \frac{2}{(k-1)r}N(kr, f) \leq \frac{2}{(k-1)r}T(kr, f) \leq \frac{2kK}{k-1},$$

hence we have $n(r, f) = O(1)$. Since we have $T(r, -f) = T(r, f) + O(1)$ by the Jensen formula, we may apply the same reasoning to see that $n(r, 1_\circ \oslash f) = O(1)$ as well. As mentioned in the above remark, the derivative of f has at most finitely many discontinuities, meaning that the graph of f consists of finitely many linear segments only, and so f is rational.

On the other hand, if f is rational, then we have for all r sufficiently large the representations

$$f(r) = \alpha_+ r + \beta_+, \qquad f(-r) = \alpha_- r + \beta_-.$$

But then

$$f^+(r) = \max(\alpha_+ r + \beta_+, 0) \leq \max\{|\alpha_+|, |\beta_+|\}r,$$

at least for $r \geq 1$. Similarly,

$$f^+(-r) = \max\{\alpha_- r + \beta_-, 0\} \leq \max\{|\alpha_-|, |\beta_-|\}r,$$

and so

$$m(r, f) \leq \frac{1}{2}\left(\max\{|\alpha_+|, |\beta_+|\} + \max\{|\alpha_-|, |\beta_-|\}\right)r.$$

Since f is rational, it has at most finitely many poles and thus there exists a positive constant κ such that $n(t, f) \leq \kappa$ holds for any $t \geq 0$. Therefore,

$$N(r, f) = \frac{1}{2}\int_0^r n(t, f)dt \leq \frac{\kappa}{2}r,$$

and we have

$$T(r, f) \leq \frac{1}{2}\left(\kappa + \max\{|\alpha_+|, |\beta_+|\} + \max\{|\alpha_-|, |\beta_-|\}\right)r,$$

and we are done. $\qquad\qquad\square$

Corollary 1.17. *A non-constant tropical rational function f is of order $\rho(f) = 1$.*

The example above, or Example 2.10 in Chapter 2 where we have $T(r, f) = O(r\log r)$ as $r \to \infty$, shows that some tropical non-rational functions can be of order one.

1.2.4 *Nevanlinna functions for tropical exponentials*

In addition to tropical polynomials and rational functions, we propose to treat here certain tropical meromorphic functions (called tropical hyper-exponential functions in literature), which are reminiscent to exponential functions $\exp(z^c)$ over the usual algebra.

Definition 1.18. Let α be a real number with $|\alpha| > 1$. Define a function $e_\alpha(x)$ on $\mathbb{R}$ by

$$e_\alpha(x) := \alpha^{[x]}(x - [x]) + \sum_{j=-\infty}^{[x]-1} \alpha^j = \alpha^{[x]}\left(x - [x] + \frac{1}{\alpha - 1}\right).$$

Similarly, if β is a real number with $|\beta| < 1$, the corresponding definition reads as

$$e_\beta(x) := \beta^{[x]}\left(\frac{1}{1 - \beta} - x + [x]\right).$$

Remark 1.19. Note that this definition has no meaning, if $\alpha = \pm 1$, resp. $\beta = \pm 1$, while the case $\beta = 0$ becomes trivial.

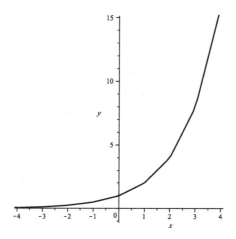

Fig. 1.5   Graph of $e_2(x)$.

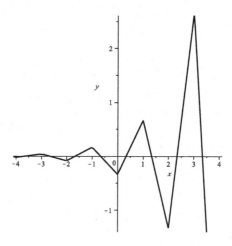

Fig. 1.6   Graph of $e_{-2}(x)$.

To get a rough idea of the behavior of these functions, one may look at the graphs of these exponential functions for $\alpha = \pm 2$, $\beta = \pm 1/2$, see the following four figures, Fig. 1.5 to Fig. 1.8.

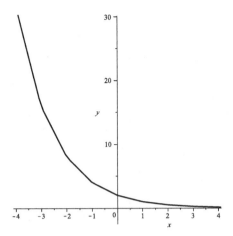

Fig. 1.7   Graph of $e_{1/2}(x)$.

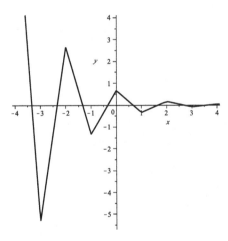

Fig. 1.8   Graph of $e_{-1/2}(x)$.

We first need to show that $e_\alpha(x)$ with $|\alpha| > 1$ determines a tropical meromorphic function:

Proposition 1.20. *The function $e_\alpha(x)$ is tropical meromorphic on $\mathbb{R}$ satisfying*

- $e_\alpha(m) = \alpha^m/(\alpha - 1)$ *for each* $m \in \mathbb{Z}$,
- $e_\alpha(x) = x + \frac{1}{\alpha-1}$ *for any* $x \in [0,1)$, *and*
- *the functional equation* $y(x+1) = y(x)^{\otimes\alpha}$ *on the whole of* $\mathbb{R}$.

Proof. The two first assertions trivially follow from Definition 1.18. The last assertion is verified by a straightforward computation:

$$e_\alpha(m) = \sum_{j=-\infty}^{m-1} \alpha^j = \alpha^{m-1} \sum_{j=-\infty}^{0} \alpha^j = \alpha^{m-1} \sum_{j=0}^{\infty} \left(\frac{1}{\alpha}\right)^j = \frac{\alpha^m}{\alpha - 1}$$

when $m \in \mathbb{Z}$,

$$e_\alpha(x) = \sum_{j=-\infty}^{-1} \alpha^j + x = x + \frac{1}{\alpha - 1}$$

when $x \in [0,1)$, and on $\mathbb{R}$

$$e_\alpha(x+1) = \sum_{j=-\infty}^{[x+1]-1} \alpha^j + \alpha^{[x+1]}(x+1-[x+1])$$

$$= \sum_{j=-\infty}^{[x]} \alpha^j + \alpha^{[x]+1}(x-[x])$$

$$= \alpha \left(\sum_{j=-\infty}^{[x]-1} \alpha^j + \alpha^{[x]}(x-[x]) \right) = \alpha\, e_\alpha(x).$$

It remains to verify that $e_\alpha(x)$ is continuous at integer points $x = m \in \mathbb{Z}$. This follows by taking $\varepsilon \in (0,1)$ and observing that

$$e_\alpha(m+\varepsilon) = \sum_{j=-\infty}^{m-1} \alpha^j + \alpha^m(m+\varepsilon - m) = \frac{\alpha^m}{\alpha - 1} + \varepsilon\alpha^m \quad \text{and}$$

$$e_\alpha(m-\varepsilon) = \sum_{j=-\infty}^{m-1-1} \alpha^j + \alpha^{m-1}(m-\varepsilon-m+1) = \frac{\alpha^m}{\alpha - 1} - \varepsilon\alpha^{m-1}.$$

$\square$

As for the case with $|\beta| < 1$, we similarly obtain

Proposition 1.21. *The function* $e_\beta(x)$ *is tropical meromorphic on* $\mathbb{R}$ *and satisfies*

- $e_\beta(m) = \beta^m/(1-\beta)$ *for each* $m \in \mathbb{Z}$,
- $e_\beta(x) = -x + \frac{1}{1-\beta}$ *for any* $x \in [0,1)$, *and*

- *the functional equation $y(x + 1) = y(x)^{\otimes \beta}$ on the whole $\mathbb{R}$.*

Proof. Again, the last assertion is the only one to be checked:

$$e_\beta(x + 1) = \sum_{j=[x+1]}^{\infty} \beta^j - \beta^{[x+1]}(x + 1 - [x + 1])$$

$$= \beta \left(\sum_{j=[x]}^{\infty} \beta^j - \beta^{[x]}(x - [x]) \right) = \beta \, e_\beta(x).$$

As for the continuity at integer points $x = m \in \mathbb{Z}$, we may take $\varepsilon \in (0, 1)$, resulting in

$$e_\beta(m + \varepsilon) = \sum_{j=m}^{\infty} \beta^j - \beta^m(m + \varepsilon - m) = \frac{\beta^m}{1 - \beta} - \varepsilon \beta^m \quad \text{and}$$

$$e_\beta(m - \varepsilon) = \sum_{j=m-1}^{\infty} \beta^j - \beta^{m-1}(m - \varepsilon - m + 1) = \frac{\beta^m}{1 - \beta} + \varepsilon \beta^{m-1}.$$

$\square$

As for the connection between $e_\alpha(x)$ with $|\alpha| > 1$ and $e_\beta(x)$ with $|\beta| < 1$, we obtain

Proposition 1.22. *Suppose $\alpha \neq \pm 1$. Then*

- $e_\alpha(-x) = \frac{1}{\alpha} e_{1/\alpha}(x)$, *and*
- $e_\alpha(0) = \frac{1}{\alpha} e_{1/\alpha}(0)$.

Proof. The first assertion immediately follows from the expressions for $e_\alpha(x)$ and $e_{1/\alpha}(x)$ and from $[-x] = -[x] - 1$. The second assertion is trivial. $\square$

We continue by giving a few elementary properties for tropical exponential functions:

Proposition 1.23. *Concerning the monotonicity, poles and roots of exponential functions, we observe that*

- *if $\alpha > 1$, then $e_\alpha(x)$ is strictly increasing, it has no poles, and it has roots exactly at $x = m \in \mathbb{Z}$, of multiplicity $\alpha^m(1 - 1/\alpha) \to \infty$ as $m \to \infty$;*

- *if $\alpha < -1$, then $e_\alpha(x)$ is locally monotone outside poles and roots, with poles of multiplicity $\alpha^{2j}(1 - \alpha)$ at odd integers $x = 2j + 1$ and roots of multiplicity $\alpha^{2j}(1 - 1/\alpha)$ at even integers $x = 2j$;*
- *if $0 < \beta < 1$, then $e_\beta(x)$ is strictly decreasing with no poles, and roots exactly at $x = m \in \mathbb{Z}$, of multiplicity $\beta^{m-1}(1 - \beta) \to 0$ as $m \to \infty$;*
- *if $-1 < \beta < 0$, then $e_\beta(x)$ is locally monotone outside poles and roots, with poles of multiplicity $\beta^{2j}(1 - 1/\beta)$ at even integers $x = 2j$, and roots of multiplicity $\beta^{2j}(1 - \beta)$ at odd integers $x = 2j + 1$.*

Proof. First suppose that $\alpha > 1$. In intervals between integers, say in $(m, m + 1)$, $m \in \mathbb{Z}$, we have $e_\alpha(x) = \alpha^m(x - m + \frac{1}{\alpha-1})$, and therefore $e'_\alpha(x) = \alpha^m > 0$, hence $e_\alpha(x)$ is increasing in this interval. Moreover, given $m \in \mathbb{Z}$ and $0 < \varepsilon < 1$, we see that

$$e_\alpha(m + \varepsilon) - e_\alpha(m) = \alpha^m \left(\varepsilon + \frac{1}{\alpha - 1} \right) - \alpha^m \frac{1}{\alpha - 1} = \varepsilon \alpha^m > 0$$

and

$$e_\alpha(m) - e_\alpha(m - \varepsilon) = \frac{\alpha^m}{\alpha - 1} - \alpha^{m-1} \left(1 - \varepsilon - \frac{1}{\alpha - 1} \right)$$
$$= \alpha^{m-1} \left(\frac{2}{\alpha - 1} + \varepsilon \right) > 0,$$

and the strict monotonicity has been verified. As to the poles and roots of $e_\alpha(x)$, they may only appear at the integer points $m \in \mathbb{Z}$. But then

$$\omega_{e_\alpha}(m) = \lim_{\varepsilon \to 0+} ((e'_\alpha(m + \varepsilon) - (e'_\alpha(m - \varepsilon)) = \alpha^m(1 - 1/\alpha) > 0,$$

and so $e_\alpha(m)$ has a root at $x = m$ of multiplicity $\alpha^m(1 - 1/\alpha)$, and $e_\alpha(x)$ has no poles.

Provided next that $\alpha < -1$, then $e_\alpha(x)$ has a zero of multiplicity $\alpha^{2j}(1 - 1/\alpha)$ at each even integer $x = 2j$ and a pole of multiplicity $\alpha^{2j}(1 - \alpha)$ at each odd integer $x = 2j + 1$, since $\omega_{e_\alpha}(m) = \alpha^m(1 - 1/\alpha)$ for each $m \in \mathbb{Z}$, completing the second assertion after we observe that local monotonicity applies between consecutive integers.

In the case of $0 < \beta < 1$, if $x \in (m, m + 1)$ for an arbitrary $m \in \mathbb{Z}$, we then have $e'_\beta(x) = -\beta^m < 0$. Therefore,

$$\omega_{e_\beta}(m) = -\beta^m - (-\beta^{m-1}) = \beta^{m-1}(1 - \beta) > 0.$$

Moreover, if $m \in \mathbb{Z}$ and $\varepsilon < 1$, then

$$e_\beta(m + \varepsilon) - e_\beta(m) = \beta^m \left(\frac{1}{1 - \beta} - \varepsilon \right) - \frac{\beta^m}{1 - \beta} = -\varepsilon \beta^m < 0$$

and

$$e_\beta(m) - e_\beta(m - \varepsilon) = \frac{\beta^m}{1-\beta} - \frac{\beta^m}{1-\beta} - \varepsilon\beta^{m-1} = -\varepsilon\beta^{m-1} < 0,$$

proving that we have a strictly decreasing case.

In the final case with $-1 < \beta < 0$, we have that $e_\beta(x)$ has a pole of multiplicity $\beta^{2j}(1 - 1/\beta)$ at each even integer $x = 2j$ and a root of multiplicity $\beta^{2j}(1 - \beta)$ at each odd integer $x = 2j + 1$. Again, local monotonicity between consecutive integers is obvious. $\qquad\square$

Proposition 1.24. *The function $e_\alpha(x)$, $\alpha \neq \pm 1$, is of infinite order and, in fact, of hyper-order one.*

Proof. If $\alpha > 1$, then $e_\alpha(x)$ is strictly positive and has no poles. Moreover, for $r = m + \varepsilon$ with $m \in \mathbb{Z}$ and $\varepsilon \in [0, 1)$, we have

$$T(r, e_\alpha) = m(r, e_\alpha) = \frac{\alpha^m\left(\varepsilon + 1/(\alpha - 1)\right) + \alpha^{-m-1}\left(1 - \varepsilon + 1/(\alpha - 1)\right)}{2}$$

$$= \frac{1}{2}\left(\frac{1}{\alpha - 1} + r - [r] + o(1)\right)\alpha^{[r]}$$

as $r \to \infty$. Therefore, $e_\alpha(x)$ is of hyper-order one:

$$\rho_2(e_\alpha(x)) := \limsup_{r\to\infty} \frac{\log\log T(r, e_\alpha)}{\log r} = \lim_{r\to\infty} \frac{\log[r] + O(1)}{\log r} = 1,$$

provided that $\alpha > 1$.

Consider next the case $\alpha < -1$. By Proposition 1.23, $e_\alpha(x)$ has a zero of multiplicity $\alpha^{2j}(1 - 1/\alpha)$ at each even integer $x = 2j$ and a pole of multiplicity $\alpha^{2j}(1 - \alpha)$ at each odd integer $x = 2j + 1$. Thus we see that when $2\ell \leq t < 2(\ell + 1)$ for some integer ℓ, that is, when $\ell = \left[\frac{t}{2}\right]$, then, by an elementary computation,

$$n(t, 1_\circ \oslash e_\alpha) = \sum_{j=-\ell}^{\ell} \alpha^{2j}\left(1 - \frac{1}{\alpha}\right) = \sum_{j=1}^{\ell}(\alpha^{2j} + \alpha^{-2j})\left(1 - \frac{1}{\alpha}\right) + 1 - \frac{1}{\alpha}$$

$$= \left(1 - \frac{1}{\alpha}\right)\left\{\frac{\alpha^2}{\alpha^2 - 1}\alpha^{2\ell} + \frac{1}{1 - \alpha^2}\alpha^{-2\ell}\right\}$$

$$= \frac{1}{\alpha(\alpha + 1)}(\alpha^2\alpha^{2\ell} - \alpha^{-2\ell}) \geq \frac{1}{|\alpha||\alpha + 1|}(|\alpha|^t - |\alpha|^{-t}).$$

Therefore,

$$T(r, 1_\circ \oslash e_\alpha) \geq N(r, 1_\circ \oslash e_\alpha) \geq \frac{1}{2|\alpha||\alpha + 1|\log|\alpha|}(|\alpha|^r - 2 - |\alpha|^{-r}), \quad (1.15)$$

resulting in $\rho_2(e_\alpha(x)) \geq 1$ by the tropical Jensen formula (1.10).

To prove the converse inequality, we first observe that

$$n(t, 1_\circ \oslash e_\alpha) = \left(1 - \frac{1}{\alpha}\right)\left\{\frac{\alpha^2}{\alpha^2 - 1}\alpha^{2\ell} + \frac{1}{1 - \alpha^2}\alpha^{-2\ell}\right\}$$

$$\leq \left(1 - \frac{1}{\alpha}\right)\frac{\alpha^2}{\alpha^2 - 1}\alpha^{2\ell} = \frac{\alpha}{\alpha + 1}|\alpha|^t,$$

so that

$$N(r, -e_\alpha) \leq \frac{\alpha}{2(\alpha + 1)\log|\alpha|}(|\alpha|^r - 1). \tag{1.16}$$

Moreover, by a simple observation,

$$-e_\alpha(r) = -\alpha^{[r]}\left(r - [r] - \frac{1}{\alpha - 1}\right) \leq \left(1 - \frac{1}{\alpha - 1}\right)|\alpha|^r,$$

while

$$|-e_\alpha(-r)| = \left|-\alpha^{[-r]}\left(-r - [-r] - \frac{1}{\alpha - 1}\right)\right| \leq \left(1 - \frac{1}{\alpha - 1}\right)|\alpha|^{-r}.$$

Therefore,

$$m(r, -e_\alpha) \leq \frac{1}{2}((-e_\alpha(r))^+ + (-e_\alpha(-r))^+)$$

$$\leq \left(1 - \frac{1}{\alpha - 1}\right)(|\alpha|^r + |\alpha|^{-r}) \leq \left(1 - \frac{1}{\alpha - 1}\right)(|\alpha|^r + 1),$$

and so

$$T(r, -e_\alpha) \leq K_1(\alpha)|\alpha|^r + K_2(\alpha),$$

where $K_1(\alpha), K_2(\alpha)$ are constants depending on α only. This implies that $\rho_2(e_\alpha(x)) \leq 1$.

The case $|\alpha| < 1$ then immediately follows by Proposition 1.22.   □

Remark 1.25. Observe that if $\alpha > 1$, resp. $0 < \beta < 1$, then $e_\alpha(x) \oplus a > 0$, resp. $e_\beta(x) \oplus a > 0$, for each $a \in \mathbb{R}$. This shows that $m\big(r, 1_\circ \oslash (e_\alpha \oplus a)\big) = m\big(r, -(e_\alpha \oplus a)\big) \equiv 0$, so that

$$T\big(r, 1_\circ \oslash (e_\alpha \oplus a)\big) = N\big(r, 1_\circ \oslash (e_\alpha \oplus a)\big),$$

and, respectively,

$$T\big(r, 1_\circ \oslash (e_\beta \oplus a)\big) = N\big(r, 1_\circ \oslash (e_\beta \oplus a)\big)$$

as well.

In order to see what happens if $\alpha \leq -1$, resp. $-1 < \beta < 0$, consider first, as an example, the case $e_\beta(x)$ with $\beta = -1/2$, see Fig. 1.8, and take $a = -1 < 0$. Then the roots of $e_\beta(x) \oplus a$ are the same as those of $e_\beta(x)$ for all $x = 2j+1 > 0$, while for $x = 2j+1 < 0$, each such root of $e_\beta(x)$, having multiplicity $\beta^{2j}(1-\beta)$, splits into two roots of $e_\beta(x) \oplus a$, with the sum of their multiplicities being equal to $\beta^{2j}(1-\beta)$, see Fig. 1.8 and Fig. 1.9. Therefore, we easily get

$$T(r, e_\beta(x)) \geq N(r, e_\beta(x)) = 2N(r, 1_\circ \oslash (e_\beta(x) \oplus a)) + O(r). \qquad (1.17)$$

More generally, the same conclusion as in (1.17) follows for all $a < 0$. In particular, this means that each $a < 0$ is a deficient value for $e_\beta(x)$ in the sense that

$$1 - \limsup_{r \to \infty} \frac{N(r, 1_\circ \oslash (e_\beta(x) \oplus a))}{T(r, e_\beta(x))} \geq 1/2 > 0.$$

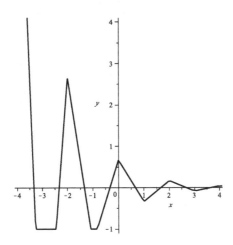

Fig. 1.9 Graph of $e_{-1/2}(x) \oplus (-1)$.

Remark 1.26. To prevent misinterpretations in what follows, let $e_\alpha(x)$ and $e_\beta(x)$ be two tropical exponential functions with $\alpha, \beta \neq 1$, $\alpha \neq \beta$ and let $s \in [0, 1)$ be a fixed real number. Then it is immediate to verify that we get for the Casoratian determinants

$$\begin{vmatrix} e_\alpha(x) & e_\alpha(x-s) \\ e_\alpha(x+1) & e_\alpha(x-s+1) \end{vmatrix} = 0,$$

and

$$\begin{vmatrix} e_\alpha(x) & e_\beta(x) \\ e_\alpha(x+1) & e_\beta(x+1) \end{vmatrix} \neq 0.$$

However, one of the tropical exponentials in the pairs $e_\alpha(x), e_\alpha(x-s)$ and $e_\alpha(x), e_\beta(x)$ is not a constant multiple of the other one. Recall, however, that $\alpha e_\alpha(-x) = e_\beta(x)$ by Proposition 1.22 whenever $\beta = 1/\alpha$. Therefore, the standard presentation of linear (in)dependence does not carry over as such to the tropical setting of meromorphic functions. See Chapter 5 below for a more detailed analysis of tropical linear independence.

Chapter 2

Tropical entire functions

This chapter is devoted to studying tropical transcendental entire functions on $\mathbb{R}$ defined by max-plus series expansions of type

$$f(x) = \bigoplus_{n=0}^{\infty} a_n \otimes x^{\otimes n} = \max_{n \in \mathbb{Z}_{\geq 0}} \{a_n + nx\}, \quad a_n \in \mathbb{R}_{\max}. \tag{2.1}$$

This is, of course, a natural extension of tropical polynomials that we have treated in Chapter 1, see also Tsai [106] for tropical polynomials and their fundamental properties. Moreover, some of tropical exponentials considered in Chapter 1 are basic examples of transcendental tropical entire functions.

The main purpose in this chapter is to estimate the order of growth and type of the function $f(x)$ by means of the coefficients a_n in the series. The formulas obtained admit similar relations as those for the logarithmic order and logarithmic type for transcendental entire function on $\mathbb{C}$ by means of their Taylor-Maclaurin series expansions. Later on, we proceed to applying these tropical formulas to some known transcendental tropical entire functions given by their series expansions. One is an ultra-discretized theta function which gives an ultra-discretization of some elliptic functions and the other one is an ultra-discrete basic hypergeometric function which forms a tropical meromorphic solution to an ultra-discretized Painlevé III equation by taking its ultra-discrete analogue of 'logarithmic shift'. We also show that both of these functions are of order two. These two functions will then be treated in more detail later in Chapters 6 and 7, respectively. Some other examples of tropical entire functions of arbitrary order are also presented as a formal ultra-discretization of complex entire functions given by q-series expansions.

2.1 Definitions and basic results

Our key problem in this chapter is now to find relations between the order of growth, resp. the type, and the coefficients of its max-plus infinite series expansion (2.1) of a tropical entire function $f(x)$ on the real line $\mathbb{R}$ corresponding to similar relations for their infinite power series expansions of entire functions on the complex plane $\mathbb{C}$. Recalling the definition of order for the Nevanlinna characteristic in (1.13), we may define the **order** (of growth) and the **type** of a function $S(r) : \mathbb{R}_{>0} \to \mathbb{R}_{>0}$ by

$$\beta = \limsup_{r\to\infty} \frac{\log S(r)}{\log r} \quad \text{and} \quad \alpha = \limsup_{r\to\infty} \frac{S(r)}{r^\beta}, \tag{2.2}$$

respectively. These definitions imply that there exists a set large enough in $\mathbb{R}_{\geq 0}$ so that an approximation behavior such as

$$S(r) = \{\alpha + o(1)\}r^\beta \qquad (r \to +\infty) \tag{2.3}$$

holds. In fact, Halburd and Korhonen [42, Corollary 3.4] show that for any given $\epsilon > 0$, a positive increasing function S on $\mathbb{R}_{\geq 0}$ of order ρ and type τ with $0 < \rho < \infty$ and $0 < \tau < \infty$ admits a set of infinite linear measure of $\mathbb{R}_{\geq 0}$ on which

$$(\tau - \epsilon)r^\rho \leq S(r) \leq (\tau + \epsilon)r^\rho$$

holds.

2.1.1 *Preliminaries*

For the purpose mentioned above, we pose two problems to be treated in this chapter:

(a) What expansion of a tropical entire function corresponds to the Taylor series expression of an entire function?

(b) How can we define the order of growth of a tropical entire function, $f(x)$ say, by means of its tropical Taylor series coefficients?

Of course, we first need to make the following

Definition 2.1. A tropical entire function $f(x)$ is a real-valued continuous piecewise linear function defined on the real line $\mathbb{R}$ such that the set $\mathcal{D}_f$ of discontinuities of the derivative $f'(x)$ has no finite limit in $\mathbb{R}$ and

$$f'(x+0) > f'(x-0) \tag{2.4}$$

holds for each $x \in \mathcal{D}_f$.

When the set $\mathcal{D}_f$ is not finite, we say $f(x)$ to be transcendental, so that the graph of $y = f(x)$ draws a (downward-convex) polygonal line in the xy-plane, which possesses infinitely many corner points exactly over each $x \in \mathcal{D}_f$. Immediate examples of transcendental tropical entire functions are tropical exponentials e_α with $\alpha > 1$ and e_β with $0 < \beta < 1$, see Proposition 1.23. For a collection of more examples, see Section 2.2 below.

Recall that an element $x \in \mathcal{D}_f$ is a root of $f(x)$ with multiplicity

$$\omega_f(x) := f'(x+0) - f'(x-0), \tag{2.5}$$

which is positive by (2.4). The convexity is a simple consequence of this property. Note that the condition (2.4) implies

$$M(r, f) := \max_{|x| \le r} |f(x)| = \max\{|f(+r)|, |f(-r)|\}.$$

This can be regarded as the maximum principle for tropical entire functions. For the sake of simplicity, we may make following normalizations to a given tropical entire function $f(x)$ while considering the two problems (a) and (b) posed above:

[N0] $f(x)$ is transcendental,
[N1] $f(x) \ge 0$ for any $x \in \mathbb{R}$,
[N2] $f(x)$ has only positive roots, that is, $\mathcal{D}_f$ locates only on the positive real ray $\mathbb{R}_{>0}$.

These normalizations enable us to concentrate our observation on the part of the graph of $y = f(x)$ inside the first quadrant of the xy-plane.
On the other hand, we do not lose generality at all with these additional assumptions [N0], [N1] and [N2]. In fact, if $f(x)$ is a tropical polynomial contrary to [N0], for a sufficiently large $r > 0$ we have

$$M(r, f) = Ar + B$$

for some constants $A \ge 0$ and $B \in \mathbb{R}$. In terms of tropical operations $a \oplus b = \max\{a, b\}$, $a \otimes b = a + b$ and $a^{\otimes b} = ba$ for $a, b \in \mathbb{R} \cup \{-\infty\}$ as well as $a \oslash b = a - b$ when $b \in \mathbb{R}$, the tropical polynomial $f(x)$ has the expression

$$f(x) = \max_{\ell \le n \le m} \{a_n x + b_n\} = \bigoplus_{n=\ell}^{m} b_n \otimes x^{\otimes a_n}$$

for integers $\ell < m$ and $a_n, b_n \in \mathbb{R}$ ($\ell \le n \le m$) with

$$a_\ell < a_{\ell+1} < \cdots < a_{m-1} < a_m.$$

Then we see $A = \max\{|a_\ell|, |a_m|\}$, which shows

$$\lim_{r\to\infty} \frac{M(r,f)}{r} = \max\{|a_\ell|, |a_m|\} = \max_{\ell \le n \le m} |a_n| = \bigoplus_{n=\ell}^{m} |a_n|.$$

Recall that this limit is replaced by

$$\lim_{r\to\infty} \frac{\log M(r,f)}{\log r} = \max\{|a_\ell|, |a_m|\}$$

for a complex Laurent polynomial $f(z) = \sum_{n=\ell}^{m} b_n z^{a_n}$ with $b_n \ne 0$. This relation is however not very interesting in the view point of analysis, since it only says that such a limit cannot be finite for any transcendental functions in each of the two cases, respectively. Indeed, we are to estimate the growth rate of a given non-constant tropical entire function $f(x)$ on $\mathbb{R}$ by the means of its order defined by

$$\limsup_{r\to\infty} \frac{\log M(r,f)}{\log r} \in [1, +\infty],$$

while, as observed in Appendix C, that is done for a non-constant entire function $f(z)$ on $\mathbb{C}$ by the means of its order or logarithmic order defined respectively by

$$\limsup_{r\to\infty} \frac{\log\log M(r,f)}{\log r} \in [0, +\infty] \quad \text{or} \quad \limsup_{r\to\infty} \frac{\log\log M(r,f)}{\log\log r} \in [1, +\infty].$$

The order of $f(x)$ is always equal to 1, if $f(x)$ is a non-constant tropical polynomial, while the order and logarithmic order attain the values 0 and 1, respectively for a non-constant complex polynomial $f(z)$. As some transcendental entire functions on $\mathbb{C}$ can be of logarithmic order 1 and therefore of order 0, certain transcendental tropical entire functions on $\mathbb{R}$ can be of order 1. Examples of such functions will be given below, see Section 2.2.

Now we assume, in particular, that $f(x)$ has a root at the origin, that is,

$$f'(-0) < f'(+0).$$

Then we can allot the root at $x = 0$ to two new roots, which are positive and negative, respectively. In fact, let us put

$$c_- := \max\{\, c < 0 \,:\, f'(c-0) < f'(c+0) \,\}$$

and

$$c_+ := \min\{\, c > 0 \,:\, f'(c-0) < f'(c+0) \,\},$$

where we put $c_- = -\infty$ if $f(x)$ has no negative root at all, while $c_+ = +\infty$ if $f(x)$ does not have any positive root. Take a suitable tropical **unit** function

$$\alpha x + \beta = \beta \otimes x^{\otimes \alpha} \quad (\alpha, \beta \in \mathbb{R})$$

so that the line $y = \alpha x + \beta$ meets two points P and Q on the graph of $y = f(x)$ having the x-segments, say P_x and Q_x respectively, and satisfying

$$c_- < P_x < 0 < Q_x < c_+.$$

In fact, the constants α and β satisfy

$$f'(-0) < \alpha < f'(+0) \quad \text{and} \quad f(0) < \beta,$$

and the numbers P_x and Q_x are the solutions of the linear equations

$$f'(-0)x + f(0) = \alpha x + \beta \quad \text{and} \quad \alpha x + \beta = f'(+0)x + f(0),$$

respectively. Therefore, we have

$$P_x = -\frac{\beta - f(0)}{\alpha - f'(-0)} < 0 \quad \text{and} \quad Q_x = \frac{\beta - f(0)}{f'(+0) - \alpha} > 0.$$

We consider the new tropical entire function

$$F(x) := \max\{f(x), \alpha x + \beta\} = f(x) \oplus \beta \otimes x^{\otimes \alpha}.$$

Two functions $f(x)$ and $F(x)$ have the same roots together with multiplicities except for $x = 0$, P_x and Q_x. Further concerning these exceptional roots, it follows that the sum of the multiplicities of the roots P_x, Q_x of $F(x)$ equals to the multiplicity of $f(x)$ at the origin. Needless to say, there is no difference in their asymptotic behavior as $x \to \pm\infty$. Furthermore, we see naturally that the graph of $y = F(x)$ locates above the line $y = \alpha x + \beta$, since

$$F(x) \oslash \left(\beta \otimes x^{\otimes \alpha}\right) = F(x) - (\alpha x + \beta) = \max\{f(x) - \alpha x - \beta, 0\} \geq 0$$

for any $x \in \mathbb{R}$. Let us emphasize again that $F(x) \equiv \beta \otimes x^{\otimes \alpha}$ on the closed interval $[P_x, Q_x] \ni 0$, and therefore $F(x)$ has no root at the origin, that is, $F'(0) = \alpha$. Then we can decompose $F(x)$ by two tropical entire functions $F_1(x)$ and $F_2(x)$ both having only positive roots and

$$F_1(x) = F_2(x) \equiv 0 \quad \text{whenever } x \leq 0$$

into the tropical sum

$$F(x) = \max\{F_1(x), F_2(-x)\} = F_1(x) \oplus F_2(-x).$$

Indeed, we simply define

$$F_1(x) = \begin{cases} f(x) & (x \geq 0) \\ 0 & (x < 0) \end{cases}, \quad F_2(x) = \begin{cases} f(-x) & (x \geq 0) \\ 0 & (x < 0) \end{cases},$$

respectively, that is, $F_1(x) = f(x \oplus 0)$ and $F_2(x) = f(-(x \oplus 0))$. Hence we may concentrate on the tropical entire function

$$g(x) := F(x) \oslash \left(\beta \otimes x^{\otimes \alpha}\right) = 0 \oplus f(x) \oslash \left(\beta \otimes x^{\otimes \alpha}\right)$$

associated to the original function $f(x)$ so that the assumptions $[N0]$, $[N1]$ and $[N2]$ are all fulfilled by $g(x)$. In what follows we will rewrite this $g(x)$ by $f(x)$.

2.1.2   *Growth order and tropical series expansions*

We now proceed to discuss the problem (a). Let $\{r_n\}_{n=1}^{\infty}$ be the increasing sequence of the roots of $f(x)$. Also let $\{s_n\}_{n=1}^{\infty}$ be the slopes of the line segments of the graph of $y = f(x)$, so that

$$s_n = \frac{f(r_{n+1}) - f(r_n)}{r_{n+1} - r_n} = f'(r_n + 0) = f'(r_{n+1} - 0), \quad n = 1, 2, \ldots.$$

Then we can express $f(x)$ by the tropical power series expansion of the form

$$f(x) = \max \left\{ 0, s_1(x - r_1), s_2(x - r_2) + s_1(r_2 - r_1), \quad \ldots, \right.$$

$$\left. \ldots, \quad s_n(x - r_n) + \sum_{j=1}^{n-1} s_j(r_{j+1} - r_j), \quad \ldots \right\}$$

$$= \max_{0 \le n < +\infty} \left\{ s_n x - \sum_{j=1}^{n-1} m_j r_j \right\},$$

where each number

$$m_n := s_n - s_{n-1} \quad (n = 1, 2, \ldots)$$

is positive as the multiplicity of the root r_n with the convention $r_0 = s_0 = m_0 = 0$. Thus the tropical power series

$$\bigoplus_{n=0}^{\infty} c_n \otimes x^{\otimes s_n} = \max_{n \in \mathbb{Z}_{\ge 0}} \{s_n x + c_n\} \tag{2.6}$$

expresses the function $f(x)$, when the coefficients c_n are given by

$$c_n := -\sum_{j=1}^{n-1} m_j r_j = 0 \oslash \left(\bigoplus_{j=1}^{n-1} r_j^{\otimes m_j} \right), \quad n = 1, 2, \ldots, \tag{2.7}$$

which tends to $-\infty$ as $n \to \infty$.

Concerning the problem (b), we see that

$$M(r, f) = \max\{|f(-r)|, |f(r)|\} = f(r)$$

and, recalling the Nevanlinna functions from Chapter 1,

$$T(r, f) = m(r, f) = \frac{1}{2}\left\{f^+(-r) + f^+(r)\right\} = \frac{1}{2}f(r) = \frac{1}{2}M(r, f), \quad r > 0, \tag{2.8}$$

for our normalized function $f(x)$. Note that $f(r) = O(r)$ as $r \to +\infty$ when and only when $f(x)$ is a tropical polynomial normalized as above, that is, the series (2.6) is finite and the sequence $\{c_n\}$ is bounded from below. Hence the order of $f(x)$ is obtained by

$$\limsup_{r\to\infty} \frac{\log T(r,f)}{\log r} = \limsup_{r\to\infty} \frac{\log M(r,f)}{\log r} = \limsup_{r\to\infty} \frac{\log f(r)}{\log r}. \qquad (2.9)$$

Of course, if this number, ρ say, is positive and finite, we also define its type as usual, that is, by

$$\limsup_{r\to\infty} \frac{M(r,f)}{r^\rho} = \limsup_{r\to\infty} \frac{f(r)}{r^\rho}.$$

This is twice of the number $\limsup_{r\to\infty} \frac{T(r,f)}{r^\rho}$, but we choose the former to define the type of tropical entire functions. Note that, in fact, $\rho = 0$ is possible only when $f(x)$ is a constant and $\rho \geq 1$ for any non-constant function $f(x)$. We have already seen that $\rho(f) = 1$ whenever $f(x)$ is any non-constant tropical polynomial. However, the converse is not true. For example, we again refer to Section 2.2.

We now see that

$$f(r) = s_n r - \sum_{j=1}^{n-1} m_j r_j = s_n r + c_n \qquad (2.10)$$

for $r \in (r_n, r_{n+1})$. As mentioned above, the sequence $c_n = -\sum_{j=1}^{n-1} m_j r_j$ tends to $-\infty$ as $n \to \infty$. From now on, we will write 0 and $-\infty$ by $1_\circ$ and $0_\circ$ as the neutral elements for tropical product $\otimes$ and tropical sum $\oplus$, respectively. Then, for example, the expression (2.7) of the coefficients c_n becomes

$$c_n := 1_\circ \oslash \left(\sum_{j=1}^{n-1} m_j r_j \right) = 1_\circ \oslash \left(\bigoplus_{j=1}^{n-1} r_j^{\otimes m_j} \right), \qquad n = 1, 2, \ldots,$$

and they have the limit $\lim_{n\to\infty} c_n = 0_\circ$ similarly as the sequence of the Taylor coefficients of a transcendental entire function tends to the limit 0. In what follows in this section, we now specify tropical entire functions to be treated:

Definition 2.2. A tropical entire function is normalized provided that its (convex) graph $\{(x, f(x)) : x \in \mathbb{R}\}$ coincides with the x-axis for all $x < 0$.

Of course, a normalized tropical entire function $f(x)$ is called transcendental, if its (infinite) series expression is of the form

$$f(x) = \bigoplus_{n=0}^{\infty} c_n \otimes x^{\otimes s_n} = \max\{ s_n x + c_n : n = 0, 1, 2, \ldots \}$$

for the coefficients $c_n (\geq 0)$ and the slopes s_n with

$$0 = s_0 < s_1 < \cdots < s_n < \cdots \rightarrow +\infty.$$

In order that the maximum at x of the family of lines $y = s_n x + c_n$ ($n = 0, 1, 2, \ldots$) describes such a curve, the y-intercepts c_n of these lines should satisfy

$$0 = c_0 > c_1 > \cdots > c_n > \cdots \rightarrow -\infty.$$

This is a necessary and sufficient condition on $f(x)$ to be normalized tropical entire and transcendental.

Example 2.3. Let $\{s_n\}_{n \in \mathbb{N}}$ be the sequence of positive numbers given by $s_{n+1} - s_n = \frac{1}{n+1}$ with $s_1 = 1$, that is,

$$s_n = \sum_{k=0}^{n-1} \frac{1}{k+1} \quad (n \in \mathbb{N}).$$

Consider an infinite tropical series

$$\bigoplus_{n=0}^{\infty} (-n + s_n) \otimes x^{\otimes s_n} = \max\{-n + s_n + s_n x : n = 0, 1, 2, \ldots\}$$

with $s_0 = 0$. Then this defines a transcendental tropical entire function, say $f(x)$, since the constants $c_n := -n + s_n$ as well as s_n ($n \in \mathbb{Z}_{\geq 0}$) satisfy the conditions required in Definition 2.2. For any r ($n \leq r < n+1$), we have $f(r) = -n + s_n + s_n r$ and thus

$$\log M(r, f) = \log r + \log \log r + O(1) \quad (r \rightarrow +\infty),$$

which implies $\lim\limits_{r \rightarrow +\infty} \dfrac{\log M(r, f)}{\log r} = 1$. On the other hand, observe that

$$c_k = -k + s_k \sim -k + \log k, \quad c_k/s_k = -k/s_k + 1 \sim -k/\log k + 1$$

and

$$\frac{\log(-c_k)}{\log(-c_k/s_k)} \rightarrow 1$$

as $k \rightarrow +\infty$.

2.1.3 Main theorem for tropical series expansions

Recall the tropical series expansion of $f(x)$ in (2.6):

$$f(x) = \bigoplus_{k=0}^{\infty} c_k \otimes x^{\otimes s_k} = \max_{k \in \mathbb{Z}_{\geq 0}} \{s_k x + c_k\}, \quad x \in \mathbb{R},$$

where the sequence of the coefficients $\{c_k\}_{k=0}^{\infty} \subset \mathbb{R}$ is given as in (2.7) by

$$c_k := -\sum_{j=0}^{k-1} m_j r_j = 1_{\circ} \oslash \left(\bigoplus_{j=0}^{k-1} r_j^{\otimes m_j} \right), \quad k = 1, 2, \ldots$$

and $c_0 = 0$, while the sequence of the exponents $\{s_k\}_{k=0}^{\infty}$ satisfies

$$s_k - s_{k-1} = m_k, \quad k = 1, 2, \ldots$$

and $s_0 = 0$ with positive integers m_j $(j \in \mathbb{N})$ and $m_0 = 0$. Then it follows that

$$0 = s_0 < s_1 < \cdots < s_k < \cdots \to +\infty$$

and

$$M(r, f) = \max_{k \in \mathbb{Z}_{\geq 0}} \{c_k + s_k r\}.$$

As our main result in this chapter, we prove the following

Theorem 2.4. *Let $f(x)$ be a transcendental tropical entire function whose tropical series expansion is $f(x) = \bigoplus_{k=0}^{\infty} c_k \otimes x^{\otimes s_k}$ with $(0 =)c_0 > c_1 > \cdots \to -\infty$ and $(0 =)s_0 < s_1 < \cdots \to +\infty$. Then we have*

$$\limsup_{r \to \infty} \frac{\log M(r, f)}{\log r} = \limsup_{k \to \infty} \frac{\log(-c_k)}{\log\{(-c_k)/s_k\}}. \tag{2.11}$$

If this limit ρ in (2.11) is finite but strictly larger than 1, then we obtain

$$\limsup_{r \to \infty} \frac{M(r, f)}{r^{\rho}} = c(\rho) \limsup_{k \to \infty} \frac{-c_k}{\{(-c_k)/s_k\}^{\rho}} \tag{2.12}$$

where

$$c(\rho) = \frac{(\rho - 1)^{\rho - 1}}{\rho^{\rho}}. \tag{2.13}$$

If two expressions on the right-hand side of (2.11) or (2.12) in this theorem are denoted by

$$\limsup_{k \to \infty} \frac{\log(1_{\circ} \oslash c_k)}{\log\{(1_{\circ} \oslash c_k)^{1/s_k}\}} \quad \text{and} \quad c(\rho) \limsup_{k \to \infty} \frac{1_{\circ} \oslash c_k}{\{(1_{\circ} \oslash c_k)^{1/s_k}\}^{\rho}},$$

respectively, then one recognizes a clear correspondence between the tropical formulas (2.11) or (2.12) and the classical complex analysis formulas (C.6) or (C.7) in Appendix C. Remark that the constant $c(\rho)$ is the same as in the case of tropical entire functions of positive and finite logarithmic order, and thus we have

$$\frac{1}{c(\rho)\rho} = \left(1 + \frac{1}{\rho - 1}\right)^{\rho - 1} \to e$$

as $\rho \to 1 + 0$ or as $\rho \to \infty$, respectively. The second relation also holds for a transcendental tropical entire function $f(x)$ of order equal to one. In fact, the type must then be infinite. See Example in Section 2.2.

Note that Theorem 2.4 answers our original problem (b). Moreover, we have found such a scaling relation that

the ratio $\dfrac{\log Y}{\log X}$ in (2.11) is replaced by the ratio $\dfrac{Y}{X^\rho}$ in (2.12).

Before proceeding to the proof of this theorem, we give some observations about the relation between the tropical series expansion of a tropical entire function on $\mathbb{R}$ and the Taylor series expansion of an entire function on $\mathbb{C}$. To this end, let us begin with the interpolation series for a given entire function on $\mathbb{C}$ and given points $\beta_n \in \mathbb{C}$ $(j = 1, 2, \ldots)$ of the form

$$f(z) = \sum_{n=0}^{\infty} \alpha_n (z - \beta_1)(z - \beta_2) \cdots (z - \beta_n), \quad \alpha_n \in \mathbb{C}, \tag{2.14}$$

which is known as Newton's interpolation series with nodes $\{\beta_n\}$. It is not difficult to see that if all these points β_j are distinct, then the coefficients α_n can be determined in terms of the β_j and the values $f(\beta_j)$ and that each α_n can be expressed as a rational function of $\beta_1, \beta_2, \ldots, \beta_{n+1}$, $f(\beta_1), f(\beta_2), \ldots, f(\beta_{n+1})$. Then the expression is a unique expansion of $f(z)$. However, the expansion exists, even if the points β_j are not all distinct. In particular, let us assume that all the β_j equal to 0. Then the formal series (2.14) reduces to the Taylor series expansion of $f(z)$ about the origin, that is, to the power series

$$f(z) = \sum_{n=0}^{\infty} \alpha_n z^n, \quad \alpha_n = \frac{f^{(n)}(0)}{n!} \quad (n \in \mathbb{Z}_{\geq 0}).$$

Returning to the tropical case, the corresponding formal series expansion to (2.14) under a formal ultra-discretization should be of the form

$$f(x) = \bigoplus_{n=0}^{\infty} \alpha_n \otimes (x \oplus \beta_1) \otimes (x \oplus \beta_2) \otimes \cdots \otimes (x \oplus \beta_n)$$

$$= \max_{n \in \mathbb{Z}_{\geq 0}} \left\{ \alpha_n + \max\{x, \beta_1\} + \max\{x, \beta_2\} + \cdots + \max\{x, \beta_n\} \right\},$$

where $\alpha_n, \beta_j \in \mathbb{R} \cup \{-\infty\}$ $(n, j \in \mathbb{Z}_{\geq 0})$. When all the β_j are equal to $0_\circ = -\infty$, it follows that $x \oplus \beta_j = x$ for every $j \in \mathbb{Z}_{\geq 0}$ and any $x \in \mathbb{R}$ and therefore

$$f(x) = \bigoplus_{n=0}^{\infty} \alpha_n \otimes x^{\otimes n} = \max_{n \in \mathbb{Z}_{\geq 0}} \{\alpha_n + nx\}.$$

Further, some of the coefficients α_n may also vanish in the tropical sense, that is, attain $0_\circ = -\infty$. Then, finally, we have the expression (2.6),

$$f(x) = \bigoplus_{k=0}^{\infty} c_k \otimes x^{\otimes s_k},$$

with the real coefficients c_k in the sense that $c_k \neq -\infty$ and the exponents s_k with the order according to size

$$s_0 < s_1 < \ldots < s_k < \ldots.$$

It could be natural to regard the expression (2.6) as the formal ultra-discretization of the Taylor series of an entire function. Then our main theorem can be also regarded as the ultra-discretization of the identities (C.6) and (C.7) presented in Appendix C. It is understood that the procedure of ultra-discretization repairs the gap of the appearance of logarithms in both cases, for example the logarithmic order of an entire function $f(z)$ and the order of a tropical entire function,

$$\limsup_{r \to \infty} \frac{\log \log M(r, f(z))}{\log \log r} \quad \text{and} \quad \limsup_{r \to \infty} \frac{\log M(r, f(x))}{\log r}.$$

Indeed, the statement of our theorem as well as its proof given below has been ultra-discretized formally in a straightforward way from the corresponding classical reasoning, giving answers to our problems (a) and (b) above. For example, our proofs in the subsequent two sections include the application of two simple estimates for a tropical entire function $f(x) = \bigoplus_{k=0}^{\infty} c_k \otimes x^{\otimes s_k}$ such as

$$\max_{|x| \leq r} |f(x)| = \max\{|f(-r)|, |f(r)|\} \quad \text{and} \quad M(r, f(x)) \geq c_k + s_k r \quad (k \in \mathbb{Z}_{\geq 0}),$$

but each of these can be considered as a sort of formal ultra-discretization of two important results in complex analysis, namely, the maximum principle and Cauchy estimates for an entire function $f(z) = \sum_{k=0}^{\infty} a_{n_k} z^{n_k}$,

$$\max_{|z| \leq r} |f(z)| = \max_{|z| = r} |f(z)| \quad \text{and} \quad M(r, f(z)) \geq |a_{n_k}| r^{n_k} \quad (k \in \mathbb{Z}_{\geq 0}),$$

respectively. Complete clarification of these connections clearly depends on further studies.

2.1.4   *Proof of Theorem 2.4, first part*

In this first part of the proof of Theorem 2.4, we start by concentrating on the following

Lemma 2.5. *We have*

$$\limsup_{r\to\infty} \frac{\log M(r,f)}{\log r} \geq \limsup_{k\to\infty} \frac{\log(-c_k)}{\log\{(-c_k)/s_k\}}$$

for a transcendental tropical entire function $f(x) = \bigoplus_{n=0}^\infty c_n \otimes x^{\otimes s_n}$. If the limit superior,

$$\rho := \limsup_{r\to\infty} \frac{\log M(r,f)}{\log r},$$

on the left is finite but strictly larger than 1, then we obtain

$$\limsup_{r\to\infty} \frac{M(r,f)}{r^\rho} \geq c(\rho) \limsup_{k\to\infty} \frac{-c_k}{\{(-c_k)/s_k\}^\rho}$$

with the constant $c(\rho)$ in (2.13).

Proof. If ρ is infinite, there is nothing to prove. Thus we now assume that ρ is finite and therefore for any given $\varepsilon > 0$, there exists an $r_0 > 0$ such that

$$\log M(r,f) \leq (\rho+\varepsilon)\log r$$

holds for all $r \geq r_0$. Here we should note that the number ρ is not smaller than 1, so that we may also assume that $\rho + \varepsilon > 1$. It follows that for any $k \in \mathbb{Z}_{\geq 0}$, and for all $r \geq r_0$,

$$\log(c_k + s_k r) \leq (\rho+\varepsilon)\log r, \quad \text{or } -c_k \geq s_k r - r^{\rho+\varepsilon},$$

is true. The function

$$\phi_k(r) := s_k r - r^{\rho+\varepsilon}$$

attains its maximum at $r = \{s_k/(\rho+\varepsilon)\}^{1/(\rho-1+\varepsilon)} =: r_*$, since

$$(d/dr)\phi_k(r) = s_k - (\rho+\varepsilon)r^{\rho-1+\varepsilon} = 0$$

if and only if $r^{\rho-1+\varepsilon} = s_k/(\rho+\varepsilon)$. The maximum is actually

$$\phi_k(r_*) = s_k\{s_k/(\rho+\varepsilon)\}^{1/(\rho-1+\varepsilon)} - \{s_k/(\rho+\varepsilon)\}^{(\rho+\varepsilon)/(\rho-1+\varepsilon)}$$
$$= s_k^{(\rho+\varepsilon)/(\rho-1+\varepsilon)} \cdot (\rho+\varepsilon)^{-(\rho+\varepsilon)/(\rho-1+\varepsilon)}(\rho-1+\varepsilon)$$

which is non-negative. Thus we have

$$-c_k \geq s_k^{(\rho+\varepsilon)/(\rho-1+\varepsilon)} \cdot (\rho-1+\varepsilon) \cdot (\rho+\varepsilon)^{-(\rho+\varepsilon)/(\rho-1+\varepsilon)}$$

for all $k \in \mathbb{Z}_{\geq 0}$, and therefore

$$\log(-c_k) \geq \frac{\rho + \varepsilon}{\rho - 1 + \varepsilon} \log s_k + \log(\rho - 1 + \varepsilon) - \frac{\rho + \varepsilon}{\rho - 1 + \varepsilon} \log(\rho + \varepsilon).$$

This implies

$$\left(1 - \frac{1}{\rho + \varepsilon}\right) \log(-c_k) \geq \log s_k + \frac{\rho - 1 + \varepsilon}{\rho + \varepsilon} \log(\rho - 1 + \varepsilon) - \log(\rho + \varepsilon)$$

$$= \log s_k + \frac{1}{\rho + \varepsilon} \log \left\{ \frac{(\rho - 1 + \varepsilon)^{\rho - 1 + \varepsilon}}{(\rho + \varepsilon)^{\rho + \varepsilon}} \right\}$$

so that

$$\log(-c_k) - \log s_k \geq \frac{1}{\rho + \varepsilon} \log(-c_k) + \frac{1}{\rho + \varepsilon} \log \left\{ \frac{(\rho - 1 + \varepsilon)^{\rho - 1 + \varepsilon}}{(\rho + \varepsilon)^{\rho + \varepsilon}} \right\}.$$

Hence,

$$\frac{\log\{(-c_k)/s_k\}}{\log(-c_k)} \geq \frac{1}{\rho + \varepsilon} + o(1)$$

as $k \to \infty$, since $-c_k \to +\infty$. This gives

$$\liminf_{k \to \infty} \frac{\log\{(-c_k)/s_k\}}{\log(-c_k)} \geq \frac{1}{\rho + \varepsilon} \quad \text{or} \quad \limsup_{k \to \infty} \frac{\log(-c_k)}{\log\{(-c_k)/s_k\}} \leq \rho + \varepsilon$$

for any $\varepsilon > 0$, so that we have

$$\limsup_{k \to \infty} \frac{\log(-c_k)}{\log\{(-c_k)/s_k\}} \leq \limsup_{r \to \infty} \frac{\log M(r, f)}{\log r}.$$

To complete the proof of the lemma, assume now that the order ρ of f satisfies $1 < \rho < +\infty$, and consider the two limit superiors

$$\limsup_{k \to \infty} \frac{-c_k}{\{(-c_k)/s_k\}^\rho} \quad \text{and} \quad \limsup_{r \to \infty} \frac{M(r, f)}{r^\rho}.$$

Moreover, suppose that the second one is finite and has the value λ, and thus for any $\varepsilon > 0$ there exists $r_0 > 0$ such that

$$c_k + s_k r \leq M(r, f) \leq (\lambda + \varepsilon) r^\rho$$

for any $r \geq r_0$ and for all $k \in \mathbb{Z}_{\geq 0}$. The function

$$\varphi_k(r) := (\lambda + \varepsilon) r^\rho - s_k r$$

of $r \geq r_0$ has its minimum at

$$r = \left(\frac{s_k}{(\lambda + \varepsilon)\rho} \right)^{1/(\rho - 1)} =: r_{**}$$

as above. Note that for a sufficiently large $k \in \mathbb{N}$, we have $r_{**} \geq r_0$. Therefore

$$-c_k \geq -\varphi_k(r_{**}) = s_k r_{**} - (\lambda + \varepsilon) r_{**}^{\rho}$$
$$= s_k^{\rho/(\rho-1)} (\lambda + \varepsilon)^{-1/(\rho-1)} \rho^{-\rho/(\rho-1)} (\rho - 1).$$

On the other hand, we have

$$\frac{-c_k}{\{(-c_k)/s_k\}^{\rho}} = (-c_k)^{1-\rho} s_k^{\rho} \leq (\lambda + \varepsilon) \frac{\rho^{\rho}}{(\rho - 1)^{\rho-1}}$$

for any $\varepsilon > 0$, so that

$$\frac{(\rho - 1)^{\rho-1}}{\rho^{\rho}} \limsup_{k \to \infty} \frac{-c_k}{\{(-c_k)/s_k\}^{\rho}} \leq \limsup_{r \to \infty} \frac{M(r, f)}{r^{\rho}}$$

as we wanted.

2.1.5   *Proof of Theorem 2.4, second part*

We now consider the opposite inequalities:

Lemma 2.6. *We have*

$$\limsup_{r \to \infty} \frac{\log M(r, f)}{\log r} \leq \limsup_{k \to \infty} \frac{\log(-c_k)}{\log\{(-c_k)/s_k\}}$$

for a transcendental tropical entire function $f(x) = \bigoplus_{n=0}^{\infty} c_n \otimes x^{\otimes s_n}$. *If the limit superior*

$$\rho := \limsup_{k \to \infty} \frac{\log(-c_k)}{\log\{(-c_k)/s_k\}},$$

on the right is finite but strictly larger than 1, then we obtain

$$\limsup_{r \to \infty} \frac{M(r, f)}{r^{\rho}} \leq c(\rho) \limsup_{k \to \infty} \frac{-c_k}{\{(-c_k)/s_k\}^{\rho}}$$

with the constant $c(\rho)$ *in* (2.13).

Proof. Assume that $1 < \rho < \infty$, since otherwise we have nothing to prove. Thus for any given $\varepsilon > 0$, there exists $k_0 \in \mathbb{N}$ such that

$$\frac{\log(-c_k)}{\log\{(-c_k)/s_k\}} < \rho + \varepsilon, \quad \text{or} \quad c_k < -s_k^{(\rho+\varepsilon)/(\rho-1+\varepsilon)},$$

holds for any integer $k \geq k_0$. Therefore, for any $r > 0$, we have

$$f(r) = M(r, f) = \max_{n \in \mathbb{Z}_{\geq 0}} \{c_n + s_n r\}$$
$$\leq O(r) + \max_{n \geq k_0} \{c_n + s_n r\}$$
$$= O(r) + \max_{n \geq k_0} \left\{ -s_n^{(\rho+\varepsilon)/(\rho-1+\varepsilon)} + s_n r \right\}.$$

Here we define the function $\phi_r(s)$ of s by

$$\phi_r(s) := -s^{(\rho+\varepsilon)/(\rho-1+\varepsilon)} + sr,$$

whose derivative is

$$(d/ds)\phi_r(s) = -\frac{\rho+\varepsilon}{\rho-1+\varepsilon}s^{1/(\rho-1+\varepsilon)} + r$$

and has the zero at

$$s = \left(\frac{\rho-1+\varepsilon}{\rho+\varepsilon}r\right)^{\rho-1+\varepsilon} =: s_* > 0.$$

Then $\phi_r(s)$ attains its maximum at

$$\phi_r(s_*) = -s_*^{(\rho+\varepsilon)/(\rho-1+\varepsilon)} + s_* r$$

$$= -\left(\frac{\rho-1+\varepsilon}{\rho+\varepsilon}r\right)^{\rho+\varepsilon} + \left(\frac{\rho-1+\varepsilon}{\rho+\varepsilon}r\right)^{\rho-1+\varepsilon}r$$

$$= \left(\frac{\rho-1+\varepsilon}{\rho+\varepsilon}\right)^{\rho-1+\varepsilon}\left(1 - \frac{\rho-1+\varepsilon}{\rho+\varepsilon}\right)r^{\rho+\varepsilon}$$

$$= \frac{(\rho-1+\varepsilon)^{\rho-1+\varepsilon}}{(\rho+\varepsilon)^{\rho+\varepsilon}}r^{\rho+\varepsilon}.$$

Thus we have

$$M(r,f) \leq \left\{\frac{(\rho-1+\varepsilon)^{\rho-1+\varepsilon}}{(\rho+\varepsilon)^{\rho+\varepsilon}} + o(1)\right\}r^{\rho+\varepsilon}$$

and therefore

$$\log M(r,f) \leq (\rho+\varepsilon)\log r + O(1)$$

as $r \to \infty$. This gives

$$\limsup_{r\to\infty}\frac{\log M(r,f)}{\log r} \leq \rho+\varepsilon$$

for any $\varepsilon > 0$, and thus our desired inequality

$$\limsup_{r\to\infty}\frac{\log M(r,f)}{\log r} \leq \limsup_{k\to\infty}\frac{\log(-c_k)}{\log\{(-c_k)/s_k\}}.$$

Next we are to prove that the second inequality

$$\limsup_{r\to\infty}\frac{M(r,f)}{r^\rho} \leq c(\rho)\limsup_{k\to\infty}\frac{-c_k}{\{(-c_k)/s_k\}^\rho}$$

with $c(\rho) = (\rho-1)^{\rho-1}/\rho^\rho$ is true, whenever $1 < \rho < \infty$ holds. For this purpose, let us put

$$\tau := \limsup_{k\to\infty}\frac{-c_k}{\{(-c_k)/s_k\}^\rho}$$

and assume $\tau < +\infty$, since the above inequality holds trivially otherwise. For an arbitrary given $\varepsilon > 0$, there exists $k_1 \in \mathbb{N}$ such that

$$\frac{-c_k}{\{(-c_k)/s_k\}^\rho} < \tau + \varepsilon, \quad \text{or} \quad c_k < -\frac{1}{(\tau + \varepsilon)^{1/(\rho-1)}} s_k^{\rho/(\rho-1)},$$

holds for any integer $k \geq k_1$. Therefore, for any $r > 0$, we have

$$M(r, f) = \max_{n \in \mathbb{Z}_{\geq 0}} \{c_n + s_n r\} \leq O(r) + \max_{n \geq k_1} \{c_n + s_n r\}$$

$$= O(r) + \max_{n \geq k_1} \left\{ -\frac{1}{(\tau + \varepsilon)^{1/(\rho-1)}} s_n^{\rho/(\rho-1)} + s_n r \right\}.$$

Here we define the function

$$\varphi_r(s) := -\frac{1}{(\tau + \varepsilon)^{1/(\rho-1)}} s^{\rho/(\rho-1)} + sr,$$

whose derivative is

$$(d/ds)\varphi_r(s) = -\frac{\rho}{(\rho - 1)(\tau + \varepsilon)^{1/(\rho-1)}} s^{1/(\rho-1)} + r.$$

This $d\varphi_r(s)/ds$ has the zero at

$$s = \left(\frac{\rho - 1}{\rho}\right)^{\rho-1} (\tau + \varepsilon) r^{\rho-1} =: s_{**} > 0,$$

where $\varphi_r(s)$ attains its maximum

$$\varphi_r(s_{**}) = -\frac{1}{(\tau + \varepsilon)^{\frac{1}{(\rho-1)}}} \left(\frac{\rho - 1}{\rho}\right)^\rho (\tau + \varepsilon)^{\frac{\rho}{(\rho-1)}} r^\rho + \left(\frac{\rho - 1}{\rho}\right)^{\rho-1} (\tau + \varepsilon) r^\rho$$

$$= (\tau + \varepsilon) \left(\frac{\rho - 1}{\rho}\right)^{\rho-1} \left(1 - \frac{\rho - 1}{\rho}\right) r^\rho = (\tau + \varepsilon) \frac{(\rho - 1)^{\rho-1}}{\rho^\rho} r^\rho.$$

Hence we have

$$M(r, f) \leq \left\{ (\tau + \varepsilon) \frac{(\rho - 1)^{\rho-1}}{\rho^\rho} + o(1) \right\} r^\rho$$

so that

$$\frac{M(r, f)}{r^\rho} \leq (\tau + \varepsilon) \frac{(\rho - 1)^{\rho-1}}{\rho^\rho} + o(1)$$

as $r \to \infty$. It follows therefore

$$\limsup_{r \to \infty} \frac{M(r, f)}{r^\rho} \leq (\lambda + \varepsilon) \frac{(\rho - 1)^{\rho-1}}{\rho^\rho}$$

for any $\varepsilon > 0$, so that we are done with the desired estimate

$$\limsup_{r \to \infty} \frac{M(r, f)}{r^\rho} \leq c(\rho) \limsup_{k \to \infty} \frac{-c_k}{\{(-c_k)/s_k\}^\rho}.$$

2.1.6 *Representations of tropical entire functions*

As we have seen in Chapter 1, there are many ways of representing a single tropical polynomial in the max-plus algebra setting. In this subsection, we proceed to point out all such notions treated in Chapter 1, as well as in Chapter 5 of the monograph [13] by Butkovič, and in the paper [106] by Tsai, are still available for tropical entire functions. To this end, we first need to consider series expansions of tropical entire functions in the compact form, and tropical conjugate entire functions related to the Legendre transformation.

To start with, consider a tropical entire function $f(x)$ of the form

$$f(x) = \max_{n \in \mathbb{Z}_{\geq 0}} \{s_n x + c_n\} = \bigoplus_{n \in \mathbb{Z}_{\geq 0}} c_n \otimes x^{\otimes s_n} \tag{2.15}$$

on $\mathbb{R}$, where the sequences $\{s_n\}$, $\{c_n\}$ of real numbers satisfy

- $0 < s_1 < s_2 < \cdots < s_n < \cdots$,

and

- $c_1 > c_2 > \cdots > c_n > \cdots (\to -\infty)$.

Then we put

$$r_j := \frac{c_{j-1} - c_j}{s_j - s_{j-1}} \quad (j \in \mathbb{N}) \tag{2.16}$$

with $c_0 = s_0 = 0$. Assume that each term $c_n \otimes x^{\otimes s_n}$ is essential in the sense by Butkovič [13], that is, there exists an $x_0 \in \mathbb{R}$ such that

$$c_n \otimes x_0^{\otimes s_n} = f(x_0) > c_j \otimes x^{\otimes s_j}$$

for any $j \neq n$. Then we have

$$0 < r_1 < r_2 < \cdots < r_j < \cdots (\to +\infty).$$

In fact, for each $j \in \mathbb{N}$, there exists $\alpha > 0$ such that

$$c_j + s_j \alpha > c_{j-1} + s_{j-1}\alpha \quad \Leftrightarrow \quad \alpha > \frac{c_{j-1} - c_j}{s_j - s_{j-1}} = r_j,$$

and

$$c_j + s_j \alpha > c_{j+1} + s_{j+1}\alpha \quad \Leftrightarrow \quad \alpha < \frac{c_j - c_{j+1}}{s_{j+1} - s_j} = r_{j+1}.$$

This sequence $\{r_j\}_{j \in \mathbb{N}}$ cannot have finite accumulation points as they are the corner points of the graph of $f(x)$ or, equivalently, the roots of the tropical transcendental entire function, so the sequence must tend to $+\infty$. Putting $r_0 = 0_\circ = -\infty$, we see that

$$f(x) = s_j x + c_j = c_j \otimes x^{\otimes s_j}$$

when and only when $x \in [r_j, r_{j+1}]$. This shows that our $f(x)$ in (2.15) is maximally represented in the sense by Tsai [106]. Moreover, the expression (2.15) is then in the compact form as defined in Chapter 1. Note that for our discussions, it is necessary that the max-plus series expansion should include only the essential terms so that they indeed contribute to the expression.

We next consider a function $f^\bullet(p)$ defined by

$$f^\bullet(p) := -\inf_{x \in \mathbb{R}}\{f(x) - px\} = \sup_{x \in \mathbb{R}}\{px - f(x)\}, \qquad (2.17)$$

which is known as the Legendre (or Legendre-Fenchel) transformation of the function $f(x)$, see e.g. [95]. It seems natural here that there might be some possible relations between $f(x)$ and $f^\bullet(p)$, when both of them are tropical entire functions. By definition, for $s_{j-1} < p < s_j$, the function $f(x) - px$ of the variable x attains its minimum, when and only when $x = r_j$, and therefore

$$f^\bullet(p) = -\{f(r_j) - pr_j\} = pr_j - f(r_j)$$
$$= pr_j - (s_j r_j + c_j) = (p - s_j)r_j - c_j.$$

When $p = s_{j-1}$, the minimum of $f(x) - px$ is attained for $x \in [r_{j-1}, r_j]$ and the value is $(s_{j-1}x + c_{j-1}) - s_{j-1}x = c_{j-1}$ and thus $f^\bullet(s_{j-1}) = -c_{j-1}$. Similarly, we have also $f^\bullet(s_j) = -c_j$. Note that if $p = s_0 = 0$,

$$\inf_{x \in \mathbb{R}}\{f(x) - px\} = \inf_{x \in \mathbb{R}} f(x) = s_0 x + c_0$$

when $x \le r_1 = 0$. For the sake of convenience, we define

$$f^\bullet(p) \equiv 0 \quad \text{when} \quad p \le 0.$$

Then we have

$$f^\bullet(p) = (p - s_j)r_j - c_j = r_j p - (s_j r_j + c_j) = -f(r_j) \otimes p^{\otimes r_j}$$

when and only when $s_{j-1} \le p \le s_j$ $(j \in \mathbb{N})$. Further note that

$$f^\bullet(s_{j-1}) = (s_{j-1} - s_j)r_j - c_j = -(c_{j-1} - c_j) - c_j = -c_{j-1}$$

by (2.17), while for $p < s_0 = 0$, $f^\bullet(p) = f^\bullet(s_0) = 0$. Hence

$$f^\bullet(p) = \max_{j \in \mathbb{Z}_{\ge 0}}\{r_j p - f(r_j)\} = \bigoplus_{j \in \mathbb{Z}_{\ge 0}} (-f(r_j)) \otimes p^{\otimes r_j}. \qquad (2.18)$$

Recall that

$$0 < r_1 < r_2 < \cdots < r_j < \cdots$$

and

$$-f(r_0) > -f(r_1) > \cdots > -f(r_j) > \cdots (\to \infty).$$

Hence $f^{\bullet}(p)$ is a tropical entire function on $\mathbb{R}$, since it is a convex function as the convex conjugate function of $f(x)$. Now we call this Legendre transform $f^{\bullet}(p)$ of our tropical entire function $f(x)$ as the **tropical conjugate entire function** of $f(x)$.

Let us next consider the representations

$$\rho = \limsup_{r \to \infty} \frac{\log M(r, f)}{\log r} = \limsup_{n \to \infty} \frac{\log(-c_n)}{\log(-c_n/n)} \tag{2.19}$$

and when $0 < \rho < \infty$,

$$\tau = \limsup_{r \to \infty} \frac{M(r, f)}{r^{\rho}} = \frac{1}{e(\rho)\rho} \limsup_{n \to \infty} \frac{-c_n}{(-c_n/n)^{\rho}} \tag{2.20}$$

for the order ρ and the type τ of f with the series expansion (2.15).

We recall that

$$M(s_j, f^{\bullet}) = f^{\bullet}(s_j) = -c_j$$

holds for $j \in \mathbb{N}$ and observe that

$$\limsup_{j \to \infty} \frac{\log(-c_j)}{\log(-c_j/j)} = \frac{1}{\liminf_{j \to \infty} \left(1 - \frac{\log j}{\log(-c_j)}\right)}$$

$$= \frac{1}{1 - \limsup_{j \to \infty} \frac{\log j}{\log(-c_j)}} = \frac{1}{1 - \frac{1}{\liminf_{j \to \infty} \frac{\log(-c_j)}{\log j}}}.$$

Then instead of studying the growth of the tropical entire function $f^{\bullet}(p)$ itself, we study the growth of its discretization

$$\gamma(j) = f^{\bullet}(s_j) \quad (j \in \mathbb{Z}_{\geq 0})$$

by defining the following growth indices:

$$\mu := \liminf_{j \to \infty} \frac{\log \gamma(j)}{\log j}$$

and

$$\sigma := \liminf_{j \to \infty} \frac{\gamma(j)}{j^{\mu}},$$

when $0 < \mu < \infty$. Then by (2.19) we have

$$\rho = \frac{1}{1 - 1/\mu}, \quad \text{or} \quad \frac{1}{\rho} + \frac{1}{\mu} = 1,$$

and thus $\mu = \frac{\rho}{\rho - 1} \in (1, \infty)$ whenever $\rho \in (1, \infty)$. Also by (2.20) we have for $\rho \in (1, \infty)$

$$\begin{aligned}
\tau &= \frac{1}{e(\rho)\rho} \limsup_{j \to \infty} \frac{-c_j}{(-c_j/j)^\rho} \\
&= \left(\frac{\rho - 1}{\rho}\right)^{\rho - 1} \frac{1}{\rho} \limsup_{j \to \infty} \frac{j^\rho}{(-c_j)^{\rho - 1}} \\
&= \left(\frac{\rho - 1}{\rho}\right)^{\rho - 1} \frac{1}{\rho} \left(\frac{1}{\liminf_{j \to \infty} \frac{-c_j}{j^{\rho/(\rho - 1)}}}\right)^{\rho - 1},
\end{aligned}$$

that is, for $\mu \in (1, \infty)$

$$\tau^{1/\rho} = \frac{1}{\mu^{1/\mu}} \cdot \frac{1}{\rho^{1/\rho}} \cdot \sigma^{1/\mu},$$

or

$$\left(\frac{1}{\tau\rho}\right)^{1/\rho} \cdot \left(\frac{1}{\sigma\mu}\right)^{1/\mu} = 1.$$

Hence we have dual relations between the order ρ and the type τ of $f(x)$ and the lower order and lower type σ of the discrete function $\gamma(j) = -c_j$ $(j \in \mathbb{Z}_{\geq 0})$ such as

$$1/\rho + 1/\mu = 1 \quad \text{and} \quad (\tau\rho)^{1/\rho} \cdot (\sigma\mu)^{1/\mu} = 1.$$

Observe that this reasoning about the growth of the discretization of the Legendre transformation $f^\bullet$ of a tropical entire function follows the reasoning presented in [106] for tropical polynomials. It seems natural that we can observe there these relations between two tropical entire functions $f(x)$ and $f^\bullet(p)$, since the convex function $f(x)$ in (2.15) is defined by the family of the linear lines $y = s_n x + c_n$ and the graph could be reproduced by the sequence of their y-intercepts c_n, which coincides with the value $-f^\bullet(s_n)$ $(n \in \mathbb{Z}_{\geq 0})$.

Tsai [106] considers the Legendre transformation $\mathcal{L}_h : \mathbb{R} \to \mathbb{R}$ of a convex function $h : \mathbb{R} \to \mathbb{R}$ and remarks that if $h(x)$ is a tropical polynomial or tropical entire function on $\mathbb{R}$, then $\mathcal{L}_h$ is a partially defined function. Namely, for the tropical polynomial $h(x) = x \oplus 2 = \max\{x, 2\}$, the value of $\mathcal{L}_h(2)$ will be infinite, since

$$\mathcal{L}_h(p) = \max_{x \in \mathbb{R}}\{px - h(x)\} = \sup_{x \in \mathbb{R}}\{px - \max\{x, 2\}\}$$

$$= \sup_{x \in \mathbb{R}} \begin{cases} px - h(x) \ (x \le 2), \\ (p-1)x \ (x > 2) \end{cases}$$

$$= \sup\left(2(p-1), \ \sup_{x>2}(p-1)x\right)$$

will be infinite if $p > 1$, 0 if $p = 1$, and $2(p-1)$ if $p < 1$ respectively, so that the so-called **effective domain** of this $\mathcal{L}_h$, given by the set $\{p \in \mathbb{R} \mid \mathcal{L}_h(p) < \infty\}$, is restricted to the interval $[1, \infty)$. Then he modified the definition of Legendre transform $\mathcal{L}_h$ as in [106, Definition 3.2], that is, its domain of definition is restricted into the set of slopes of the function $h(x) = \bigoplus_{n \in \mathbb{Z}_{\ge 0}}(c_n \otimes x^{\otimes s_n})$. Concretely, $\mathcal{L}_h$ is restricted to the sequence $\{s_n\}_{n \in \mathbb{Z}_{\ge 0}}$ and therefore

$$\mathcal{L}_h(s_n) = \sup_{x \in \mathbb{R}}\{s_n x - h(x)\} \quad (n \in \mathbb{Z}_{\ge 0}).$$

For example, if $h(x)$ is a tropical monomial

$$x \oplus 2 = 1 \otimes x^{\otimes 1} \oplus 2 \otimes x^{\otimes 0}$$

as above,

$$\mathcal{L}_h(1) = \sup_{x \in \mathbb{R}}\{x - \max\{x, 2\}\}$$

$$= \sup\left\{0, \max_{x<2}\{x - 2\}\right\} = 0,$$

$$\mathcal{L}_h(0) = \sup_{x \in \mathbb{R}}\{0 - \max\{x, 2\}\}$$

$$= \sup\left\{\max_{x \ge 2}\{-x\}, -2\right\} = -2.$$

Returning to our case when $h(x)$ is $f(x)$ given by (2.15), we have followed the idea of Tsai to consider $\mathcal{L}_h(s_j) = f^\bullet(s_j)$ which was denoted by $\gamma(j)$ as the discretization of $f^\bullet(p)$.

Butkovič [13, Chapter 5] also studies tropical polynomials of the form

$$p(x) = \bigoplus_{r=0}^{p} c_r \otimes x^{\otimes j_r} = \max_{r=0,\dots,p}\{c_r + j_r x\},$$

where $c_r, j_r \in \mathbb{R}$ and $p \in \mathbb{Z}_{\ge 0}$, which he calls max-algebraic polynomials and abbreviates them as max-polynomials. In his notation, the exponents j_r, $j_0 < j_1 < \cdots < j_p$, are not restricted to integers. He calls a max-polynomial $p(x)$ a standard max-polynomial if $c_0 = 0 = j_0$, since every max-polynomial can be written as a tropical multiplication of a tropical monomial $c \otimes x^{\otimes j}$ and a standard max-polynomial. Although this terminology is, of course, usable for tropical entire functions by taking $p = +\infty$, such a simple expression by means of a single tropical monomial does not hold, since we are to study Laurent-type infinite series expressions of type

$$\bigoplus_{r=-\infty}^{+\infty} c_r \otimes x^{\otimes j_r} = \max_{-\infty < r < +\infty} \{c_r + j_r x\}.$$

This is then a product of $q_-(x) \otimes q_+(x)$ with standard tropical entire functions $q_+(x)$ and $q_-(-x)$. In spite of this, it is natural for us to assume that our tropical entire functions are of a **standard form**:

$$\bigoplus_{r=0}^{+\infty} c_r \otimes x^{\otimes j_r} = \max_{0 \le r < +\infty} \{c_r + j_r x\},$$

$j_0 < j_1 < \ldots < j_r < \ldots \to +\infty.$

As pointed out in Chapter 1 and in [106], there is still a similar ambiguity in expressions for tropical entire functions as is the case for tropical polynomials. Tsai introduced [106] the notion of maximal representation for tropical polynomials. His key idea was to apply the (modified) Legendre transformation $\mathcal{L}_f$ of a convex function $f(x)$ on $\mathbb{R}$ defined by

$$\mathcal{L}_f(p) = \max_{x \in \mathbb{R}} \{px - f(x)\}.$$

This idea works for tropical entire functions as well, since tropical entire functions are convex on $\mathbb{R}$. Following Definition 3.3 in [106], we say that a tropical entire function

$$f(x) = \bigoplus_{r=0}^{+\infty} c_r \otimes x^{\otimes j_r} \qquad (2.21)$$

is **maximally represented** if

$$c_r = -\mathcal{L}_f(j_r),$$

for all $r \in \mathbb{Z}_{\ge 0}$. Then the observations in [106, Theorem 3.6, Corollaries 3.7 and 3.8] remain true in the case of tropical entire functions as well:

- Every tropical entire function has a unique maximal representation.
- Two tropical entire functions have the same graph if and only if their maximally represented polynomials are the same.

- An expression (2.21) of a tropical entire function $f(x)$ is maximally represented if and only if every term $c_r \otimes x^{\otimes j_r}$ $(r \in \mathbb{Z}_{\geq 0})$ in the expression contributes to the function $f(x)$ in the sense that the value of the term $c_r \otimes x_0^{\otimes j_r}$ at $x = x_0$ attains the same value $f(x_0)$ of the tropical entire function, that is, it contributes to the maximum

$$\max_{r \in \mathbb{Z}_{\geq 0}} \{c_r + j_r x_0\}.$$

This gives a geometric way to make the representation (2.21) maximal according to the following two cases:

(i) When a piece of the line $j_{r_0} x + c_{r_0}$ appears in the graph of $f(x)$, then one can find that the term is necessary for the maximal representation.

(ii) If the line is completely below the graph of $f(x)$, one needs to move the line $j_{r_0} x + c_{r_0}$ up to intersect the graph of $f(x)$ at exactly one point. The new value α for c_{r_0} is actually obtained as $-\mathcal{L}_f(j_{r_0})$ from the graph of $f(x)$ and the touching point is indeed the intersection of two lines defined by certain two other terms $j_{r_-} x + c_{r_-}$ and $j_{r_+} x + c_{r_+}$ with $j_{r_-} < j_{r_0} < j_{r_+}$. In fact, we have

$$\alpha = \max \left\{ \frac{j_{r_+} - j_{r_0}}{j_{r_+} - j_{r_-}} (c_{r_-} - c_{r_+}) + c_{r_+} \right\},$$

where the maximum runs over all distinct r_- and r_+ in $\mathbb{Z}_{\geq 0} \setminus \{r_0\}$.

On the other hand, similarly as in [106, Definition 5.1] to tropical polynomials, one can find the compact form of a given tropical entire function by omitting such terms in (2.21) that have no contribution to the component of the graph of $f(x)$. In fact, let us say that a tropical entire function $f(x)$ given by (2.21) is in the **compact form** if $f(x)$ and any tropical entire function $g(x)$ obtained as $f(x)$ replaced its term $c_{r_*} \otimes x^{\otimes j_{r_*}}$ $(r_* \in \mathbb{Z}_{\geq 0})$ by $\alpha \otimes x^{\otimes j_{r_*}}$ with $\alpha < c_{r_*}$ are different functions. Intuitively, the maximal (and standard) representation of a tropical entire function $f(x)$ is the compact form when all the terms contribute to the graph of $f(x)$. For this purpose, one should have deleted those terms $c_r \otimes x^{\otimes j_r}$ completely from (2.21) if the line $j_r x + c_r$ remains below of the graph of $f(x)$ except possible for one point where it touches the graph.

Returning to Butkovič's observations in [13], we can say also that a term $c_s \otimes x^{\otimes j_s}$ of a tropical entire function $f(x)$ in (2.21) is **inessential** or **essential** according to the case whether $c_s \otimes x^{\otimes j_s} \leq f(x)$ holds for every $x \in$

$\mathbb{R}$ or not. We obtain the compact form of $f(x)$ by removing all inessential terms from the representation. Then an essential term can be characterized as follows, see Lemma 5.0.1 in [13].

Proposition 2.7 ([13, Lemma 5.0.1]). *If a term $c_s \otimes x^{\otimes j_s}$, $s \in \mathbb{Z}_{\geq 0}$, is essential in the max-plus series*

$$\bigoplus_{r=0}^{\infty} c_r \otimes x^{\otimes j_r}, \quad j_0 < j_1 < \ldots < j_r < \ldots,$$

then

$$\frac{c_s - c_{s+1}}{j_{s+1} - j_s} > \frac{c_{s-1} - c_s}{j_s - j_{s-1}}.$$

Proof. Since the term $c_s \otimes x^{\otimes j_s}$ is essential and the sequence $\{j_r\}_{r=0}^{\infty}$ is increasing, there is an $\alpha \in \mathbb{R}$ such that

$$c_s + j_s \alpha > c_{s-1} + j_{s-1} \alpha$$

and

$$c_s + j_s \alpha > c_{s+1} + j_{s+1} \alpha.$$

Hence

$$\frac{c_s - c_{s+1}}{j_{s+1} - j_s} > \alpha > \frac{c_{s-1} - c_s}{j_s - j_{s-1}}.$$

That is,

$$\bigoplus_{r=0}^{\infty} c_r \otimes x^{\otimes j_r} = c_s \otimes x^{\otimes j_s}$$

when and only when

$$x \in \left[\frac{c_{s-1} - c_s}{j_s - j_{s-1}}, \frac{c_s - c_{s+1}}{j_{s+1} - j_s} \right].$$

The corner points of the graph of the max-plus infinite series

$$f(x) = \bigoplus_{r=0}^{\infty} c_r \otimes x^{j_r} = \max_{r=0}^{\infty} \{c_r + j_r x\}$$

are

$$\left\{ \frac{c_r - c_{r+1}}{j_{r+1} - j_r} \right\}_{r=0}^{\infty},$$

which form an increasing sequence. $\qquad\square$

Remark 2.8. With the same notation, $f(x)$ has also a max-plus infinite product form

$$f(x) = \bigotimes_{r=0}^{\infty} \left\{ x \oplus \left(\frac{c_r - c_{r+1}}{j_{r+1} - j_r} \right) \right\}^{\otimes (j_{r+1} - j_r)}$$

$$= \sum_{r=0}^{\infty} \max\{(j_{r+1} - j_r)x, c_r - c_{r+1}\}.$$

This is parallel to the case of tropical polynomials, see Lemma 1.6 in Chapter 1.

2.2 Examples of tropical entire functions

This section is devoted to giving some non-trivial examples of tropical entire functions, adding to our previous examples given by tropical exponential functions.

2.2.1 *Tropical entire functions of arbitrary order*

To construct tropical entire functions of arbitrary order, we first recall three entire functions on $\mathbb{C}$ given in [9, Examples 5.2] with q-series expansion, obtaining the anticipated tropical functions by a formal ultra-discretization. This simply means that we convert the following three q-series into our desired three max-plus series by a formal replacement determined as below.

Example 2.9 ([9]). (1) Let q be a complex number with $0 < |q| < 1$. For any fixed number $\alpha > 1$,

$$f_\alpha(z) = \sum_{n=0}^{\infty} q^{n^{\alpha/(\alpha-1)}} z^n$$

is an entire function which has the logarithmic order

$$\limsup_{r \to \infty} \frac{\log \log M(r, f)}{\log \log r} = \alpha$$

and logarithmic type

$$\limsup_{r \to \infty} \frac{\log M(r, f)}{(\log r)^\alpha} = \frac{(\alpha - 1)^{\alpha-1}}{\alpha^\alpha} \cdot \frac{1}{(\log(1/|q|))^{\alpha-1}}.$$

(2) The entire function

$$f_1(z) = \sum_{n=0}^{\infty} q^{e^n} z^n$$

is of logarithmic order 1 and of infinite logarithmic type.

(3) The entire function

$$f_\infty(z) = \sum_{n=1}^{\infty} q^{n(\log n)^2} z^n$$

is of infinite logarithmic order.

Each of these entire functions can now be formally ultra-discretized respectively as follows:

Example 2.10. Let Q be a positive real number, which is to correspond to the number $\log(1/|q|)$ for the above $q \in \mathbb{C}$ with $0 < |q| < 1$.

(1) For any fixed number $\alpha > 1$,

$$f_\alpha(x) = \bigoplus_{n=0}^{\infty} \left\{ (-Q)^{\otimes n^{\alpha/(\alpha-1)}} \otimes x^{\otimes n} \right\} = \max_{n \in \mathbb{Z}_{\geq 0}} \left\{ -n^{\alpha/(\alpha-1)} Q + nx \right\}$$

is a tropical entire function which has the order α and type

$$\frac{(\alpha-1)^{\alpha-1}}{\alpha^\alpha} \cdot \frac{1}{Q^{\alpha-1}}.$$

In fact, putting $c_n := -n^{\alpha/(\alpha-1)} Q$, which tends to $-\infty$ as $n \to \infty$, we have

$$\limsup_{n \to \infty} \frac{\log(-c_n)}{\log\{(-c_n)/n\}} = \frac{\alpha/(\alpha-1)}{1/(\alpha-1)} = \alpha$$

and

$$\limsup_{n \to \infty} \frac{-c_n}{\{(-c_n)/n\}^\alpha} = \lim_{n \to \infty} \frac{n^{\alpha/(\alpha-1)} Q}{\{n^{\alpha/(\alpha-1)-1} Q\}^\alpha} = \frac{1}{Q^{\alpha-1}}.$$

(2) The tropical entire function

$$f_1(x) = \bigoplus_{n=0}^{\infty} \left\{ (-Q)^{\otimes e^n} \otimes x^{\otimes n} \right\} = \max_{n \in \mathbb{Z}_{\geq 0}} \left\{ -e^n Q + nx \right\}$$

is of order 1 and infinite type. In fact, putting $c_n := -e^n Q$ which tends to $-\infty$ as $n \to \infty$, we have

$$\limsup_{n \to \infty} \frac{\log(-c_n)}{\log\{(-c_n)/n\}} = 1.$$

Further, a direct estimate concludes that the type of $f_1(x)$ is infinite. Indeed, by definition, we obtain

$$\limsup_{r \to \infty} \frac{M(r, f_1)}{r} \geq n$$

for any positive integer n, since when $r > 0$,

$$M(r, f_1) \geq c_n + nr$$

is always true for an arbitrary fixed n. In addition,

$$\limsup_{n \to \infty} \frac{-c_n}{(-c_n)/n} = \lim_{n \to \infty} \frac{e^n Q}{e^n Q/n} = \lim_{n \to \infty} n = +\infty.$$

(3) The tropical entire function

$$f_\infty(x) = \bigoplus_{n=1}^{\infty} \left\{ (-Q)^{\otimes n(\log n)^2} \otimes x^{\otimes n} \right\} = \max_{n \in \mathbb{N}} \left\{ -n(\log n)^2 Q + nx \right\}$$

has infinite order. In fact, with $c_n := -n(\log n)^2 Q$ tending to $-\infty$ as $n \to \infty$, two identities

$$\log(-c_n) = \log n + 2 \log \log n + \log Q \quad \text{and} \quad \log\{(-c_n)/n\} = 2 \log \log n + \log Q$$

imply our order estimate

$$\limsup_{n \to \infty} \frac{\log(-c_n)}{\log\{(-c_n)/n\}} = \lim_{n \to \infty} \frac{\log n}{2 \log \log n} = +\infty.$$

2.2.2 *A q-analogue of the exponential function*

Ismail [53, p. 356–357] mentions that a q-analogue of the exponential function

$$E^{(\alpha)}(z; q) := \sum_{n=0}^{\infty} \frac{z^n}{(q; q)_n} q^{\alpha n^2}, \quad 0 < \alpha, \ 0 < q < 1, \tag{2.22}$$

due to Atakishiyev, is an entire function satisfying

$$\lim_{r \to \infty} \frac{\log M\left(r, E^{(\alpha)}(\cdot, q)\right)}{(\log r)^2} = \frac{1}{4\alpha \log q^{-1}}. \tag{2.23}$$

Here, as in [53], we denote

$$(q; q)_n = \prod_{k=1}^{n} (1 - q^k), \quad n = 1, 2, \ldots.$$

As one may easily see by a direct computation, the entire function $E^{(\alpha)}(z; q)$ is of logarithmic order two and of logarithmic type $\sigma_q = 1/(4\alpha \log q^{-1})$.

As a formal ultra-discrete analogue of $E^{(\alpha)}(z;q)$ we obtain

$$uE^{(A)}(x;Q) := \bigoplus_{n=0}^{\infty} x^{\otimes n} \otimes (-Q)^{\otimes(An^2)} \oslash [-Q;-Q]_n$$

$$= \max_{n \in \mathbb{Z}_{\geq 0}} \left\{ -n^2 AQ + nx \right\}, \quad A > 0, \ Q > 0.$$

Here, as an ultra-discrete analogue of $(q;q)_n$, we denote

$$[Q;Q]_n = \bigotimes_{k=1}^{n} (1_\circ \oplus Q^{\otimes k}) = \sum_{k=1}^{n} kQ = \frac{n(n+1)}{2} Q, \quad n = 1, 2, \ldots,$$

and

$$[-Q;-Q]_n = \bigotimes_{k=1}^{n} (1_\circ \oplus (-Q)^{\otimes k}) = \sum_{k=1}^{n} 1_\circ = 1_\circ = 0, \quad n = 1, 2, \ldots,$$

respectively, for $Q > 0$.

The function $uE^{(A)}(x;Q)$ is a transcendental tropical entire function by Definition 2.2. Moreover, the function $uE^{(A)}(x;Q)$ is of order 2 and of type $1/(4AQ)$, which is a result to be obtained by applying Theorem 2.4. Now we see such a natural correspondence as $Q = \log q^{-1}$ concerning the parameters of the two functions $E^{(\alpha)}(z;q)$ and $uE^{(A)}(x;Q)$.

In [53], Section 21.1 and on p. 533, Ismail notes that many of the examples studied in the chapter are entire functions f having the property

$$M(r,f) = \exp\big(c(\log r)^2\big), \tag{2.24}$$

and proposes to define q-**order** ρ_q of f by

$$\rho_q = \limsup_{r \to \infty} \frac{\log \log M(r,f)}{\log \log r}, \tag{2.25}$$

and its q-**type** σ_q by

$$\sigma_q = \inf\Big\{ K \ : \ M(r,f) < \exp\big(K(\log r)^\rho\big) \Big\}, \tag{2.26}$$

if the q-order ρ_q of f is finite. It follows from (2.23) that the entire function $E^{(\alpha)}(z;q)$ is of q-order $\rho_q = 2$ and q-type $\sigma_q = 1/(4\alpha \log q^{-1})$. This should be compared with the fact that the tropical entire function $uE^{(A)}(x;Q)$ is of order $\rho = 2$ and of type $1/(4AQ)$ due to our Theorem 2.4. Therefore, the translation of q into $\log q^{-1}$ might suit the names 'logarithmic order/type' and q-order/q-type. The reason is that the former is about differences and the latter is about q-differences. In fact, they are transformed to each other with the translation $z = q^t$, or $t = \log_q z$. We may also note that

$$x \otimes y = \log_q(q^x \times q^y), \quad |x \oplus y - \log_q(q^x + q^y)| \leq \log_q 2, \quad x^{\otimes y} = \log_q\{(q^x)^y\},$$

respectively.

Following the procedure of formal ultra-discretization that we have used in Section 2.2.1 in order to transform three entire functions in Example 2.9 into three tropical entire functions in Example 2.10, we may observe Nobe's ultra-discrete theta function $\Theta_0(x)$ studied below in Chapter 6 from the q-difference view point. The θ-function

$$\theta(z, q) = \sum_{n=-\infty}^{+\infty} q^{n^2} z^n$$

is defined for $|q| < 1$ and $z \in \mathbb{C} \setminus \{0\}$. Let us consider the following tropical series expression,

$$\bigoplus_{n=-\infty}^{+\infty} (-Q)^{\otimes n^2} \otimes x^{\otimes n} = \max_{-\infty < n < +\infty} \left\{ -n^2 Q + nx \right\}$$

for $Q > 0$. This is essentially the same function as $\Theta_0(x) + x^2$ that Nobe has derived. We should remark here that Ormerod has obtained a basic ultra-discrete hypergeometric function as the application of the valuation to a basic hypergeometric function by applying the lift $x \to t^{-x}$. These two procedures might be based on the same root of thoughts.

2.2.3 A tropical entire function related to q-difference equations

First recall a familiar example of q-difference equations in the plane $\mathbb{C}$. Consider a transcendental entire function

$$f(z) = \prod_{j=0}^{\infty} (1 - q^j z) \tag{2.27}$$

for a constant $q \in \mathbb{C}$ with $0 < |q| < 1$. This function, known as the q-shifted factorial $(z; q)_\infty$, see [27] or [53], satisfies the q-difference equation

$$(1 - z)f(qz) = f(z) \quad (z \in \mathbb{C}) \tag{2.28}$$

and has the growth property (see Bergweiler, Ishizaki and Yanagihara [7,8])

$$\log M(r, f) = -\frac{1}{2 \log |q|} (\log r)^2 (1 + o(1)) \tag{2.29}$$

as $r \to \infty$. On the other hand, for the Taylor series expansion

$$f(z) = \sum_{n=0}^{\infty} a_n z^n, \tag{2.30}$$

we have the recurrence formula

$$(1 - q^n)a_n = -q^{n-1}a_{n-1}, \qquad n \geq 1, \quad a_0 = 1, \tag{2.31}$$

concerning the coefficients a_n in the series (2.30), and therefore

$$a_n = q^{n(n-1)/2} \bigg/ \prod_{k=1}^{n} (q^k - 1), \quad n \geq 1. \tag{2.32}$$

We now proceed to look at a formal ultra-discretization of this example. Again, this means that we simply observe the following formal max-plus infinite product as the tropical counterpart of the function in (2.27):

$$f(x) = \bigotimes_{j=0}^{\infty} \left(1_\circ \oplus (-q)^{\otimes j} \otimes x \right) = \sum_{j=0}^{\infty} \max\{0, x - jq\} \tag{2.33}$$

for a real constant $q > 0$ which is equivalent to $0_\circ < -q = 1_\circ \oslash q < 1_\circ$. Note that for any fixed value of x the tropical product in (2.33) is finite, since

$$f(x) = \sum_{j=0}^{[x/q]} (x - jq),$$

where the **Gaussian** symbol $[r]$ stands for the largest integer $\leq r$. It is not difficult to see that $f(x)$ is a normalized transcendental tropical entire function in the sense of Definition 2.2 and the set of its roots is exactly the sequence $\{jq = q^{\otimes j}\}_{j=0}^{\infty}$. We may also take the sum $\sum_{j=1}^{\infty}$ instead of $\sum_{j=0}^{\infty}$, that is, to take $f(x - q)$ instead of $f(x)$, in order to make our $f(x)$ satisfy the assumption $[N0]$ as well as $[N1]$ and $[N2]$. This function $f(x)$, and of course $g(x) := f(x - q)$ as well, satisfy the ultra-discrete equation

$$(1_\circ \oplus x) \otimes f(x - q) = f(x) \quad (x \in \mathbb{R}), \tag{2.34}$$

which is an ultra-discrete counterpart of the q-difference equation (2.28). In fact,

$$f(x - q) = \sum_{j=0}^{\infty} \max\{0, (x - q) - jq\} = \sum_{j=0}^{\infty} \max\{0, x - (j + 1)q\}$$

$$= \sum_{j=0}^{[x/q]-1} \max\{0, x - (j + 1)q\} = \sum_{j=0}^{[x/q]} \max\{0, x - jq\} - \max\{0, x\}$$

$$= f(x) - \max\{0, x\}.$$

For the purpose of obtaining the ultra-discrete counterpart of the Taylor expansion (2.30), we first note that there exists an integer m such as $mq \leq x < (m + 1)q$, that is, $m = [x/q]$, and when $m \leq -1$,

$$x - jq < (m - j + 1)q < 0$$

for $j = 0, 1, 2, \ldots$, we have $f(x) = 0$. When $m \in \mathbb{Z}_{\geq 0}$,

$$(m - j)q \leq x - jq < (m - j + 1)q$$

for $j = 0, 1, 2, \ldots$, we have

$$\max\{0, x - jq\} = \begin{cases} x - jq & (0 \leq j \leq m), \\ 0 & (m + 1 \leq j < +\infty), \end{cases}$$

so that we obtain

$$f(x) = \sum_{j=0}^{m} (x - jq) = (m + 1)x - \frac{m(m + 1)}{2}q$$

$$= \max\left\{0, x, 2x - 3q, \ldots, (m + 1)x - \frac{m(m + 1)}{2}q, \ldots\right\}$$

$$= \bigoplus_{n=0}^{\infty} (-q)^{\otimes \frac{(n-1)n}{2}} \otimes x^{\otimes n},$$

for any $x \in \mathbb{R}$. In fact, for $x \in [mq, (m + 1)q)$, that is, for $m = [x/q]$, we have

$$(m + 1)x - \frac{m(m + 1)}{2}q - \left\{(k + 1)x - \frac{k(k + 1)}{2}q\right\}$$

$$= (m - k)x - \frac{(m - k)(m + k + 1)}{2}q \qquad (2.35)$$

$$= (m - k)\left(x - \frac{m + k + 1}{2}q\right) \qquad (k = 0, 1, 2, \ldots).$$

It follows that when $0 \leq k < m$, so $k + 1 \leq m$, then

$$x - \frac{m + k + 1}{2}q = (x - mq) + \frac{m - (k + 1)}{2}q \geq 0$$

and when $m < k$, so $m + 1 \leq k$, then

$$x - \frac{m + k + 1}{2}q = -\{(m + 1)q - x\} - \frac{k - (m + 1)}{2}q \leq 0.$$

Hence, the quantity in (2.35) is always non-negative and therefore

$$\max_{0 \leq n < \infty}\left\{nx - \frac{n(n - 1)}{2}q\right\} = ([x/q] + 1)x - \frac{[x/q]([x/q] + 1)}{2}q.$$

Then by putting

$$a_n := (-q)^{\otimes(n-1)n/2}, \quad n \geq 0, \qquad (2.36)$$

the expression given above becomes the tropical series expansion

$$f(x) = \bigotimes_{n=0}^{\infty} a_n \otimes x^{\otimes n} \qquad (2.37)$$

which corresponds to the Taylor series expansion (2.30). Observe the relation

$$a_n = \frac{n(n-1)}{2}(-q) = \frac{(n-1)(n-2)}{2}(-q) + (n-1)(-q)$$

which gives

$$a_n = a_{n-1} \otimes (-q)^{\otimes(n-1)}, \quad (n \geq 1). \tag{2.38}$$

Note that there is a difference in the n by 1, but they are the ultra-discrete analogues of (2.32) and (2.31) for the coefficients a_n in the series (2.30). In fact, we see that they are equal to

$$a_n = \frac{(1/q)^n}{\prod_{k=1}^{n}(1 - (1/q)^k)} \quad \text{and} \quad a_n = a_{n-1} \times \frac{1/q}{1 - (1/q)^n},$$

respectively. Moreover,

$$q^{\otimes n} \oslash \bigotimes_{k=1}^{n}(1_\circ \oplus q^k) = \left(n - \frac{n(n+1)}{2}\right)q = (-q)^{\frac{n(n-1)}{2}},$$

since $q > 0$. Equation

$$(1_\circ \oplus x) \otimes f(x - q) = f(x) \tag{2.39}$$

becomes

$$\max\{x, 0\} + f(x - q) = f(x)$$

in the conventional algebra. The tropical entire function $f(x)$ given by the tropical series expansion

$$f(x) = \bigoplus_{n=0}^{\infty} a_n \otimes x^{\otimes n} = \max_{0 \leq n < \infty}\{nx + a_n\} = ([x/q] + 1)x + a_{[x/q]+1},$$

where the coefficients a_n are defined as in (2.36), satisfies

$$f(x - q) = ([(x - q)/q] + 1)(x - q) + a_{[(x-q)/q]+1} = [x/q](x - q) + a_{[x/q]}.$$

Thus for $x \geq 0$ we have the identity

$$x + [x/q](x - q) + a_{[x/q]} = ([x/q] + 1)x + a_{[x/q]+1},$$

which is equivalent to the recurrence relation

$$a_n = a_{n-1} - (n-1)q \quad \text{for } n := [x/q] + 1 \in \mathbb{N}.$$

Note that for $x < 0$ we have $f(x) = f(x - q) = 1_\circ \oplus x \equiv 0$.
For the growth estimate of $f(x)$, we first remark that

$$M(r, f) = |f(r)| = f(r) = q([r/q]+1)(r/q-[r/q]/2) = \frac{q}{2}\left(\frac{r}{q}\right)^2\{1+o(1/r)\}$$

and therefore

$$\log M(r, f) = 2\log r - \log q - \log 2 + o(1) = (\log r)^{\otimes 2} \oslash (\log q^{\otimes 2}) \otimes \{1_\circ \oplus o(1)\}$$

as $r \to \infty$, which corresponds to the estimate (2.29). Hence the order $\rho(f) = 2$. In particular, f is of regular growth in the sense that

$$\lim_{r \to \infty} \frac{\log M(r, f)}{\log r} = 2,$$

while its type equals to

$$\lim_{r \to \infty} \frac{M(r, f)}{r^2} = \frac{1}{2q}.$$

Hence, the order of $f(x)$ given by (2.33) coincides with the logarithmic order of the entire function $f(z)$ in (2.27) and the type $1/(2q)$ of $f(x)$ corresponds to the logarithmic type of $f(z)$, that is,

$$-\frac{1}{2\log|q|} = \frac{1}{2\log(1/|q|)}.$$

Let us proceed to study the behavior of the coefficients a_n given by (2.36),

$$a_n = (-q)^{\otimes n(n-1)/2} = -\frac{n(n-1)}{2}q = \frac{n^2}{2}\left(1 - \frac{1}{n}\right)(-q).$$

This implies

$$\log(-a_n) = 2\log n + O(1) \quad \text{as } n \to \infty,$$

and thus

$$\lim_{n \to \infty} \frac{\log(-a_n)}{\log n} = 2.$$

Moreover, as in the case of the above type estimate, we have a comparison of

$$\frac{M(r, f)}{r^2} = \frac{1}{2q}(1 + o(1/r))$$

and

$$\frac{-a_n}{(qn)^2} = \frac{n^2}{2}\left(1 - \frac{1}{n}\right)\frac{q}{(qn)^2} = \frac{1}{2q}\left(1 - \frac{1}{n}\right),$$

both of which tend to $1/(2q)$ as $r \to \infty$ and $n \to \infty$, respectively. Observe here the relation $n = [r/q] + 1$ between two variables $r \in \mathbb{R}_{\geq 0}$ and $n \in \mathbb{Z}_{\geq 0}$. The function

$$f(x) = \bigoplus_{n=0}^{\infty} a_n \otimes x^{\otimes n} = ([x/q] + 1)x + a_{[x/q]+1}$$

possesses the maximum modulus

$$M(r, f) = ([r/q] + 1)r + a_{[r/q]+1} = n \cdot r + a_n.$$

By this example, one might have predicted the pair of relations such as

$$\limsup_{r \to \infty} \frac{\log M(r, f)}{\log r} = \limsup_{n \to \infty} \frac{\log(-a_n)}{\log n} = \rho, \text{ say,}$$

and

$$\limsup_{r \to \infty} \frac{M(r, f)}{r^\rho} = \frac{1}{q^\rho} \limsup_{n \to \infty} \frac{-a_n}{n^\rho} \quad \text{when } \rho \in (0, +\infty)$$

to be true, in similarity to the order (C.2) and type (C.3) described in Appendix C for entire functions. However, we want to claim the following pair of relations:

$$\limsup_{r \to \infty} \frac{\log M(r, f)}{\log r} = \limsup_{n \to \infty} \frac{\log(-a_n)}{\log(-a_n/n)} = \rho, \text{ say,}$$

and

$$\limsup_{r \to \infty} \frac{M(r, f)}{r^\rho} = c(\rho) \limsup_{n \to \infty} \frac{-a_n}{(-a_n/n)^\rho} \quad \text{when } \rho \in (0, +\infty),$$

by introducing a certain suitable constant $c(\rho)$ depending on the value of ρ only. This possibility can be supported by the scaling relation between logarithmic order (C.6) and logarithmic type (C.7) of entire functions described in Appendix C. Concerning the example observed above, we see that

$$-a_n = \frac{n^2}{2}\left(1 - \frac{1}{n}\right)q \quad \text{and} \quad -\frac{a_n}{n} = \frac{n}{2}\left(1 - \frac{1}{n}\right)q$$

so that

$$\lim_{n \to \infty} \frac{\log(-a_n)}{\log(-a_n/n)} = \lim_{n \to \infty} \frac{2\log n}{\log n} = 2 \quad \text{and} \quad \lim_{n \to \infty} \frac{-a_n}{(-a_n/n)^2} = \frac{q/2}{(q/2)^2} = \frac{2}{q}.$$

2.2.4   *A concluding remark*

Concerning entire solutions to some linear q-difference equations with polynomial coefficients, Hayman and Bergweiler proved in [6, Theorem 1]

Theorem 2.11 ([6]). *Suppose that f is a transcendental entire solution of the equation*

$$f(z) - a(z)f(cz) = 0, \tag{2.40}$$

where $0 < |c| < 1$ and a is a polynomial of degree d. Then there exists a non-negative integer p and $A, z_1, \ldots, z_d$ in $\mathbb{C} \setminus \{0\}$ such that

$$f(z) = Az^p \prod_{\mu=1}^{d} \prod_{j=0}^{\infty} \left(1 - \frac{c^j z}{z_\mu}\right). \tag{2.41}$$

Hayman and Bergweiler note that the coefficient $a(z)$ should be given by

$$a(z) = c^{-p} \prod_{\mu=1}^{d} \left(1 - \frac{z}{z_\mu}\right). \tag{2.42}$$

Wen [108] applies the idea of Hayman and Bergweiler and shows the way to construct an entire function of an arbitrarily given integer logarithmic order (see [108, Example 2]):

Example 2.12 ([108]). Denote $W_1(z) = 1 - z$ and $W_2(z) = \prod_{n=1}^{\infty} W_1(zq^n)$, where $0 < |q| < 1$. As we have observed, $W_2(z) = (z; q)_\infty$ is an entire function of logarithmic order 2 and solves the q-difference equation

$$W_2(z/q) = (1 - z)W_2(z) = W_1(z)W_2(z).$$

The entire function $W_3(z) = \prod_{n=1}^{\infty} W_2(zq^n)$ satisfies

$$W_3(z/q) = W_2(z)W_3(z),$$

and hence it has logarithmic order 3 by applying his theorem [108, Theorem 1.1]. Proceeding with this construction inductively, for each positive integer m, the entire function

$$W_m(z) = \prod_{n=1}^{\infty} W_{m-1}(zq^n)$$

satisfies the q-difference equation

$$W_m(z/q) = W_{m-1}(z)W_m(z)$$

and thus has logarithmic order $\rho_{\log}(W_m) = m$.

It is worth mentioning that this discussion could be translated into the tropical situation so that we would have a tropical entire function $\mathcal{W}_m$ of an arbitrarily given integer order m which satisfies the ultra-discrete equation of the form $\mathcal{W}_m(x - 1) = \mathcal{W}_{m-1}(x) \otimes \mathcal{W}_m(x)$. In fact, we have observed above that this is the case for $m = 2$ by putting

$$\mathcal{W}_1(x) = 1_\circ \oplus x = \max\{0, x\}$$

and

$$\mathcal{W}_2(x) = \bigoplus_{n=0}^{\infty} (-q)^{\otimes n(n-1)/2} \otimes x^{\otimes n} = \bigotimes_{n=0}^{\infty} \mathcal{W}_1(x - nq).$$

For each positive integer m, the function

$$\mathcal{W}_m(x) = \bigotimes_{n=0}^{\infty} \mathcal{W}_{m-1}(x - nq)$$

would be our desired solution.

We conclude this chapter by posing the following

Problem. Can we find a q-difference or ultra-discrete analogue of the summability theorem due to Whittaker (see [109, Theorem 3] and [110, Theorem 5]) on the linear difference equation $g(z + 1) = f(z)g(z)$ with an entire or meromorphic coefficient $f(z)$? In other words, are all the following statements proved in the same way by a formal ultra-discrete translation?

(W) If $f(z)$ is an entire or meromorphic function of order ρ, there is a meromorphic function $g(z)$, of order less than or equal to $\rho + 1$, such that $g(z + 1) = f(z)g(z)$.

(q-W) If $f(z)$ is an entire or meromorphic function of logarithmic order ρ_q, there is a meromorphic function $g(z)$, of logarithmic order less than or equal to $\rho_q + 1$, such that $g(qz) = f(z)g(z)$.

(u-W) If $f(x)$ is a tropical entire or meromorphic function of order ρ_u, there is a tropical meromorphic function $g(x)$, of order less than or equal to $\rho_u + 1$, such that $g(x + 1) = f(x) \otimes g(x)$.

Chapter 3

One-dimensional tropical Nevanlinna theory

In Chapter 1 we introduced Nevanlinna functions for tropical meromorphic functions and looked at a few examples of their behavior in the case of tropical polynomials and rational functions. In this chapter we present the full details of the one-dimensional tropical Nevanlinna theory, including the first and the second main theorems.

3.1 Poisson–Jensen formula in the tropical setting

In what follows in this chapter, a tropical meromorphic function f is to be understood in the sense of Definition 1.1 above, unless otherwise specified. We may also call f to be restricted meromorphic, whenever all of its one-sided derivatives (slopes) are integers.

The Poisson–Jensen formula in the extended tropical setting is formally as in the restricted meromorphic case, see [46], Lemma 3.1. The same proof applies.

Theorem 3.1. *Suppose f is a meromorphic function on $[-r, r]$, for some $r > 0$ and denote the distinct zeros, resp. poles, of f in this interval by a_μ, resp. by b_ν, with their corresponding multiplicities τ_f, see Definition 1.3, attached. Then for any $x \in (-r, r)$ we get the Poisson–Jensen formula*

$$f(x) = \frac{1}{2}(f(r) + f(-r)) + \frac{x}{2r}(f(r) - f(-r))$$

$$-\frac{1}{2r}\sum_{|a_\mu| < r}\tau_f(a_\mu)(r^2 - |a_\mu - x|r - a_\mu x) + \frac{1}{2r}\sum_{|b_\nu| < r}\tau_f(b_\nu)(r^2 - |b_\nu - x|r - b_\nu x).$$

In the particular case of $x = 0$ we obtain the tropical Jensen formula

$$f(0) = \frac{1}{2}(f(r) + f(-r)) - \frac{1}{2}\sum_{|a_\mu|<r}\tau_f(a_\mu)(r - |a_\mu|) + \frac{1}{2}\sum_{|b_\nu|<r}\tau_f(b_\nu)(r - |b_\nu|).$$

Proof. As in [46], we define an increasing sequence $(c_j), j = -p, \ldots, q$ in $(-r, r)$ in the following way. Let $c_0 = x$, and let the other points in this sequence be the points in $(-r, r)$ at which the derivative of f does not exist, i.e., f has either a zero or a pole at these points. Further, we denote by m_j slopes of the line segments in the graph of f. In particular, we define $m_{j-1} := \lim_{x \to c_j^-} f'(x)$ for $j = -p, \ldots, 0$, resp. $m_{j+1} := \lim_{x \to c_j^+} f'(x)$ for $j = 0, \ldots, q$. Elementary geometric observation implies

$$f(r) - f(x) = m_1(c_1 - x) + m_2(c_2 - c_1) + \cdots + m_q(c_q - c_{q-1}) + m_{q+1}(r - c_q)$$
$$= -m_1 x + m_{q+1} r + c_1(m_1 - m_2) + \cdots + c_q(m_q - m_{q+1})$$
$$= m_1(r - x) - \sum_{j=1}^{q}(m_j - m_{j+1})(r - c_j).$$

By a parallel reasoning,

$$f(x) - f(-r) = m_{-1}(r + x) - \sum_{j=1}^{p}(m_{-j-1} - m_{-j})(r + c_{-j}).$$

Multiplying the above two equalities by $(r + x)$ and $(r - x)$, respectively, and subtracting, we obtain

$$2rf(x) = r(f(r) + f(-r)) + x(f(r) - f(-r)) + (m_{-1} - m_1)(r^2 - x^2)$$
$$+ \sum_{j=1}^{p}(m_{-j-1} - m_{-j})(r^2 - (x - c_{-j})r - c_{-j}x)$$
$$+ \sum_{j=1}^{q}(m_j - m_{j+1})(r^2 - (c_j - x)r - c_j x)$$
$$= r(f(r) + f(-r)) + x(f(r) - f(-r))$$
$$+ \sum_{j=-p}^{q} -\omega_f(c_j)(r^2 - |c_j - x|r - c_j x).$$

Recalling the definition of the multiplicity τ_f for roots and poles of f, the claim is an immediate consequence of this equality. $\qquad \square$

3.2 Basic properties of Nevanlinna functions

It is easy to verify that several basic inequalities, see [46], for the proximity function and the characteristic function hold in our present setting as well. In particular, the following simple observations are immediately proved by the corresponding definitions:

Lemma 3.2. *(i) If $f \leq g$, then $m(r, f) \leq m(r, g)$.*

(ii) Given f, g, then $m(r, \max\{f, g\}) \geq m(r, f)$.

(iii) Given a positive number α, then

$$m(r, f^{\otimes \alpha}) = m(r, \alpha f) = \alpha m(r, f),$$
$$N(r, f^{\otimes \alpha}) = N(r, \alpha f) = \alpha N(r, f),$$
$$T(r, f^{\otimes \alpha}) = T(r, \alpha f) = \alpha T(r, f).$$

(iv) Given tropical meromorphic functions f, g, then

$$m(r, f \otimes g) \leq m(r, f) + m(r, g),$$
$$N(r, f \otimes g) \leq N(r, f) + N(r, g),$$
$$T(r, f \otimes g) \leq T(r, f) + T(r, g).$$

(v)

$$T(r, f \oplus g) \leq T(r, f) + T(r, g).$$

Proof. Most of the assertions in this lemma are just obvious consequences of the corresponding definitions. However, the incquality $N(r, f \otimes g) \leq N(r, f) + N(r, g)$ perhaps needs a short argument. Indeed, by the definition of the (negative) pole multiplicity $\omega_f(x) := \lim_{\varepsilon \to 0+} (f'(x + \varepsilon) - f'(x - \varepsilon))$, it is clear that $\omega_f(x)$ is linear at x with respect to f:

$$\omega_{f+g}(x) = \omega_f(x) + \omega_g(x).$$

But then

$$\max\{\omega_{f+g}(x), 0\} \leq \max\{\omega_f(x), 0\} + \max\{\omega_g(x), 0\}$$

for each $x \in \mathbb{R}$. Hence $n(t, f + g) \leq n(t, f) + n(t, g)$ for all $t > 0$, and the asserted inequality follows by integration. $\square$

Remark 3.3. Observe that whenever $f \leq g$, the inequality $N(r, f) \leq N(r, g)$ is not necessarily true. Similarly, the inequality

$$N(r, f \oplus g) = N(r, \max\{f, g\}) \leq \max\{N(r, f), N(r, g)\}$$

may fail. Indeed, as for the case $f \leq g$, take f, g satisfying this inequality so that the graph of f is constant outside of $[-1, 1]$ and is $\wedge\wedge$-shaped in $[-1, 1]$, and let g be defined correspondingly as $\wedge$-shaped. Then f has two poles, while g has only one. If the slopes are suitably defined, then $N(r, f) > N(r, g)$. As for the case of $\max\{f, g\}$, a corresponding example is easily constructed. The corresponding observations are true for the characteristic function as well, provided just that the proximity functions are small enough.

We next proceed to show that characteristic functions $T(r, f)$ are nondecreasing. As a preparation to this important, elementary result, observe that counting functions $N(r, f)$ are positive, continuous and non-decreasing piecewise linear functions of r as well. This is an immediate consequence of their definition (1.8).

Theorem 3.4. *The characteristic function $T(r, f)$ is a positive, continuous, non-decreasing piecewise linear function of r.*

Proof. By its definition (1.9), the characteristic function $T(r, f)$ is positive and continuous. Moreover, outside of the slope discontinuities of f, both $m(r, f)$ and $N(r, f)$ are locally linear, hence $T(r, f)$ as well. To prove the monotonicity of $T(r, f)$, observe that we may consider $T(r, -f)$ equally well by the Jensen formula (1.10)

$$T(r, f) = T(r, -f) + f(0).$$

Suppose first that $r, -r$ are not slope discontinuity points of f. Then we may differentiate $T(r, f)$ to obtain

$$f'(r) + n(r, f) = f'(-r) + n(r, -f).$$

If first $f(-r) < 0, f(r) < 0$, then $m(r, f) = 0$, and by continuity of f, we have $m(\rho, f) = 0$ as ρ increases from r to $r + \varepsilon$ for ε sufficiently small. In this interval, $T(\rho, f) = N(\rho, f)$, and as $N(r, f)$ is increasing, the same holds for $T(r, f)$ as r increases.

If then $f(-r) \geq 0$ and $f(r) \geq 0$, then $-f(-r) \leq 0, -f(r) \leq 0$, and so $m(r, -f) = 0$. Clearly, $m(r, -f)$ is non-decreasing as r increases from r to $r + \varepsilon$, hence $T(r, -f)$ and $T(r, f)$ are non-decreasing as well.

Assuming next that $f(-r) < 0$, while $f(r) \geq 0$, we have $f'(-r) = 0$ around $-r$, and

$$T'(r, f) = \frac{1}{2}(f'(r) + n(r, f)) = \frac{1}{2}(f'(-r) + n(r, -f)) = \frac{1}{2}n(r, -f) \geq 0,$$

hence $T(r, f)$ is increasing at r.

Finally, if $f(-r) \geq 0$ and $f(r) < 0$, then $f'(r) = 0$ and

$$T'(r, f) = T'(r, -f) = \frac{1}{2}(f'(-r) + n(r, -f)) = \frac{1}{2}(f'(r) + n(r, f))$$
$$= \frac{1}{2}n(r, f) \geq 0.$$

It remains to consider points of slope discontinuity of f. Considering $-f$, if needed, we may assume that r is a pole of f, of multiplicity $\tau(r, f) = f'(r - \varepsilon) - f'(r + \varepsilon)$ for an ε sufficiently small. Suppose, contrary to the claim, that $T(\rho, f) < T(r, f)$ for $\rho > r$ close to r, say for $\rho = r + \varepsilon$. Since the characteristic function is piecewise linear, and r is the only pole of f close to r, it is geometrically obvious that we must have $T'(r - \varepsilon, f) > T'(r + \varepsilon, f)$. But then

$$f'(r - \varepsilon) + n(r - \varepsilon, f) > f'(r + \varepsilon) + n(r + \varepsilon, f)$$
$$= f'(r + \varepsilon) + n(r - \varepsilon, f) + \tau(r, f)$$
$$= f'(r + \varepsilon) + n(r - \varepsilon, f) + f'(r - \varepsilon) - f'(r + \varepsilon)$$
$$= n(r - \varepsilon, f) + f'(r - \varepsilon),$$

a contradiction. $\qquad\square$

3.2.1 *First main theorem*

As usual in the Nevanlinna theory, the next step from the Poisson–Jensen formula is to formulate the first main theorem. To this end, we recall the notation $L_f := \inf\{f(b)\}$ over all poles b of f, i.e.

$$L_f := \inf\{f(b) : \omega_f(b) < 0\}.$$

In particular, if f has no poles (and so f is said to be tropical entire), then we have $L_f = \inf \emptyset = +\infty$.

Theorem 3.5. *Let f be tropical meromorphic. Then*

$$T(r, 1_\circ \oslash (f \oplus a)) = T(r, -\max\{f, a\}) \leq T(r, f) + \max\{a, 0\} - \max\{f(0), a\}$$

for any $a \in \mathbb{R}$ and any $r > 0$. Moreover, an asymptotic equality

$$T(r, 1_\circ \oslash (f \oplus a)) = T(r, -\max\{f, a\}) = T(r, f) - \max\{f(0), a\} + \varepsilon(r, a)$$

holds for any $r > 0$ with $0 \leq \varepsilon(r, a) \leq \max\{a, 0\}$, provided that $-\infty < a < L_f$.

Proof. Making use of the tropical Jensen formula (1.10), we immediately conclude that

$$T(r, 1_\circ \oslash (f \oplus a)) = T(r, -\max\{f, a\})$$
$$= T(r, \max\{f, a\}) - \max\{f(0), a\}$$
$$\leq T(r, f) + \max\{a, 0\} - \max\{f(0), a\}$$

for any $a \in \mathbb{R}$ and for any $r > 0$. Here we also used the inequality $T(r, g \oplus h) = T(r, \max\{g, h\}) \leq T(r, g) + T(r, h)$ and the simple observation that $T(r, a) = a^+ = \max\{a, 0\}$. Further,

$$\max\{a, 0\} - \max\{f(0), a\} = \left\{ \begin{array}{lll} a - f(0) \leq 0; & a > 0, f(0) \geq a \\ a - a & = 0; & a > 0, f(0) < a \\ 0 - f(0) \leq |a|; & a \leq 0, f(0) \geq a \\ 0 - a & = |a|; & a \leq 0, f(0) < a \end{array} \right\} \leq |a|.$$

To obtain the asserted asymptotic equality, suppose first that f has at least one pole and that $-\infty < a < L_f$. In this case, we have $N(r, \max\{f, a\}) = N(r, f)$. Therefore,

$$T(r, 1_\circ \oslash (f \oplus a)) = T(r, -\max\{f, a\})$$
$$= T(r, \max\{f, a\}) - \max\{f(0), a\}$$
$$= m(r, \max\{f, a\}) + N(r, \max\{f, a\}) - \max\{f(0), a\}$$
$$\geq m(r, f) + N(r, f) - \max\{f(0), a\}$$
$$\geq T(r, f) - \max\{f(0), a\},$$

according to the monotonicity of $m(r, f)$, Lemma 3.2, with respect to f.

Finally, if $L_f = +\infty$, that is, if f has no poles, the asymptotic equality holds as well. In fact, because of $T(r, f) = m(r, f)$ then and $f \oplus a \geq f$ for any $a \in \mathbb{R}$, we have

$$T(r, 1_\circ \oslash (f \oplus a)) = T(r, -\max\{f, a\})$$
$$= T(r, \max\{f, a\}) - \max\{f(0), a\}$$
$$\geq m(r, \max\{f, a\}) - \max\{f(0), a\}$$
$$\geq m(r, f) - \max\{f(0), a\}$$
$$= T(r, f) - \max\{f(0), a\}.$$

$\square$

Example 3.6. A non-constant linear function $f(x) = \alpha x + \beta$ with $\alpha > 0$ and $\beta > 0$ satisfies

$$T(r, f) = m(r, f) = \begin{cases} \beta & \left(0 \le r < \frac{\beta}{\alpha}\right) \\ \frac{\alpha}{2}r + \frac{\beta}{2} & \left(\frac{\beta}{\alpha} \le r\right) \end{cases}.$$

It is a simple exercise to verify by this example that the error term $\varepsilon(r, a)$ in Theorem 3.5, may run over the whole interval $\left[0, \max\{f(0), a\}\right)$.

Remark 3.7. Given a tropical meromorphic function f, observe that $1_\circ \oslash (f \oplus 0) = -\max\{f, 0\} \le 0$. Therefore, $m(r, 1_\circ \oslash \max\{f, 0\}) = 0$ and so, $N(r, 1_\circ \oslash \max\{f, 0\}) = T(r, 1_\circ \oslash \max\{f, 0\}) = T(r, f) - \max\{f(0), 0\}$ by the first main theorem above. Moreover, $1_\circ \oslash f = 1_\circ \oslash (f \oplus (-\infty))$ and $1_\circ \oslash (f \oplus 0)$ do not coincide in general. Therefore, in particular, $N(r, 1_\circ \oslash f)$ and $N(r, 1_\circ \oslash (f \oplus 0))$ also do not coincide in general. As an example, suppose that $f(0) \ge 0$, and suppose that $N(r, 1_\circ \oslash f) = N(r, 1_\circ \oslash (f \oplus 0))$. Then, by the Jensen formula,

$$N(r, 1_\circ \oslash \max\{f, 0\}) = T(r, f) - \max\{f(0), 0\} = T(r, f) - f(0) = T(r, -f)$$
$$= N(r, 1_\circ \oslash f) + m(r, 1_\circ \oslash f),$$

hence $m(r, 1_\circ \oslash f) = m(r, -f) = 0$, and we conclude that $f \ge 0$.

3.2.2 *Tropical Cartan identity*

To formulate the tropical variant of the classical Cartan identity

$$T(r, f) = \frac{1}{2\pi} \int_0^{2\pi} N\left(r, e^{i\varphi}, f\right) d\varphi + \log^+ |f(0)|, \qquad (3.1)$$

see (A.4) in Appendix A, we first observe by the First Main Theorem that

$$N(r, -\max(f, 0)) \le T(r, f) + \max\{1_\circ, 0\} - \max\{f(0), 1_\circ\} = T(r, f) - f^+(0).$$

By the tropical Jensen formula, we also have

$$N(r, -\max(-f, 0)) \le T(r, -f) - (-f(0))^+ = T(r, f) - f(0) - (-f(0))^+.$$

Since $f(0) + (-f(0))^+ = f^+(0)$, we obtain

$$T(r, f) \ge \max\{N(r, -\max(f, 0)), N(r, -\max(-f, 0))\} + f^+(0).$$

Actually, equality holds here and the tropical Cartan identity follows:

Theorem 3.8 (Tropical Cartan identity). *Let $f(x)$ be a tropical meromorphic function, $f \not\equiv -\infty$, on $\mathbb{R}$. Then we have*

$$T(r, f) = \bigoplus_{\sigma=\pm 1} N(r, -\max(\sigma f, 0)) + f^+(0)$$

$$= \bigoplus_{\sigma=\pm 1} N\left(r, 1_\circ \oslash ((\sigma f) \oplus 1_\circ)\right) + f^+(0). \tag{3.2}$$

Remark 3.9. Note that by the definition of tropical counting functions, we have

$$N(r, -\max(-f, 0)) \oplus N(r, -\max(f, 0))$$

$$= \frac{1}{2} \int_0^r \{n(t, -\max(-f, 0)) \oplus n(t, -\max(f, 0))\} dt.$$

By Theorem 3.8,

$$\frac{d}{dr} T(r, f) = \frac{1}{2}\{n(t, -\max(-f, 0)) \oplus n(t, -\max(f, 0))\},$$

is non-decreasing.

Corollary 3.10. $T(r, f)$ *is an increasing convex function of r.*

Proof. This is an immediate consequence of the preceding remark, completely parallel to the classical result in Nevanlinna theory that $T(r, f)$ is an increasing convex function of $\log r$, see Theorem A.3. $\qquad\square$

Proof of Theorem 3.8. An application of the tropical Jensen formula to the function $f^+(x) = f(x) \oplus 1_\circ$ gives

$$f^+(0) = \frac{1}{2} \sum_{\sigma=\pm 1} f^+(\sigma r) + N(r, f^+) - N(r, 1_\circ \oslash f^+). \tag{3.3}$$

By definition, the first sum on the right-hand side equals to the tropical proximity function $m(r, f)$, hence

$$f^+(0) = m(r, f) + N(r, f^+) - N(r, 1_\circ \oslash f^+).$$

Therefore, the key part in this proof consists of analyzing $N(r, f^+) - N(r, 1_\circ \oslash f^+)$. This will be done by the following lemma:

Lemma 3.11. *Let f be tropical meromorphic function. Then*

$$N(r, f \oplus 1_\circ) - N\left(r, 1_\circ \oslash (f \oplus 1_\circ)\right)$$

$$= N(r, f) - \bigoplus_{\sigma=\pm 1} N\left(r, 1_\circ \oslash ((\sigma f) \oplus 1_\circ)\right). \tag{3.4}$$

Once this has been proved, the identity (3.3) may easily be written in the form

$$f^+(0) = T(r, f) - \bigoplus_{\sigma = \pm 1} N\left(r, 1_\circ \oslash \left((\sigma f) \oplus 1_\circ\right)\right),$$

proving the tropical Cartan identity. □

Therefore, it remains to prove the lemma:

Proof of Lemma 3.11. By the definition of the integrated counting function for poles, it is sufficient to prove that the point-wise identity

$$\tau_{f \oplus 1_\circ}(x_0) - \tau_{1_\circ \oslash (f \oplus 1_\circ)}(x_0) = \tau_f(x_0) - \max_{\sigma = \pm 1}\left\{\tau_{1_\circ \oslash ((\sigma f) \oplus 1_\circ)}(x_0)\right\} \quad (3.5)$$

holds at each point $x_0 \in \mathbb{R}$. Of course, multiplicities for poles only are to be counted here. Recall from Definition 1.3 that $\tau_f(x) = -\omega_f(x)$, if f has a pole at x and $\tau_f(x) = \omega_f(x)$, if f has a root at x. Of course, $\tau_f(x)$ is non-negative everywhere. To prove this local identity, we distinguish our considerations into three cases depending on the behavior of $f(x)$ at x_0, i.e. according to whether $\omega_f(x_0) < 0$, that is, whether f has a pole of order $-\omega_f(x_0)$ at x_0, or $\omega_f(x_0) \geq 0$, meaning that f has at most a root of order $\omega_f(x_0)$ at x_0.

<u>Case 1-i:</u> Suppose first that $\omega_f(x_0) < 0$ while $f(x_0) > 0$. Then $f(x) > 0$ for all $x \in \mathbb{R}$ sufficiently near to x_0, and we have

$$f \oplus 1_\circ = \max(f, 0) = f,$$
$$1_\circ \oslash (f \oplus 1_\circ) = -\max(f, 0) = -f, \quad \text{and}$$
$$1_\circ \oslash \left((-f) \oplus 1_\circ\right) = -\max(-f, 0) = 0$$

near x_0. Therefore, $\tau_{f \oplus 1_\circ}(x_0) = \tau_f(x_0)$. Moreover, $1_\circ \oslash (f \oplus 1_\circ)$ has a root at x_0. Since $1_\circ \oslash ((-f) \oplus 1_\circ)$ vanishes near x_0, $1_\circ \oslash (f \oplus 1_\circ)$ and $1_\circ \oslash ((-f) \oplus 1_\circ)$ have no contributions in computing the point-wise identity (3.5) at x_0. Hence, the identity (3.5) holds at x_0.

<u>Case 1-ii:</u> Suppose next that $\omega_f(x_0) < 0$ while $f(x_0) < 0$. Then $f(x) < 0$ near x_0, and so

$$f \oplus 1_\circ = \max(f, 0) = 0,$$
$$1_\circ \oslash (f \oplus 1_\circ) = -\max(f, 0) = 0, \quad \text{and}$$
$$1_\circ \oslash \left((-f) \oplus 1_\circ\right) = -\max(-f, 0) = f$$

near x_0. Clearly, the left-hand side of (3.5) vanishes at x_0. On the other hand,

$$\max_{\sigma=\pm 1}\left\{\tau_{1_\circ \oslash((\sigma f)\oplus 1_\circ)}(x_0)\right\} = \max\{0, \tau_f(x_0)\},$$

and the right-hand side of (3.5) vanishes as well.

<u>Case 1-iii:</u> Suppose now that $\omega_f(x_0) < 0$ while $f(x_0) = 0$. If $f(x) \leq 0$ near x_0, the reasoning in Case 1-ii applies. To consider the cases when the sign of $f(x)$ changes as x goes through the point x_0, suppose first that $f(x) \geq 0$ for $x < x_0$, and $f(x) < 0$ for $x > x_0$. Then, looking at the graphs of $f \oplus 1_\circ, 1_\circ \oslash (f \oplus 1_\circ)$ and $1_\circ \oslash ((-f) \oplus 1_\circ)$ near x_0, we see that $f \oplus 1_\circ$ has a root at x_0, thus having no contribution to (3.5) at x_0. To compute the pole contributions in (3.5) for the remaining terms, observe first that $f'(x_0 - 0) > f'(x_0 + 0)$, since f has a pole at x_0. But then $\tau_f(x_0) = f'(x_0 - 0) - f'(x_0 + 0)$, $\tau_{1_\circ \oslash(f \oplus 1_\circ)}(x_0) = -f'(x_0 - 0)$ and $\tau_{1_\circ \oslash((-f)\oplus 1_\circ)}(x_0) = -f'(x_0 + 0)$. The identity (3.5) then follows at once. Finally, if we have $f(x) < 0$ for $x < x_0$ and $f(x) \geq 0$ for $x > x_0$ near x_0, a completely parallel reasoning applies; we leave this as an easy exercise.

<u>Case 2-i:</u> Suppose now that we have $\omega_f(x_0) \geq 0$ while $f(x_0) > 0$. Then $f(x) > 0$ near $x = x_0$, so we have

$$f \oplus 1_\circ = \max(f, 0) = f,$$
$$1_\circ \oslash (f \oplus 1_\circ) = -\max(f, 0) = -f, \quad \text{and}$$
$$1_\circ \oslash ((-f) \oplus 1_\circ) = -\max(-f, 0) = 0,$$

there. Then $f, f \oplus 1_\circ$ and $1_\circ \oslash ((-f) \oplus 1_\circ)$ have no contribution to (3.5) at x_0 and the identity follows since $-\tau_{1_\circ \oslash(f \oplus 1_\circ)}(x_0)$ only remains in both sides of (3.5).

<u>Case 2-ii:</u> Suppose next that $\omega_f(x_0) \geq 0$ while $f(x_0) < 0$. Then $f(x) < 0$ near $x = x_0$, and we have

$$f \oplus 1_\circ = \max(f, 0) = 0,$$
$$1_\circ \oslash (f \oplus 1_\circ) = -\max(f, 0) = 0, \quad \text{and}$$
$$1_\circ \oslash ((-f) \oplus 1_\circ) = -\max(-f, 0) = f$$

around x_0. This means that all terms towards (3.5) vanish, and we may continue to

<u>Case 2-iii:</u> where $\omega_f(x_0) \geq 0$ with $f(x_0) = 0$. If now $f(x) \geq 0$ near $x = x_0$, the reasoning in the Case 2-i applies, and we may proceed to considering

the situation when $f(x_0) < 0$ for $x < x_0$ and $f(x) \geq 0$ for $x > x_0$. Similarly as in Case 1-iii, we may look at the graphs of f, $f \oplus 1_\circ = \max(f, 0)$, $1_\circ \oslash (f \oplus 1_\circ) = \min(-f, 0)$ and $1_\circ \oslash ((-f) \oplus 1_\circ) = \min(f, 0)$ to see that $1_\circ \oslash (f \oplus 1_\circ)$ and $1_\circ \oslash ((-f) \oplus 1_\circ)$ only have a pole contribution to (3.5). But now $\tau_{1_\circ \oslash (f \oplus 1_\circ)}(x_0) = f'(x_0 + 0)$ and $\tau_{1_\circ \oslash ((-f) \oplus 1_\circ)}(x_0) = f'(x_0 - 0)$. Since $f'(x_0 + 0) \geq f'(x_0 - 0)$ by the assumption that $\omega_f(x_0) \geq 0$, the identity (3.5) follows at once. It remains to treat the case when $f(x_0) \geq 0$ for $x < x_0$ and $f(x) < 0$ for $x > x_0$. This is again nothing but an exercise, completing the proof of the lemma. □

3.3 Auxiliary results from real analysis

3.3.1 *Borel-type theorems*

Borel-type results have appeared to be a key tool to proving the lemma on logarithmic derivatives in the classical Nevanlinna theory. Similarly they are unavoidable while proceeding to the tropical counterpart of that lemma, as well as to its various variants. For the convenience of the reader, we include here several of results of Borel type. Keeping in mind possible future applications in the tropical theory, we recall here more of these results than what is actually needed in what immediately follows. Since all these results are of pure real analysis, we omit their proofs here, giving exact references for those readers who might be interested to study the proofs as well.

We start with the most classical result, originally due to Borel in 1897. Our reference here is [48], Lemma 2.4.

Theorem 3.12. *Given r_0, suppose $S : [r_0, \infty) \to [1, \infty)$ is a continuous, non-decreasing function. Then*

$$S\left(r + \frac{1}{S(r)}\right) < 2S(r)$$

for all $r \geq r_0$ outside a possible exceptional set of linear measure ≤ 2.

However, in tropical considerations, Borel type results with an exceptional set of finite logarithmic measure are typically needed. Here a set $E \subset [1, \infty)$ is said to be of finite logarithmic measure if $\int_E dt/t$ is finite. As the key result of this type is perhaps the following theorem due to Halburd and

Korhonen, see [41], Lemma 2.2:

Theorem 3.13. *Let $T : (0, \infty) \to (0, \infty)$ be a non-decreasing continuous function, $s > 0$, $\beta < 1$, and let $F \subset [0, \infty)$ be the set of all r such that $T(r) \leq \beta T(r + s)$. If F is of infinite logarithmic measure, then T must be of infinite order, i.e. $\limsup_{r \to \infty} (\log T(r) / \log r) = \infty$.*

Corollary 3.14. *Let $T : (0, \infty) \to (0, \infty)$ be a non-decreasing continuous function of finite order $\limsup_{r \to \infty} (\log T(r) / \log r) < \infty$, and let $s > 0$, $\alpha > 1$. Then $T(r + s) \leq \alpha T(r, f)$ outside an exceptional set of finite logarithmic measure.*

Below, we frequently need to consider tropical meromorphic functions that are of infinite order, while being, however, of restricted growth in the sense that their hyper-order $\limsup_{r \to \infty} (\log \log T(r, f) / \log r) < 1$. As to this situation, we recall a variant of Theorem 3.13, due to Halburd, Korhonen and Tohge, see [45], Lemma 8.3:

Theorem 3.15. *Let $T : [0, \infty) \to [0, \infty)$ be a non-decreasing continuous function of hyper-order $\rho_2(f) < 1$, let $s > 0$ and let $0 < \delta < \gamma < 1 - \rho_2(f)$. Then*

$$T(r + s) = T(r) + O\left(\frac{T(r)}{r^\gamma}\right) = T(r) + o\left(\frac{T(r)}{r^\delta}\right),$$

whenever r runs to infinity outside of a set of finite logarithmic measure.

Remark 3.16. In the preceding theorem, we trivially conclude that $T(r + s) \leq 2T(r)$, provided that r is large enough and outside an exceptional set of finite logarithmic measure.

For completeness and possible further applications, we include a few related Borel type results. As for the following one, usually called as a Borel–Nevanlinna theorem, see [26], p. 90:

Theorem 3.17. *Let $u(r)$ be a continuous, non-decreasing function on $[r_0, \infty)$, tending to $+\infty$ as $r \to \infty$. Let $\phi(u)$ be a continuous positive non-increasing function on $[u_0, \infty)$, with $u_0 = u(r_0)$, tending to zero as $u \to \infty$ and satisfying the integral condition*

$$\int_{u_0}^{\infty} \phi(u) du < \infty.$$

Then for all $r \geq r_0$ outside an exceptional a set of finite linear measure, the inequality

$$u(r + \phi(u(r))) < u(r) + 1$$

holds.

If you use $\sqrt{\log u(r)}$ instead of $u(r)$ and $1/u^2$ instead of $\phi(u)$ in the preceding Theorem 3.17, see [26], p. 91, we obtain

Theorem 3.18. *Let $u(r)$ be a function satisfying the conditions in Theorem 3.17, and fix $\varepsilon > 0$. Then for all $r \geq r_0$ outside an exceptional set of finite linear measure, the inequality*

$$u \left(r + \frac{1}{\log u(r)} \right) < u(r)^{1+\varepsilon}$$

holds.

Recalling that a set $E \subset [1, \infty)$ is called to be a set of finite logarithmic measure, provided that the integral $\int_E dt/t$ converges, we may use $u(e^r)$ instead of $u(r)$ in Theorem 3.18 to obtain

Theorem 3.19. *Let $u(r)$ be a function satisfying the conditions in Theorem 3.17, and fix $\varepsilon > 0$. Then for all $r \geq r_0$ outside an exceptional set of finite logarithmic measure, the inequality*

$$u \left(r + \frac{r}{\log u(r)} \right) < u(r)^{1+\varepsilon}$$

holds.

Finally, we add the following slightly more complicated Borel-type results, to be found in [16], Lemma 3.3.1:

Theorem 3.20. *Suppose $F(r), \phi(r)$ are positive, non-decreasing, continuous functions in $[r_0, \infty)$ and $F(r) \geq e$. Moreover, let $\xi(r)$ be a positive, non-decreasing, continuous function in $[e, \infty)$, $C > 1$ be a constant and*

$$E := \left\{ r \in [r_0, \infty) \, \Big| \, F \left(r + \frac{\phi(r)}{\xi(F(r))} \right) \geq CF(r) \right\}$$

be a closed set in $[r_0, \infty)$. Then, for all $R < \infty$, we have the inequality

$$\int_{E \cap [r_0, R]} \frac{dr}{\phi(r)} \leq \frac{1}{\xi(e)} + \frac{1}{\log C} \int_e^{F(r)} \frac{dx}{x\xi(x)}.$$

3.4   Variants of the lemma on tropical quotients

A typical application of Borel-type theorems is the lemma on tropical quotients, corresponding to the classical lemma on logarithmic derivatives in complex analysis. For possible future applications, we offer here several variants of this important result. However, contrary to the classical complex analysis, the lemma on tropical quotients is not needed to prove what is a tropical counterpart of the second main theorem.

Before proceeding to these crucial lemmas, we give two observations concerning the relation between the integrated and non-integrated counting functions.

Lemma 3.21. *Let f be tropical meromorphic. Then, for all $c \neq 0, \alpha > 1, r > 0$ and for $\rho = \frac{1}{2}(\alpha + 1)(r + |c|)$,*

$$n(\rho, f) \leq \frac{4}{\alpha - 1} \frac{1}{r + |c|} N(\alpha(r + |c|), f).$$

Proof. First observe that $\frac{1}{2}(\alpha + 1)(r + |c|) = \rho < \alpha(r + |c|)$. Then

$$N(\alpha(r + |c|), f) \geq \frac{1}{2} \int_{\rho}^{\alpha(r+|c|)} n(t, f) dt \geq \frac{\alpha - 1}{4}(r + |c|)n(\rho, f),$$

and the assertion follows. $\square$

The next lemma is an application of Borel reasoning:

Lemma 3.22. *Given $\varepsilon > 0$ and $R > 0$, we have*

$$n(r, f) \leq \frac{4}{r} N(r, f)^{1+\varepsilon}$$

for all $r > R$ outside an exceptional set of finite logarithmic measure.

Proof. There is nothing to prove, if $n(r, f) = 0$ for all r. Hence, suppose f admits at least one pole. By the proof of Theorem 1.14, we have $n(r, f) \leq \frac{2}{(k-1)r} N(kr, f)$ whenever $k > 1$. Take $k = 1 + 1/N(r, f)^{\varepsilon}$. Then we have

$$n(r, f) \leq \frac{2}{r} N(r, f)^{\varepsilon} N\left(r + \frac{r}{N(r, f)^{\varepsilon}}, f\right). \tag{3.6}$$

Apply now Theorem 3.20 to $N(r, f)$, taking $\phi(r) := r$ and $\xi(x) = x^{\varepsilon}$. We may also assume that $N(r_0, f) \geq e$, and we restrict us to considering values of $r \geq r_0$ only. By Theorem 3.20, we conclude that

$$N\left(r + \frac{r}{N(r, f)^{\varepsilon}}, f\right) \leq 2N(r, f) \tag{3.7}$$

outside an exceptional set of finite logarithmic measure. The claim now follows by combining (3.6) and (3.7). □

Theorem 3.23. *Let f be tropical meromorphic. Then, for any $\varepsilon > 0$,*

$$m(r, f(x+c) \oslash f(x)) \le \frac{2^{1+\varepsilon}10|c|}{r}(T(r+|c|,f)^{1+\varepsilon} + o(T(r+|c|,f)))$$

holds outside an exceptional set of finite linear measure.

Proof. We repeat the proof given in [46], Lemma 3.8, only slightly modified. Suppose we have $\rho > r + |c|$ and $x \in [-r, r]$. Denote by a_μ, resp. b_ν, the roots, resp. the poles in f in $[-r, r]$. By the Poisson–Jensen formula in Theorem 3.1, we obtain that

$$f(x+c) - f(x) = \frac{c}{2\rho}(f(\rho) - f(-\rho))$$

$$+ \frac{1}{2\rho}\sum_\mu (|a_\mu - x - c| - |a_\mu - x|)\rho + ca_\mu$$

$$- \frac{1}{2\rho}\sum_\nu (|b_\nu - x - c| - |b_\nu - x|)\rho + cb_\nu.$$

Since for a constant a, $m(r, a) = a^+$, we conclude by the monotonicity of the proximity function that

$$m(r, f(x+c) - f(x)) = \left(\frac{c}{2\rho}(f(\rho) - f(-\rho))\right)^+$$

$$+ \sum_\mu m\left(r, \frac{1}{2\rho}\sum_\mu (|a_\mu - x - c| - |a_\mu - x|)\rho + ca_\mu\right)$$

$$+ \sum_\nu m\left(r, \frac{1}{2\rho}\sum_\nu (|b_\nu - x - c| - |b_\nu - x|)\rho + cb_\nu\right).$$

Next observe that

$$\left(\frac{c}{2\rho}(f(\rho) - f(-\rho))\right)^+ \le \left|\frac{c}{2\rho}(f(\rho) - f(-\rho))\right|$$

$$\le \frac{|c|}{2\rho}\left(f^+(\rho) + f^+(-\rho) + (-f)^+(\rho) + (-f)^+(-\rho)\right)$$

$$\le \frac{|c|}{2\rho}(m(\rho, f) + m(\rho, -f)).$$

Moreover, since

$$|a_\mu - x - c| - |a_\mu - x| \le ||a_\mu - x - c| - |a_\mu - x|| \le |(a_\mu - x - c) - (a_\mu - x)| = |c|,$$

we obtain, by monotonicity of the proximity function again that

$$m\left(r, \frac{1}{2\rho}(|a_\mu - x - c| - |a_\mu - x|)\rho + ca_\mu\right) \le \frac{|c|}{2} + \frac{1}{2\rho}|ca_\mu| \le |c|.$$

Similarly,

$$m\left(r, \frac{1}{2\rho}(|b_\nu - x - c| - |b_\nu - x|)\rho + cb_\nu\right) \le |c|.$$

Combining now the preceding estimates with the Jensen formula (or the first main theorem), we get

$$m(r, f(x+c) - f(x)) \le |c|\left(\frac{1}{\rho}(m(\rho, f) + m(\rho, -f)) + (n(\rho, f) + n(\rho, -f))\right)$$

$$\le \frac{|c|}{\rho}\left(T(\rho, f) + T(\rho, -f) + 4T(\rho, f)^{1+\varepsilon} + 4T(\rho, -f)^{1+\varepsilon}\right)$$

$$\le \frac{5|c|}{\rho}\left(T(\rho, f)^{1+\varepsilon} + T(\rho, -f)^{1+\varepsilon}\right)$$

$$\le \frac{10|c|}{\rho}\left(T(\rho, f)^{1+\varepsilon} + o(T(\rho, f))\right).$$

To complete the proof, we may now choose $\rho = r + |c| + \frac{1}{T(r+|c|,f)}$. Applying Theorem 3.12, the assertion readily follows. $\square$

Theorem 3.24. *Let f be tropical meromorphic. Then, for all $\alpha > 1$ and all $r > 0$, we have*

$$m(r, f(x+c) \oslash f(x)) \le \frac{16|c|}{r + |c|}\frac{1}{\alpha - 1}(T(\alpha(r + |c|), f) + |f(0)|/2.$$

Proof. First fix $\rho = \frac{1}{2}(\alpha + 1)(r + |c|) < \alpha(r + |c|)$ as in Lemma 3.21 and observe that $\frac{1}{\alpha+1} < \frac{2}{\alpha-1}$. Next pick from the proof of Theorem 3.23 the inequality

$$m(r, f(x+c) - f(x)) \le |c|\left(\frac{1}{\rho}(m(\rho, f) + m(\rho, -f)) + (n(\rho, f) + n(\rho, -f))\right).$$

Now making use of Lemma 3.21 first to estimate the non-integrated counting functions here with the integrated ones, and then applying the definition and monotonicity of the characteristic function, we obtain

$$m(r, f(x+c) - f(x))$$

$$\leq |c| \left(\frac{2}{(\alpha+1)(r+|c|)} (T(\alpha(r+|c|), f) + T(\alpha(r+|c|), -f)) \right)$$

$$+ \frac{4}{\alpha-1} \frac{1}{r+|c|} (T(\alpha(r+|c|), f) + T(\alpha(r+|c|), -f)).$$

Since $\frac{1}{\alpha+1} \leq \frac{2}{\alpha-1}$, the Jensen formula implies that

$$m(r, f(x+c) - f(x)) \leq \frac{16|c|}{r+|c|} \frac{1}{\alpha-1} T(\alpha(r+|c|), f) + |f(0)|/2.$$

$\square$

We complete this subsection by giving some corollaries in the case of tropical meromorphic functions of finite order, resp. of hyper-order less than one.

Corollary 3.25. *Let f be tropical meromorphic of finite order ρ, and let $\varepsilon > 0$. Then*

$$m(r, f(x+c) \oslash f(x)) = O(r^{\rho+\varepsilon-1})$$

holds outside an exceptional set of finite linear measure.

Proof. Since f is of finite order ρ, we have $T(r, f) \leq r^{\rho+\tilde\varepsilon}$ for all r large enough. Denoting $K := 2^{1+\tilde\varepsilon} 10|c|$, Theorem 3.23 implies that for some δ, $0 < \delta < 1$, say,

$$m(r, f(x+c) \oslash f(x)) \leq \frac{K+\delta}{r} (r+|c|)^{(\rho+\tilde\varepsilon)(1+\tilde\varepsilon)}$$

holds outside an exceptional set of finite linear measure. The claim now immediately follows by choosing $\tilde\varepsilon = \tilde\varepsilon(\varepsilon) > 0$ small enough. $\square$

Corollary 3.26. *Let f be tropical meromorphic of finite order ρ, and take $\tau < 1$, $r_0 > 0$. Then*

$$m(r, f(x+c) \oslash f(x)) = O(T(r, f)/r^\tau)$$

holds for all $r > r_0$ outside an exceptional set of finite logarithmic measure.

Proof. First observe by Remark following Theorem 3.15 that $T(r+c, f) \leq 2T(r, f)$ outside an exceptional set of finite logarithmic measure. Since f is of finite order, we may take σ so that $T(r, f) \leq r^\sigma$ for all $r > r_0$. Choosing $\varepsilon > 0$ in Theorem 3.23 so that $\varepsilon\sigma < 1 - \tau$ yields the assertion. $\square$

A similar estimate of type $m(r, f(x + c) \oslash f(x)) = O\left(e^{(1+\varepsilon)r^\delta}/r\right)$, where $\rho_2 + \delta < 1$, easily follows in the case of hyper-order $\rho_2 < 1$. However, making use of Theorem 3.20, we prefer to formulate this situation as

Theorem 3.27. *Let f be tropical meromorphic of finite hyper-order ρ_2. If $0 < \beta < 1 - \rho_2$ and $\gamma \in (\beta, 1 - \rho_2)$, then*

$$m(r, f(x + c) \oslash f(x)) = O(T(r, f)/r^\gamma) = o(T(r, f)/r^\beta)$$

holds outside an exceptional set of finite logarithmic measure.

Proof. If f happens to be of finite order, then we may take $\beta \in (0, 1)$ arbitrarily. Since by Corollary 3.26

$$m(r, f(x + c) \oslash f(x)) = O(T(r, f)/r^\gamma)$$

and $\gamma > \beta$, it follows that

$$m(r, f(x + c) \oslash f(x)) = o(T(r, f)/r^\beta)$$

as r goes to infinity outside an exceptional set of finite logarithmic measure, proving the claim in this case.

Therefore, we may assume that f is of infinite order. To apply the generalized Borel lemma, Theorem 3.20, we define $\phi(r) := r, \xi(x) := (\log x)^{1+\varepsilon}$ and

$$\alpha = 1 + \frac{\phi(r + |c|)}{(r + |c|)\xi(T(r + |c|, f))} = 1 + \frac{1}{(\log T(r + |c|, f))^{1+\varepsilon}}.$$

By Theorem 3.24, we have

$$m(r, f(x + c) - f(x)) \leq \frac{16|c|}{r + |c|}\frac{1}{\alpha - 1}(T(\alpha(r + |c|), f) + |f(0)|/2$$

$$= \frac{16|c|}{r + |c|}(\log T(r + |c|, f))^{1+\varepsilon}(T(\alpha(r + |c|), f) + |f(0)|/2$$

outside of an exceptional set of finite logarithmic measure. To proceed, fix now $\varepsilon > 0$ so that $(\rho_2(f) + \varepsilon)(1 + \varepsilon) \leq 1 - \gamma < 1 - \beta$. Then, since $(\log T(r + |c|, f))^{1+\varepsilon} \leq (r + |c|)^{1-\gamma}$, we get

$$\frac{(\log T(r + |c|, f))^{1+\varepsilon}}{r + |c|} \leq (r + |c|)^{-\gamma} \to 0$$

as r approaches to infinity. Concerning the term

$$T(\alpha(r + |c|), f) = T\left(r + |c| + \frac{r + |c|}{(\log T(r + |c|, f))^{1+\varepsilon}}, f\right)$$

above, we may apply Theorem 3.20 to conclude that

$$T(\alpha(r + |c|), f) \leq CT(r + |c|, f)$$

outside an exceptional set of finite logarithmic measure, and further by Remark to Theorem 3.15

$$T(\alpha(r + |c|), f) \leq 2T(r, f).$$

Therefore we do have, outside an exceptional set of finite logarithmic measure,

$$m(r, f(x + c) - f(x)) \leq 32|c|C(r + |c|)^{-\gamma}T(r, f)$$
$$= O(r^{\beta-\gamma})\frac{T(r, f)}{r^\beta} = o(1)\frac{T(r, f)}{r^\beta},$$

completing the proof. ◻

3.5 Second main theorem

3.5.1 *General form of the second main theorem*

For the convenience of the reader, we offer two different presentations for the second main theorem in these notes: This section consists of presenting the second main theorem in the basic setting of tropical meromorphic functions $f : \mathbb{R} \to \mathbb{R}$, while Chapter 5 below offers the more general framework of tropical holomorphic curves, including an analogue of the classical Cartan version of the second main theorem (Theorem 5.21). In this section, we first offer a sequence of preparatory lemmas. These lemmas as well as their proofs are first given, in some cases, by using classical notation and reasoning, completing then the lemmas by their corresponding tropical versions. These lemmas form the necessary machinery to the second main theorem, see Theorem 3.40 below. The section will then be completed by some rather immediate consequences and remarks related to the second main theorem.

We first give three elementary observations to invoke the Nevanlinna functions in a moment.

Lemma 3.28. *For any $p \in \mathbb{N}$ and any $a_k \in \mathbb{R}$, $1 \leq k \leq p$, we have*

$$1_\circ \oslash \left(\bigotimes_{k=1}^p (f \oplus a_k)\right) = \bigotimes_{k=1}^p (1_\circ \oslash (f \oplus a_k)).$$

Proof. This lemma is nothing but writing in the tropical notation the following trivial identity:

$$-\sum_{k=1}^{p}\max\{f,a_k\} = \sum_{k=1}^{p}(-\max\{f,a_k\}).$$

$\square$

Lemma 3.29. *For each $p \in \mathbb{N}$, and each $a_k \in \mathbb{R}$, $1 \le k \le p$, we have*

$$\max\left\{\sum_{k=1}^{p}\max\{f,a_k\}, p\max\{a_1,\dots,a_p\}\right\} = p\max\{f,\max\{a_1,\dots,a_p\}\},$$

(3.8)

hence in tropical notation

$$\left(\bigotimes_{k=1}^{p}(f\oplus a_k)\right)\oplus\left(\bigoplus_{k=1}^{p}a_k\right)^{\otimes p} = \left(f\oplus\left(\bigoplus_{k=1}^{p}a_k\right)\right)^{\otimes p}.$$

Proof. Suppose first that $f \le \max\{a_1,\dots,a_p\}$. Looking at the right-hand side of (3.8), we have

$$p\max\{f,\max(a_1,\dots,a_p)\} = p\max\{a_1,\dots,a_p\}.$$

But then

$$\sum_{k=1}^{p}\max\{f,a_k\} \le p\max\{f,\max\{a_1,\dots,a_p\}\} = p\max\{a_1,\dots,a_p\}$$

and so

$$\max\left\{\sum_{k=1}^{p}\max(f,a_k), p\max(a_1,\dots,a_p)\right\} = p\max\{a_1,\dots,a_p\})$$

$$= p\max\{f,\max\{a_1,\dots,a_p\}\}.$$

On the other hand, if $f > \max\{a_1,\dots,a_p\}$, then

$$\sum_{k=1}^{p}\max\{f,a_k\} \ge \sum_{k=1}^{p}\max\{\max\{a_1,\dots,a_p\},a_k\} = p\max\{a_1,\dots,a_p\}.$$

Therefore,

$$\max\left\{\sum_{k=1}^{p}\max\{f,a_k\}, p\max\{a_1,\dots,a_p\}\right\} = \sum_{k=1}^{p}\max\{f,a_k\}$$

$$\le \sum_{k=1}^{p}\max\{f,\max\{a_1,\dots,a_p\}\} = pf$$

and

$$pf = p \max\{f, \max\{a_1, \ldots, a_p\}\} \leq \sum_{k=1}^{p} \max\{f, a_k\}$$

$$= \max \left\{ \sum_{k=1}^{p} \max\{f, a_k\}, p \max\{a_1, \ldots, a_p\} \right\},$$

proving the assertion. $\qquad\square$

Lemma 3.30. *For any* $p \in \mathbb{N}$ *and any* $a_k \in \mathbb{R}$, $1 \leq k \leq p$, *and for* $b_k := -(p-1)a_k$, $1 \leq k \leq p$, *we have*

$$\sum_{k=1}^{p} \max\{x, a_k\} \geq \min_{1 \leq k \leq p} \{\max\{x, a_k\} + (p-1)a_k\} \qquad (3.9)$$

for each $x \in \mathbb{R}$.

Remark 3.31. Clearly, the assertion (3.9) is equivalent to

$$-\sum_{k=1}^{p} \max\{x, a_k\} \leq - \min_{1 \leq k \leq p} \{\max\{x, a_k\} + (p-1)a_k\}$$

$$= - \min_{1 \leq k \leq p} \{\max\{x, a_k\} - b_k\} = \max_{1 \leq k \leq p} \{b_k - \max\{x, a_k\}\}.$$

Therefore, in tropical notation, we obtain the inequality

$$1_\circ \oslash \left(\bigotimes_{k=1}^{p} (f \oplus a_k) \right) \leq \bigoplus_{k=1}^{p} (b_k \oslash (x \oplus a_k)). \qquad (3.10)$$

Proof. To prove Lemma 3.30, we may assume that $a_1 \leq a_2 \leq \cdots \leq a_p$.

Assume first that $x \leq a_1$. Then we have

$$\sum_{k=1}^{p} \max\{x, a_k\} = \sum_{k=1}^{p} a_k \geq pa_1,$$

while

$$\min_{1 \leq k \leq p} \{\max\{x, a_k\} + (p-1)a_k\} = \min_{1 \leq k \leq p} \{a_k + (p-1)a_k\} = p \min_{1 \leq k \leq p} a_k = pa_1.$$

Therefore, the assertion (3.9) follows (with equality in this case).

Next assume that $a_p \leq x$. Then

$$\sum_{k=1}^{p} \max\{x, a_k\} = px \geq x + (p-1)a_1$$

and, on the other hand,

$$\min_{1\le k\le p}\{\max\{x,a_k\}+(p-1)a_k\}=x+(p-1)a_1,$$

and the assertion (3.9) again follows with equality.

To complete the proof, we now have to deal with the case $a_1\le x\le a_p$. Then we must have $a_{j-1}\le x\le a_j$ for some j with $2\le j\le p$. This means that $\max\{x,a_k\}=x$ for all k with $1\le k\le j-1$ and that $\max\{x,a_k\}=a_k$ for all k with $j\le k\le p$. Then, for the left-hand side of (3.9), we obtain

$$\sum_{k=1}^{p}\max\{x,a_k\}=(j-1)x+\sum_{k=j}^{p}a_k\ge x+(j-2)a_{j-1}+\sum_{k=j}^{p}a_k\ge x+(p-1)a_{j-1},$$

and, for the right-hand side of (3.9),

$$\min_{1\le k\le p}\{\max\{x,a_k\}+(p-1)a_k\}\le \max\{x,a_{j-1}\}+(p-1)a_{j-1}=x+(p-1)a_{j-1},$$

completing the proof. $\square$

We now proceed by proving some propositions on Nevanlinna functions needed to prove the second main theorem.

Proposition 3.32. *For each $p\in\mathbb{N}$, each $a_k\in\mathbb{R}$, $1\le k\le p$, and each $c\in\mathbb{R}, c\neq 0$, we have*

$$m\left(r,1_\circ\oslash\left(\bigotimes_{k=1}^{p}(f\oplus a_k)\right)\right)$$

$$\le m(r,1_\circ\oslash f(x+c))+m\left(r,f(x+c)\oslash\left(\bigotimes_{k=1}^{p}(f\oplus a_k)\right)\right)$$

$$=T(r,f(x+c))-N(r,1_\circ\oslash f(x+c))+m\left(r,f(x+c)\oslash\left(\bigotimes_{k=1}^{p}(f\oplus a_k)\right)\right)$$

$$-f(c).$$

Proof. The inequality in the assertion immediately follows from

$$1_\circ\oslash\left(\bigotimes_{k=1}^{p}(f\oplus a_k)\right)=(1_\circ\oslash f(x+c))\otimes\left(f(x+c)\oslash\left(\bigotimes_{k=1}^{p}(f\oplus a_k)\right)\right)$$

by recalling the basic inequality $m(r,g\otimes h)\le m(r,g)+m(r,h)$. To obtain the last part of the assertion, observe that

$$m(r,1_\circ\oslash f(x+c))=T(r,f(x+c))-f(c)-N(r,1_\circ\oslash f(x+c))$$

by the Jensen formula. $\square$

Proposition 3.33. *For each $p \in \mathbb{N}$, each $a_k \in \mathbb{R}$, $1 \leq k \leq p$, and each $c \in \mathbb{R}, c \neq 0$, we have*

$$m\left(r, f(x+c) \oslash \left(\bigotimes_{k=1}^{p}(f \oplus a_k)\right)\right)$$

$$\leq m\left(r, \bigoplus_{k=1}^{p}(f(x+c) \oslash (f(x) \oplus a_k))\right) + (p-1) \max_{1 \leq j \leq p}\{-a_1, \ldots, -a_p, 0\}.$$

Proof. Recalling the notation $b_k := -(p-1)a_k$ in Lemma 3.30, and the attached remark, we first observe that

$$f(x+c) - \sum_{k=1}^{p} \max\{f(x), a_k\} \leq f(x+c) + \max_{k=1}^{p}\{b_k - \max\{f(x), a_k\}\}.$$

But then

$$f(x+c) - \sum_{k=1}^{p} \max\{f(x), a_k\} \leq \max_{k=1}^{p}\{f(x+c) + b_k - \max\{f(x), a_k\}\}$$

$$= \max_{k=1}^{p}\{b_k + (f(x+c) - \max\{f(x), a_k\})\}$$

$$\leq (p-1)\max_{k=1}^{p}\{-a_k\} + \max_{k=1}^{p}\{f(x+c) - \max\{f(x), a_k\}\}$$

$$\leq (p-1)\max_{k=1}^{p}\{-a_k, 0\} + \max_{k=1}^{p}\{f(x+c) - \max\{f(x), a_k\}\}.$$

In tropical notation, this amounts to

$$f(x+c) \oslash \left(\bigotimes_{k=1}^{p}(f \oplus a_k)\right) \leq \bigoplus_{k=1}^{p}(f(x+c) \oslash (f(x) \oplus a_k)) + (p-1)\max\{a_1, \ldots, a_p, 0\}.$$

The asserted inequality now follows by applying the monotonicity of the proximity function. □

Proposition 3.34. *For each $p \in \mathbb{N}$ and each $a_k \in \mathbb{R}$, we have*

$$T\left(r, 1_\circ \oslash \left(\bigotimes_{k=1}^{p}(f \oplus a_k)\right)\right) = m\left(r, 1_\circ \oslash \left(\bigotimes_{k=1}^{p}(f \oplus a_k)\right)\right)$$

$$+ N\left(r, 1_\circ \left(\bigotimes_{k=1}^{p}(f \oplus a_k)\right)\right)$$

$$= T\left(r, \bigotimes_{k=1}^{p}(f \oplus a_k)\right) - \bigotimes_{k=1}^{p}(f(0) \oplus a_k).$$

Proof. To obtain the asserted identity, it is sufficient to apply the Jensen formula to the function $F(x) := \bigotimes_{k=1}^{p}(f \oplus a_k)$. $\qquad\square$

Remark 3.35. In what follows below, we need to apply the observation

$$F(0) = \bigotimes_{k=1}^{p}(f(0) \oplus a_k) = \sum_{k=1}^{p}\max\{f(0), a_k\}.$$

Proposition 3.36. *For each $p \in \mathbb{N}$ and each $a_k \in \mathbb{R}$, we have*

$$N\left(r, 1_\circ \oslash \left(\bigotimes_{k=1}^{p}(f \oplus a_k)\right)\right) \leq \sum_{k=1}^{p} N(r, 1_\circ \oslash (f \oplus a_k)).$$

Proof. First observe that the identity in Lemma 3.28 may be written in the form

$$1_\circ \oslash \left(\bigotimes_{k=1}^{p}(f \oplus a_k)\right) = \sum_{k=1}^{p}(1_\circ \oslash (f \oplus a_k)).$$

The claim now follows by recalling the monotonicity of the counting function from Lemma 3.2(iii). $\qquad\square$

Proposition 3.37. *For each $p \in \mathbb{N}$ and each $a_k \in \mathbb{R}$, we have*

$$T\left(r, \bigotimes_{k=1}^{p}(f \oplus a_k)\right) \leq pT(r, f) + \sum_{k=1}^{p}\max\{a_k, 0\}.$$

Proof. Recalling the monotonicity properties of the characteristic function from Lemma 3.2(iii) and Lemma 3.2(iv), we immediately obtain

$$T\left(r, \bigotimes_{k=1}^{p}(f \oplus a_k)\right) \leq \sum_{k=1}^{p}T(r, f \oplus a_k)$$

$$\leq \sum_{k=1}^{p}(T(r, f) + T(r, a_k))$$

$$= pT(r, f) + \sum_{k=1}^{p}\max\{a_k, 0\}.$$

As to the last equality here, recall that $T(r, a) = \max\{a, 0\}$ for each constant $a \in \mathbb{R}$. $\qquad\square$

Proposition 3.38. *For each $p \in \mathbb{N}$ and each $a_k \in \mathbb{R}$ such that $a_k < L_f := \inf(f(b) : \omega_f(b) < 0)$, we have*

$$T\left(r, \bigotimes_{k=1}^{p}(f \oplus a_k)\right) \geq pT(r, f) - p\max\{a_1, \ldots, a_p, 0\}.$$

Proof. Making use of the tropical notation of Lemma 3.29, and the simple fact that $T(r, g^{\otimes p}) = pT(r, g)$, we first see that

$$T\left(r, \left(\bigotimes_{k=1}^{p}(f \oplus a_k)\right) \oplus \left(\bigoplus_{k=1}^{p} a_k\right)^{\otimes p}\right) = pT\left(r, f \oplus \left(\bigoplus_{k=1}^{p} a_k\right)\right).$$

We next observe that

$$pT\left(r, \left(\bigotimes_{k=1}^{p}(f \oplus a_k)\right) \oplus \left(\bigoplus_{k=1}^{p} a_k\right)^{\otimes p}\right) \geq pT(r, f).$$

This needs a short reasoning. Indeed, we clearly have $f \oplus (\bigoplus_{k=1}^{p} a_k) = \max\{f, \bigoplus_{k=1}^{p} a_k\} \geq f$, and therefore

$$m\left(r, f \oplus \left(\bigoplus_{k=1}^{p} a_k\right)\right) \geq m(r, f)$$

by Remark to Lemma 3.2. On the other hand,

$$N\left(r, f \oplus \left(\bigoplus_{k=1}^{p} a_k\right)\right) = N(r, f),$$

since we have $\bigoplus_{k=1}^{p} a_k = \max\{a_1, \ldots, a_p\} < L_f$ by assumption, and therefore the set of poles of $f \oplus (\bigoplus_{k=1}^{p} a_k)$ equals to the set of poles of f.

Invoking again the monotonicity of the characteristic function, we next conclude that

$$pT\left(r, f \oplus \left(\bigoplus_{k=1}^{p} a_k\right)\right) = T\left(r, \left(\bigotimes_{k=1}^{p}(f \oplus a_k)\right) \oplus \left(\bigoplus_{k=1}^{p} a_k\right)^{\otimes p}\right)$$

$$\leq T\left(r, \bigotimes_{k=1}^{p}(f \oplus a_k)\right) + pT\left(r, \bigoplus_{k=1}^{p} a_k\right).$$

But now

$$T\left(r, \bigoplus_{k=1}^{p} a_k\right) = \max\left\{\left(\max_{k=1}^{p} a_k\right), 0\right\} = \max\{a_1, \ldots, a_p, 0\},$$

and so

$$T\left(r, \bigotimes_{k=1}^{p}(f \oplus a_k)\right) + p\max\{a_1, \ldots, a_p, 0\} \geq pT(r, f),$$

completing the proof. $\qquad\square$

Remark 3.39. Assuming that we have $\max\{a_1, \ldots, a_p\} < L_f$, and combining Proposition 3.37 and Proposition 3.38, we obtain the estimate

$$\left| T\left(r, \bigotimes_{k=1}^{p}(f \oplus a_k)\right) - pT(r, f) \right| \le p\max\{a_1, \ldots, a_p, 0\},$$

i.e.

$$T\left(r, \bigotimes_{k=1}^{p}(f \oplus a_k)\right) = pT(r, f) + O(1).$$

We are now ready formulate and prove the second main theorem in the tropical setting:

Theorem 3.40. *Let f be a tropical meromorphic function and let $c > 0$, $q \in \mathbb{N}$ and $a_1, \ldots, a_q \in \mathbb{R}$ be distinct such that $\max\{a_1, \ldots, a_q\} < L_f = \inf(f(b) : \omega(b) < 0)$. Then*

$$qT(r, f) \le \sum_{j=1}^{q} N(r, 1_\circ \oslash (f \oplus a_j)) + T(r, f(x+c)) - N(r, 1_\circ \oslash f(x+c))$$

$$+ m(r, f(x+c) \oslash f(x)) - f(c) + q\max\{a_1, \ldots, a_q, 0\}$$

$$+ (q-1)\max\{-a_1, \ldots, -a_q, 0\} + \sum_{j=1}^{q}\max\{f(0), a_j\}.$$

Proof. To prove the asserted inequality, we need to apply the preceding propositions from Proposition 3.32 to Proposition 3.38. To make understanding of the following chain of inequalities easier for the reader, we point out the corresponding proposition applied at each phase:

$$qT(r, f) - \sum_{j=1}^{q} N(r, 1_\circ \oslash (f \oplus a_j)) - q\max\{a_1, \ldots, a_q, 0\}$$

$$- (q-1)\max\{-a_1, \ldots, -a_q, 0\}$$

(Proposition 3.36)

$$\le qT(r, f) - N\left(r, 1_\circ \oslash \left(\bigotimes_{k=1}^{q}(f \oplus a_k)\right)\right) - q\max\{a_1, \ldots, a_q, 0\}$$

$$- (q-1)\max\{-a_1, \ldots, -a_q, 0\}$$

(Proposition 3.33)

$$\leq qT(r,f) - N\left(r, 1_\circ \oslash \left(\bigotimes_{k=1}^{q}(f \oplus a_k)\right)\right) - q\max\{a_1, \ldots, a_q, 0\}$$

$$+ m\left(r, \bigoplus_{k=1}^{q}(f(x+c) \oslash (f(x) \oplus a_k))\right) - m\left(r, f(x+c) \oslash \left(\bigotimes_{k=1}^{q}(f \oplus a_k)\right)\right)$$

(Proposition 3.38)

$$\leq T\left(r, \bigotimes_{k=1}^{q}(f \oplus a_k)\right) - N\left(r, 1_\circ \oslash \left(\bigotimes_{k=1}^{q}(f \oplus a_k)\right)\right)$$

$$+ m\left(r, \bigoplus_{k=1}^{q}(f(x+c) \oslash (f(x) \oplus a_k))\right) - m\left(r, f(x+c) \oslash \left(\bigotimes_{k=1}^{q}(f \oplus a_k)\right)\right)$$

(Proposition 3.34)

$$\leq T\left(r, 1_\circ \oslash \left(\bigotimes_{k=1}^{q}(f \oplus a_k)\right)\right) + \max_{k=1}^{q}\{f(0) \oplus a_k\} - N\left(r, 1_\circ \oslash \left(\bigotimes_{k=1}^{q}(f \oplus a_k)\right)\right)$$

$$+ m\left(r, \bigoplus_{k=1}^{q}(f(x+c) \oslash (f(x) \oplus a_k))\right) - m\left(r, f(x+c) \oslash \left(\bigotimes_{k=1}^{q}(f \oplus a_k)\right)\right)$$

(Definition of the characteristic function)

$$= m\left(r, 1_\circ \oslash \left(\bigotimes_{k=1}^{q}(f \oplus a_k)\right)\right) + \max_{k=1}^{q}\{f(0) \oplus a_k\}$$

$$+ m\left(r, \bigoplus_{k=1}^{q}(f(x+c) \oslash (f(x) \oplus a_k))\right) - m\left(r, f(x+c) \oslash \left(\bigotimes_{k=1}^{q}(f \oplus a_k)\right)\right)$$

(Proposition 3.32)

$$\leq T(r, f(x+c)) - N(r, 1_\circ \oslash f(x+c)) - f(c) + \max_{k=1}^{q}\{f(0) \oplus a_k\}$$

$$+ m\left(r, \bigoplus_{k=1}^{q}(f(x+c) \oslash (f(x) \oplus a_k))\right)$$

$$\leq T(r, f(x+c)) - N(r, 1_\circ \oslash f(x+c)) - f(c) + \max_{k=1}^{q}\{f(0) \oplus a_k\}$$

$$+ m(f(x+c) \oslash f(x))).$$

Here the last inequality follows from the fact that $-\max\{f(x), a_k\} \leq f(x)$, and therefore

$$\bigoplus_{k=1}^{q}(f(x+c) \oslash (f(x) \oplus a_k)) = \max_{k=1}^{q}\{f(x+c) - \max\{f(x), a_k\}\}$$

$$\leq f(x+c) - f(x) = f(x+c) \oslash f(x),$$

recalling the monotonicity of the proximity function. The asserted claim now follows by rearranging the inequality

$$qT(r,f) - \sum_{j=1}^{q} N(r, 1_\circ \oslash (f \oplus a_j)) - q\max\{a_1, \ldots, a_q, 0\}$$

$$- (q-1)\max\{-a_1, \ldots, -a_q, 0\}$$

$$\leq T(r, f(x+c)) - N(r, 1_\circ \oslash f(x+c)) - f(c) + \max_{k=1}^{q}\{f(0) \oplus a_k\}$$

$$+ m(r, f(x+c) \oslash f(x))$$

we here obtained. □

3.5.2   *Variants of the second main theorem and deficiencies*

As noted before, tropical meromorphic functions of hyper-order $\rho_2 < 1$ are of special importance in the tropical Nevanlinna theory and its applications. In particular, the second main theorem takes a special, more simple form in this situation:

Theorem 3.41. *Suppose f is a non-constant tropical meromorphic function of hyper-order $\rho_2 < 1$, and let $a_1, \ldots, a_q, q \geq 1$ be distinct real values that satisfy $\max\{a_1, \ldots, a_q\} < L_f$. If $0 < \delta < 1 - \rho_2$, then*

$$(q-1)T(r,f) \leq \sum_{j=1}^{q} N(r, 1_\circ \oslash (f \oplus a_j)) - N(r, 1_\circ \oslash f) + o(T(r,f)/r^\delta)$$

outside an exceptional set of finite logarithmic measure.

Proof. To prove the claim, we need to invoke Theorem 3.15 and Theorem 3.27 to simplify the right hand side of the second main theorem. Indeed, first observe that

$$m(r, f(x+c) \oslash f(x)) = o(T(r,f)/r^\delta)$$

by Theorem 3.27. But then

$$T(r, f(x+c)) \leq m(r, f(x+c) \oslash f(x)) + m(r, f(x)) + N(r, f(x+c))$$

$$\leq m(r, f) + N(r, f(x+c)) + o(T(r,f)/r^\delta)$$

$$= T(r, f) + N(r, f(x+c)) - N(r, f) + o(T(r,f)/r^\delta).$$

By Theorem 3.15, we now conclude that

$$N(r, 1_\circ \oslash f) = N((r - |c|) + |c|, 1_\circ \oslash f)$$

$$= N(r - |c|, 1_\circ \oslash f) + o\left(\frac{T(r - |c|, f)}{(r - |c|)^\delta}\right)$$

$$= N(r - |c|, 1_\circ \oslash f) + o(T(r,f)/r^\delta),$$

and therefore

$$N(r, 1_\circ \oslash f(x + c)) \geq N(r - |c|, 1_\circ \oslash f) = N(r, 1_\circ \oslash f) + o(T(r, f)/r^\delta).$$

Observing that the remaining terms in the second main theorem are constants, we may now combine the previous inequalities to obtain

$$qT(r, f) \leq \sum_{j=1}^{q} N(r, 1_\circ \oslash (f \oplus a_j)) + T(r, f) + N(r, f(x + c)) - N(r, f)$$
$$+ N(r, 1_\circ \oslash f) + o(T(r, f)/r^\delta).$$

It now remains to observe that

$$N(r, f(x + c)) \leq N(r + |c|, f) = N(r, f) + o(T(r, f)/r^\delta)$$

by applying Theorem 3.15 again. □

Corollary 3.42. *Suppose f is a non-constant tropical meromorphic function of hyper-order $\rho_2 < 1$, and suppose that $a < L_f$. If $0 < \delta < 1 - \rho_2$, then*

$$N(r, 1_\circ \oslash f) \leq N(r, 1_\circ \oslash (f \oplus a)) + o(T(r, f)/r^\delta).$$

Proof. Take $q = 1$ in Theorem 3.41. □

Remark 3.43. Recalling the notion of defect from the classical Nevanlinna theory, we may define

$$\delta(\infty, f) := 1 - \limsup_{r \to \infty} \frac{N(r, f)}{T(r, f)}$$

and

$$\delta(a, f) := 1 - \limsup_{r \to \infty} \frac{N(r, 1_\circ \oslash (f \oplus a))}{T(r, f)}$$

for $a \in \mathbb{R}$. Since $1_\circ \oslash f = 1_\circ \oslash (f \oplus (-\infty))$, we may also define

$$\delta(-\infty, f) := \delta(\infty, 1_\circ \oslash f).$$

Therefore, Corollary 3.42 tells that whenever $\delta(\infty, 1_\circ \oslash f) = 0$ and $a < L_f$, then $\delta(a, f) = 0$, provided that $\rho_2(f) < 1$. A bit of additional analysis brings us to the somewhat surprising conclusion that tropical meromorphic functions of hyper-order $\rho_2 < 1$ have, in some sense, no deficient values, except possibly roots and poles. To this end, we prove

Theorem 3.44. *Suppose f is a non-constant tropical meromorphic function of hyper-order $\rho_2 < 1$, and let $a_1, \ldots, a_q$, $q \geq 1$ be distinct real values that satisfy*

$$\max\{a_1, \ldots, a_q\} < L_f. \tag{3.11}$$

Moreover, suppose that

$$\ell_f := \inf\{f(a) \,:\, \omega_f(a) > 0\} > -\infty. \tag{3.12}$$

If now $0 < \delta < 1 - \rho_2$, then

$$qT(r, f) \leq \sum_{j=1}^{q} N(r, 1_\circ \oslash (f \oplus a_j)) + o(T(r, f)/r^\delta) \tag{3.13}$$

outside an exceptional set of finite logarithmic measure.

Proof. Fix now a real number $\mu < \min\{\max\{a_1, \ldots, a_q\}, \ell_f\}$, and define $g := f - \mu$, $b_j := a_j - \mu$ for $j = 1, \ldots, q$ and $b_0 := 0$. Then we may immediately apply Theorem 3.41 to the function g and the $q + 1$ distinct values $b_0, \ldots, b_q$ to obtain

$$qT(r, g) \leq \sum_{j=0}^{q} N(r, 1_\circ \oslash (g \oplus b_j)) - N(r, 1_\circ \oslash g) + o(T(r, g)/r^\delta)$$

outside an exceptional set of finite logarithmic measure. Clearly, we note that $\ell_g = \ell_f - \mu > 0$. Therefore, $1_\circ \oslash (g \oplus 0) = -\max\{g, 0\}$ and $1_\circ \oslash g = -g$ have the same roots. Hence,

$$N(r, 1_\circ \oslash (g \oplus b_0)) = N(r, 1_\circ \oslash g).$$

This now first results in

$$qT(r, g) \leq \sum_{j=1}^{q} N(r, 1_\circ \oslash (g \oplus b_j)) + o(T(r, g)/r^\delta)$$

outside an exceptional set of finite logarithmic measure. Since $f \oplus a_j - \mu = \max\{f, a_j\} - \mu = \max\{f, b_j + \mu\} - \mu = \max\{f - \mu, b_j\} = g \oplus b_j$, we next get

$$qT(r, f - \mu) \leq \sum_{j=1}^{q} N(r, 1_\circ \oslash (f \oplus a_j - \mu)) + o(T(r, f)/r^\delta).$$

But now $T(r, f) \leq T(r, f - \mu) + T(r, \mu) = T(r, f - \mu) + \mu^+$, and

$$N(r, 1_\circ \oslash (g \oplus b_j)) = N(r, -(f \oplus a_j) - \mu) = N(r, -f \oplus a_j + \mu)$$
$$= N(r, -f \oplus a_j) = N(r, 1_\circ \oslash (f \oplus a_j)).$$

The claim now readily follows. $\qquad\qquad\qquad\qquad\qquad\qquad\square$

Corollary 3.45. *Suppose f is a non-constant tropical meromorphic function of hyper-order $\rho_2 < 1$, and take $0 < \delta < 1 - \rho_2$. If now $\ell_f > -\infty$ and $a < L_f$, then*

$$T(r, f) \leq N(r, 1_\circ \oslash (f \oplus a)) + o(T(r, f)/r^\delta)$$

outside an exceptional set of finite logarithmic measure.

Remark 3.46. Observe that by this result, a non-constant tropical meromorphic function of hyper-order $\rho_2 < 1$ and such that $\ell_f > -\infty$ has no deficient values $a < L_f$ in the sense that $\delta(a, f) = 1$. However, one should keep in mind that completely omitted values may well appear, as one can see by looking at linear functions that have neither roots nor poles, resp. suitable rational functions that may well have no roots, resp. no poles. Moreover, the assumption that $\rho_2 < 1$ cannot be deleted. As an example, recall that for each tropical exponential function e_β with $|\beta| < 1$, all values $a < 0$ are deficient: $\delta(a, e_\beta) \geq 1/2$, see Remark after Proposition 1.24.

Chapter 4

Clunie and Mohon'ko type theorems

In this chapter, we prove tropical counterparts to three lemmas from classical complex analysis, frequently used in applications of Nevanlinna theory. As to these classical results, the reader may look at a lemma due to Valiron and Mohon'ko, see [68], Theorem 2.2.5, another one due to Mohon'ko, see [68], Proposition 9.2.3, and the original Clunie lemma in [18]. Previous tropical versions for these lemmas may be found in [46], [71] and [69]. The first two of these references are treating the restricted case of integer slopes only.

4.1 Valiron–Mohon'ko and Mohon'ko lemmas in tropical setting

Before proceeding to formulate these results, we need to recall what are tropical difference polynomials in tropical meromorphic functions and their shifts.

To determine the necessary notations, let $\lambda = (\lambda_0, \lambda_1, \ldots, \lambda_m)$ be a multi-index of real numbers, and consider

$$P_\lambda(x, f) := \bigotimes_{j=0}^{p} f(x + c_j)^{\otimes \lambda_j}$$
$$= \lambda_0 f(x) + \lambda_1 f(x + c_1) + \cdots + \lambda_m f(x + c_m)$$

with given real shifts $c_1, \ldots, c_p$ in $(-\infty, +\infty)$. Then, an expression of the form

$$P(x, f) = \bigoplus_{\lambda \in \Lambda} a_\lambda(x) \otimes P_\lambda(x, f) = \bigoplus_{\lambda \in \Lambda} a_\lambda(x) \otimes \bigotimes_{j=0}^{p} f(x + c_j)^{\otimes \lambda_j},$$

i.e.

$$P(x, f) = \max_{\lambda \in \Lambda} \left\{ a_\lambda(x) + \sum_{j=0}^{p} \lambda_j f(x + c_j) \right\}$$

with tropical meromorphic coefficients $a_\lambda(x)$ ($\lambda \in \Lambda$) over a finite index set $\Lambda = \Lambda[P]$ is called a **tropical difference polynomial** of total degree

$$\deg(P) := \max_{\lambda \in \Lambda[P]} \|\lambda\| \ (\in \mathbb{R})$$

in f and its shifts, with $\|\lambda\| := \lambda_0 + \cdots + \lambda_p$. Observe that the degree $\|\lambda\|$ may well be $= 0$ or even negative. This is an aspect to be carefully checked in the subsequent considerations.

In what follows, we frequently need to consider the maximum $\max_\lambda a_\lambda(x)$ of the coefficients $a_\lambda(x)$ of a tropical difference polynomial as well as the maximum $\max_\lambda \{-a_\lambda(x)\}$ of $-a_\lambda(x)$. One should perhaps remark that $\max_\lambda \{-a_\lambda(x)\}$ does not coincide with the maximum of the coefficients of $1_\circ \oslash P$. Indeed,

$$1_\circ \oslash P(x, f) = - \bigoplus_{\lambda \in \Lambda} a_\lambda(x) \otimes \bigotimes_{j=0}^{p} f(x + c_j)^{\otimes \lambda_j}$$

$$= - \max_{\lambda \in \Lambda} \left\{ a_\lambda(x) + \sum_{j=0}^{p} \lambda_j f(x + c_j) \right\}$$

$$= \min_{\lambda \in \Lambda} \left\{ -a_\lambda(x) - \sum_{j=0}^{p} \lambda_j f(x + c_j) \right\}.$$

Later on, in our considerations of the Clunie lemma, we also need to consider the maximum of the leading terms of a tropical difference polynomial P, say $M(P) := \max_{\|\lambda\|=\deg(P)} a_\lambda(x)$. In that connection, i.e., in the proof of Theorem 4.7, we need to separate these maxima for different tropical difference polynomials. To this end, we may apply the notations such as $\max_P a_\lambda(x), \max_Q a_\lambda$ etc.

In this section we consider tropical difference polynomials of tropical meromorphic functions f under the assumption that f is of hyper-order $\rho_2(f) < 1$. In such a situation, we say that a function $\alpha(x)$ is **small** (in the tropical sense) with respect to f, if $T(r, \alpha) = S_\delta(r, f)$ holds with a quantity $S_\delta(r, f) = o(T(r, f)/r^\delta)$ outside of a set of finite logarithmic measure, where $0 < \delta < 1 - \rho_2(f)$. Similarly, we say that α is **proximity small** (in the tropical sense) with respect to f, if $m(r, \alpha) = S_\delta(r, f)$ outside of a set of finite logarithmic measure. In particular, these notions are being applied to coefficients of tropical difference polynomials. Observe, however, that all key results in this subsection are first presented without the above assumption for the hyper-order of f; the cases with hyper-order $\rho_2(f) < 1$ then follow as immediate corollaries.

The proofs in this subsection essentially rely on the notion of the proximity function only, in addition to some elementary analysis. We first offer a preliminary proposition:

Proposition 4.1. *Suppose that f is tropical meromorphic and a tropical difference polynomial $P(x, f)$ contains a term $P_\lambda(x, f)$ with $\|\lambda\| > 0$. Then*

$$\|\lambda\| m(r, f) \leq \sum_{j=1}^{p} \lambda_j m(r, f(x) \oslash f(x + c_j)) + m(r, P(x, f)) + m(r, 1_\circ \oslash a_\lambda)$$

$$(4.1)$$

Proof. First observe that

$$P(x, f) = \max_{\lambda \in \Lambda} \left\{ a_\lambda(x) + \sum_{j=0}^{p} \lambda_j f(x + c_j) \right\} \geq a_\lambda(x) + \sum_{j=0}^{p} \lambda_j f(x + c_j)$$

for each $\lambda \in \Lambda$. Therefore,

$$m\left(r, a_\lambda(x) + \sum_{j=0}^{p} \lambda_j f(x + c_j)\right) \leq m(r, P(x, f))$$

holds for all $\lambda \in \Lambda$. Recalling that $\|\lambda\| = \lambda_0 + \cdots + \lambda_p$, we may apply the monotonicity of the proximity function to obtain

$$\|\lambda\| m(r, f) = m(r, \|\lambda\| f)$$

$$= m\left(r, \sum_{j=0}^{p} \lambda_j \big(f(x) - f(x + c_j)\big) + \sum_{j=0}^{p} \lambda_j f(x + c_j) + a_\lambda(x) - a_\lambda(x)\right)$$

$$\leq m\left(r, \sum_{j=0}^{p} \lambda_j \big(f(x) - f(x + c_j)\big)\right) + m\left(r, a_\lambda(x) + \sum_{j=0}^{p} \lambda_j f(x + c_j)\right)$$

$$+ m(r, -a_\lambda)$$

$$\leq m\left(r, \sum_{j=0}^{p} \lambda_j \big(f(x) - f(x + c_j)\big)\right) + m(r, P(x, f)) + m(r, -a_\lambda),$$

completing the proof. □

Corollary 4.2. *Suppose that the assumptions of Proposition 4.1 are satisfied. If, in addition, f is of hyper-order $\rho_2(f) < 1$, $\deg(P) > 0$ and $m(r, P(x, f)) = S_\delta(r, f)$, then*

$$m(r, f) = S_\delta(r, f).$$

Proof. The claim immediately follows from Theorem 3.27.   □

The following theorem may now be understood as a partial tropical counterpart to the classical Valiron–Mohon'ko result, see Theorem A.12 in Appendix A, [71], Theorem 2.3 and [69], Theorem 6.2:

Theorem 4.3. *Given a tropical meromorphic function f and a tropical difference polynomial*

$$P(x, f) = \bigoplus_{\lambda \in \Lambda} \left(a_\lambda(x) \otimes \bigotimes_{j=1}^{p} f(x + c_j)^{\otimes \lambda_j} \right),$$

we have

$$\left| m(r, P(x, f)) - m(r, \deg(P)f(x)) \right|$$

$$\leq \max \left\{ m(r, \max_\lambda a_\lambda(x)) + m\left(r, \max_{\lambda \in \Lambda} \bigotimes_{j=1}^{p}(f(x + c_j) - f(x))^{\otimes \lambda_j}\right), \right.$$

$$\left. m(r, \max_\lambda\{-a_\lambda(x)\}) + m\left(r, \max_{\lambda \in \Lambda} \bigotimes_{j=1}^{p}(f(x + c_j) - f(x))^{\otimes(-\lambda_j)}\right) \right\}.$$

Proof. First observing that

$$P(x, f) \leq \max_{\lambda \in \Lambda} a_\lambda(x) + \max_{\lambda \in \Lambda} \sum_{j=0}^{p} \lambda_j f(x + c_j)$$

$$\leq \max_{\lambda \in \Lambda} a_\lambda(x) + \max_{\lambda \in \Lambda} \sum_{j=0}^{p} \lambda_j \{f(x + c_j) - f(x)\} + \deg(P)f(x),$$

we obtain

$$m(r, P(x, f)) \leq m(r, \max_\lambda a_\lambda(x)) + m\left(r, \max_{\lambda \in \Lambda} \bigotimes_{j=1}^{p}(f(x + c_j) - f(x))^{\otimes \lambda_j}\right)$$

$$+ m(r, \deg(P)f(x)).$$

On the other hand, for any $\lambda \in \Lambda$,

$$P(x, f) \geq a_\lambda(x) + \sum_{j=0}^{p} \lambda_j \{f(x + c_j) - f(x)\} + \|\lambda\| f(x),$$

that is,

$$\|\lambda\| f(x) \leq P(x, f) - a_\lambda(x) - \sum_{j=0}^{p} \lambda_j \{f(x + c_j) - f(x)\}$$

holds, and therefore

$$\deg(P)f(x) \leq P(x, f) + \max_\lambda \{-a_\lambda(x)\} + \max_{\lambda \in \Lambda} \sum_{j=0}^{p} (-\lambda_j)\{f(x + c_j) - f(x)\}.$$

This now implies, by monotonicity of the proximity function again,

$$m(r, \deg(P)f(x)) \leq m(r, P(x, f)) + m(r, \max_\lambda \{-a_\lambda(x)\})$$

$$+ m\left(r, \max_{\lambda \in \Lambda}\left\{\bigotimes_{j=1}^{p}(f(x + c_j) - f(x))^{\otimes(-\lambda_j)}\right\}\right),$$

completing the proof. □

By Theorem 3.27, we now conclude, as a special case of Theorem 4.3,

Corollary 4.4. *Suppose that the assumptions of Theorem 4.3 are satisfied. If, in addition, f is of hyper-order $\rho_2(f) < 1$, $\deg(P) > 0$ and the coefficients of $P(x, f)$ are all small with respect to f, then*

$$m(r, P(x, f)) = \deg(P)m(r, f) + S_\delta(r, f).$$

The next theorem is related, in spirit, to the Mohon'ko lemma in complex analysis, see Theorem A.17 in Appendix A. For a slightly different previous version in the tropical setting, see [69], Theorem 6.4. An essential difference while compared to the classical complex analysis is that our assumptions below mean that $P(x, 0) = 0$, while in complex analysis $P(x, 0) \neq 0$.

Theorem 4.5. *Let f be a tropical meromorphic solution of a tropical difference polynomial equation*

$$P(x, f) = \bigoplus_{\lambda \in \Lambda} a_\lambda(x) \otimes \bigotimes_{j=0}^{p} f(x + c_j)^{\otimes \lambda_j} = 0$$

such that $\|\lambda\| \neq 0$ for all $\lambda \in \Lambda$. Then both

$$m(r, f) \leq 2\left(\max_{\lambda \in \Lambda}\left|\frac{1}{\|\lambda\|}\right|\right)\left(m(r, \max_\lambda |a_\lambda(x)|)\right.$$

$$\left. + \sum_{j=0}^{p} \max_{\lambda \in \Lambda}\left|\frac{\lambda_j}{\|\lambda\|}\right| m(r, |f(x + c_j) \oslash f(x)|)\right),$$

(4.2)

and

$$m\big(r, 1_\circ \oslash (f \oplus a)\big) \le 2 \left(\max_{\lambda \in \Lambda} \left| \frac{1}{\|\lambda\|} \right| \right) \left(m\big(r, \max_\lambda |a_\lambda(x)|\big) \right.$$

$$\left. + \sum_{j=0}^p \max_{\lambda \in \Lambda} \left| \frac{\lambda_j}{\|\lambda\|} \right| m\big(r, |f(x+c_j) \oslash f(x)|\big) \right) \tag{4.3}$$

hold for each $a \in \mathbb{R}$.

Proof. Writing equation $P(x, f) = 0$ in the form

$$P(x, f) = \max_{\lambda \in \Lambda} \left\{ a_\lambda(x) + \sum_{j=0}^p \lambda_j f(x + c_j) \right\} = 0,$$

we first see that

$$a_\lambda(x) + \sum_{j=0}^p \lambda_j f(x + c_j) \le 0 \quad (x \in \mathbb{R})$$

for each $\lambda \in \Lambda$.

Fixing now $r \in \mathbb{R}$ for a while, there exists $\lambda_r = (\lambda_{r,1}, \ldots, \lambda_{r,p}) \in \Lambda$ such that

$$a_{\lambda_r}(r) + \sum_{j=0}^p \lambda_{r,j}(f(r + c_j) - f(r)) + \|\lambda_r\| f(r) = 0.$$

Therefore,

$$a_{\lambda_r}(r) + \sum_{j=0}^p \lambda_{r,j}(f(x + c_j) - f(x)) = -\|\lambda_r\| f(r)$$

holds. Since $\|\lambda_r\| \ne 0$ by assumption, we have

$$f(r) = -\frac{1}{\|\lambda_r\|} a_{\lambda_r}(r) - \sum_{j=0}^p \frac{\lambda_{r,j}}{\|\lambda_r\|}(f(r + c_j) - f(r)).$$

This now implies

$$f^+(r) \le |f(r)| \le \left| \frac{1}{\|\lambda_r\|} \right| |a_{\lambda_r}(r)| + \sum_{j=0}^p \left| \frac{\lambda_{r,j}}{\|\lambda_r\|} \right| |f(r + c_j) - f(r)|$$

$$\le \max_{\lambda \in \Lambda} \left| \frac{1}{\|\lambda\|} \right| (|a_{\lambda_r}(r)| + |a_{\lambda_r}(-r)|) + \sum_{j=0}^p \max_{\lambda \in \Lambda} \left| \frac{\lambda_j}{\|\lambda\|} \right| |f(r + c_j) - f(r)|.$$

Similarly,

$$f(-r) = -\frac{1}{\|\lambda_{-r}\|} a_{\lambda_{-r}}(-r) - \sum_{j=0}^{p} \frac{\lambda_{-r,j}}{\|\lambda_{-r}\|} (f(-r+c_j) - f(-r))$$

and therefore

$$f^+(-r) \leq |f(-r)| \leq \max_{\lambda \in \Lambda} \left| \frac{1}{\|\lambda\|} \right| (|a_{\lambda_{-r}}(r)| + |a_{\lambda_{-r}}(-r)|)$$

$$+ \sum_{j=0}^{p} \max_{\lambda \in \Lambda} \left| \frac{\lambda_j}{\|\lambda\|} \right| |f(-r+c_j) - f(-r)|.$$

Averaging the estimates for $f^+(r)$ and $f^+(-r)$, we conclude that

$$m(r,f) \leq \max_{\lambda \in \Lambda} \left| \frac{1}{\|\lambda\|} \right| (m(r, |a_{\lambda_r}|) + m(r, |a_{\lambda_{-r}}|))$$

$$+ \frac{1}{2} \sum_{j=0}^{p} \max_{\lambda \in \Lambda} \left| \frac{\lambda_j}{\|\lambda\|} \right| (|f(r+c_j) - f(r)| + |f(-r+c_j) - f(-r)|)$$

$$\leq 2 \max_{\lambda \in \Lambda} \left| \frac{1}{\|\lambda\|} \right| m(r, \max_{\lambda \in \Lambda} |a_\lambda|)$$

$$+ \sum_{j=0}^{p} \max_{\lambda \in \Lambda} \left| \frac{\lambda_j}{\|\lambda\|} \right| m(r, |f(r+c_j) - f(r)|),$$

and the first assertion now immediately follows.

To prove the second assertion, we first estimate $m(r, 1_\circ \oslash f)$. Similarly as above, we first see that

$$-f(r) = \frac{1}{\|\lambda_r\|} a_{\lambda_r}(r) + \sum_{j=0}^{p} \frac{\lambda_{r,j}}{\|\lambda_r\|} (f(r+c_j) - f(r)),$$

and

$$-f(-r) = \frac{1}{\|\lambda_{-r}\|} a_{\lambda_{-r}}(-r) + \sum_{j=0}^{p} \frac{\lambda_{-r,j}}{\|\lambda_{-r}\|} (f(-r+c_j) - f(-r)).$$

Therefore, we readily obtain for $(-f(r))^+$, resp. for $(-f(-r))^+$, the estimates corresponding to what we had for $(f(r))^+$, resp. for $(f(-r))^+$ above, and the second assertion now immediately follows for $m(r, 1_\circ \oslash f)$.

Finally, since $-\max\{f(x), a\} \leq -f(x)$ holds for any $a \in \mathbb{R}$,

$$m(r, 1_\circ \oslash (f \oplus a)) \leq m(r, 1_\circ \oslash f),$$

and the second assertion has been completely proved. $\qquad \square$

Corollary 4.6. *If, in addition to the assumptions of the previous theorem, f is of hyper-order $\rho_2(f) < 1$ and f is a solution of a tropical difference polynomial equation with small coefficients:*

$$P(x, f) = \bigoplus_{\lambda \in \Lambda} a_\lambda(x) \otimes f^{\otimes \lambda}(x) = 0$$

such that $\|\lambda\| \not= 0$ for all $\lambda \in \Lambda$, then

$$m(r, f) = S_\delta(r, f), \qquad m(r, -f) = S_\delta(r, f)$$

and for any $a \in \mathbb{R} \setminus \{0\}$,

$$m\big(r, 1_\circ \oslash (f \oplus a)\big) = m(r, -\max\{f, a\}) \leq m(r, -f) = m(r, 1_\circ \oslash f) = S_\delta(r, f).$$

Proof. This is an immediate consequence of Theorem 3.27 combined with the conclusion of Theorem 4.5. □

4.2 Tropical Clunie lemma

The rest of this chapter is devoted to proving the Clunie lemma, see Theorem 4.9 below. This lemma, see [18], Lemma 1, for the original version of the result, later on appeared to be a most powerful tool in the field of ordinary differential equations in the complex plane. The basic idea in these applications has been to show that the proximity function $m(r, P(z, f))$ of a differential polynomial is small whenever a meromorphic function f satisfies an ordinary differential equation of type $f^n P(z, f) = Q(z, f)$, provided that the coefficients of $P(z, f), Q(z, f)$ are small, and the total degree of $Q(z, f)$ in f and its derivatives is $\leq n$, see e.g. [68], Lemma 2.4.2, and the numerous applications therein. One may find several versions of the Clunie lemma in subsequent complex analytic literature. As to the most general one, we may refer to [113], Theorem 1. In addition to the differential versions of the Clunie lemma, several corresponding difference versions have recently appeared as well, see e.g. [40], Theorem 3.1. The most general version in the difference setting may be found in [70], Theorem 2.3.

As to the Clunie lemma in the tropical setting, recall first the basic form due to Halburd and Southall in [46], Theorem 4.5:

Theorem 4.7. *Let $P(x, f), Q(x, f)$ be two tropical difference polynomials with small coefficients. If f is a tropical meromorphic function satisfying equation*

$$f(x)^{\otimes n} \otimes P(x, f) = Q(x, f)$$

such that the degree of Q in f and its shifts is at most n, then for any $\varepsilon > 0$,

$$m\big(r, P(x, f)\big) = O\left\{ r^{-1}\Big(T(r + |c|, f)^{1+\varepsilon} + o\{T(r + |c|, f)\}\Big)\right\},$$

holds outside of an exceptional set of finite logarithmic measure.

Restricting us to the case of finite order, an easy application of Lemma 3.25 proves

Corollary 4.8. *Let $P(x, f), Q(x, f)$ be two tropical difference polynomials with small coefficients. If f is a tropical meromorphic function of finite order ρ satisfying equation*

$$f^{\otimes n}(x) \otimes P(x, f) = Q(x, f)$$

such that the degree of Q in f and its shifts is at most n, then

$$m(r, f) = S_\rho(r, f).$$

Here we skip proving the two preceding tropical Clunie variants, as they are just special cases of the following version, which may be understood to corresponding [70], Theorem 2.3. This theorem is essentially as [69], Theorem 6.6; for a slightly restricted version in the integer slope case, see [71], Theorem 3.3. Concerning the proof below, observe that proving the complex analytic versions, see [113] and [70], extensive applications of Hölder-type inequalities are needed, but this is not the case in the tropical setting.

To simplify the notations for the tropical Clunie theorem below, we denote $\Delta_c f$ for the c-difference operator of a tropical meromorphic function f, hence

$$(\Delta_c f)(x) = f(x + c) - f(x).$$

Observe that

$$|\Delta_c f| = \max\{\Delta_c f, -\Delta_c f\}$$

and

$$|(\Delta_c f)^{\otimes \lambda_j}| = \max\{(\Delta_c f)^{\otimes \lambda_j}, -(\Delta_c f)^{\otimes \lambda_j}\} = \max\{\lambda_j \Delta_c f, -\lambda_j \Delta_c f\}.$$

Moreover, by the definition of the proximity function,

$$|\Delta_c f|(r) = |\Delta_c f|^+(r) \le |\Delta_c f|^+(r) + |\Delta_c f|^+(-r) = 2m(r, |\Delta_c f|)$$

and

$$|(\Delta_c f)^{\otimes \lambda_j}| \le 2m(r, |(\Delta_c f) \otimes \lambda_j|).$$

By a similar reasoning, we obtain

$$|f|(r) \le 2m(r, |f|)$$

for any tropical meromorphic function.

Theorem 4.9. *Let $P(x, f), Q(x, f), H(x, f)$ be tropical difference polynomials in f and its shifts,*

$$P(x, f) = \bigoplus_{\lambda \in \Lambda[P]} a_\lambda(x) \bigotimes_{j=1}^{p} f(x + c_j)^{\otimes \lambda_j},$$

$$Q(x, f) = \bigoplus_{\mu \in \Lambda[Q]} b_\mu(x) \bigotimes_{j=1}^{p} f(x + c_j)^{\otimes \mu_j},$$

$$H(x, f) = \bigoplus_{\nu \in \Lambda[H]} d_\nu(x) \bigotimes_{j=1}^{p} f(x + c_j)^{\otimes \nu_j}.$$

If f is a tropical meromorphic function satisfying equation

$$H(x, f) \otimes P(x, f) = Q(x, f), \tag{4.4}$$

if $\deg(P) \ge 0$ and if $\deg(Q) \le \deg(H)$ for the total degrees in f and its shifts, then

$$m(r, P(x, f)) \le 2\Big\{ \max_{\lambda \in \Lambda[P]} m(r, |a_\lambda|) + \max_{\mu \in \Lambda[Q]} m(r, |b_\mu|) + \max_{\nu \in \Lambda[H]} m(r, |d_\nu|) \Big\}$$

$$+ 2\Big\{ \max_{\lambda \in \Lambda[P]} m\Big(r, \sum_{j=0}^{p} |(\Delta_{c_j} f)^{\otimes \lambda_j}| \Big) + \max_{\mu \in \Lambda[Q]} m\Big(r, \sum_{j=0}^{p} |(\Delta_{c_j} f)^{\otimes \mu_j}| \Big)$$

$$+ \max_{\nu \in \Lambda[H]} m\Big(r, \sum_{j=0}^{p} |(\Delta_{c_j} f)^{\otimes \nu_j}| \Big) \Big\}.$$

Proof. Given a fixed $r > 0$, we put

$$S_+ := \{ s : f(s) \ge 0, |s| = r \} \quad \text{and} \quad S_- := \{ s : f(s) < 0, |s| = r \}.$$

Clearly, $S_+ \cap S_- = \emptyset$ and $S_+ \cup S_- = \{\pm r\}$. Then

$$m(r, P(x, f)) = \frac{1}{2}(P(r, f)^+ + P(-r, f)^+)$$

may formally be written as

$$m(r, P(x, f)) = \frac{1}{2} \left(\sum_{s \in S_+} P(s, f)^+ + \sum_{s \in S_-} P(s, f)^+ \right). \qquad (4.5)$$

If now $s \in S_-$, then we may use the notations and estimates above to obtain

$$P(s, f) = \max_{\lambda \in \Lambda[P]} \left\{ a_\lambda(s) + \sum_{j=0}^{p} \lambda_j \big(f(s + c_j) - f(s) \big) + \|\lambda\| f(s) \right\}$$

$$\leq \max_{\lambda \in \Lambda[P]} \left\{ a_\lambda(s) + \max_{\lambda \in \Lambda[P]} \sum_{j=0}^{p} \lambda_j \big(f(s + c_j) - f(s) \big) \right\} + \deg(P) f(s)$$

$$\leq \max_{\lambda \in \Lambda[P]} |a_\lambda|(s) + \max_{\lambda \in \Lambda[P]} \left\{ \sum_{j=0}^{p} \big| (f(s + c_j) - f(s))^{\otimes \lambda_j} \big| \right\}$$

$$\leq 2 \max_{\lambda \in \Lambda[P]} m(r, |a_\lambda|) + 2 \max_{\lambda \in \Lambda[P]} \sum_{j=0}^{p} m \big(r, |(\Delta_{c_j} f)^{\otimes \lambda_j}| \big).$$

We applied here the observation that $\deg(P) f(s) \leq 0$, due to our present assumptions that $\deg(P) \geq 0$ and $f(s) < 0$. Thus we immediately get for $s \in S_-$,

$$P(s, f)^+ \leq 2 \max_{\lambda \in \Lambda[P]} m(r, |a_\lambda|) + 2 \max_{\lambda \in \Lambda[P]} \sum_{j=0}^{p} m \big(r, |(\Delta_{c_j} f)^{\otimes \lambda_j}| \big). \qquad (4.6)$$

When $s \in S_+$, we similarly obtain

$$Q(s, f) \leq \max_{\mu \in \Lambda[Q]} |b_\mu(s)| + \max_{\mu \in \Lambda[Q]} \sum_{j=0}^{p} \big| (f(s + c_j) - f(s))^{\otimes \mu_j} \big| + \deg(Q) f(s),$$
$$(4.7)$$

while

$$Q(s, f) = H(s, f) \otimes P(s, f)$$

$$= P(s, f) + \max_{\nu \in \Lambda[H]} \left\{ d_\nu(s) + \sum_{j=0}^{p} \nu_j \big(f(s + c_j) - f(s) \big) + \|\nu\| f(s) \right\}$$

$$\geq P(s, f) + \max_{\nu \in \hat{\Lambda}[H]} \left\{ d_\nu(s) + \sum_{j=0}^{p} \nu_j \big(f(s + c_j) - f(s) \big) + \deg(H) f(s) \right\}$$

$$= P(s, f) + \max_{\nu \in \hat{\Lambda}[H]} \left\{ d_\nu(s) + \sum_{j=0}^{p} \nu_j \big(f(s + c_j) - f(s) \big) \right\} + \deg(H) f(s),$$

where $\nu \in \hat{\Lambda}[H]$ stands for those multi-indices of H only that are of maximal total degree in f and its shifts. Taking now $\tau \in \hat{\Lambda}[H]$ such that

$$d_\tau(s) + \sum_{j=0}^p \tau_j(f(s+c_j) - f(s)) = \max_{\nu \in \hat{\Lambda}[H]} \left\{ d_\nu(s) + \sum_{j=0}^p \nu_j(f(s+c_j) - f(s)) \right\},$$

we therefore conclude that

$$P(s,f) + \deg(H)f(s)$$

$$\leq Q(s,f) - \left\{ d_\tau(s) + \sum_{j=0}^p \tau_j\big(f(s+c_j) - f(s)\big) \right\}$$

$$\leq Q(s,f) + |d_\tau(s)| + \sum_{j=0}^p \tau_j\big(f(s+c_j) - f(s)\big)$$

$$\leq Q(s,f) + \max_{\nu \in \Lambda[H]} |d_\nu(s)| + \max_{\nu \in \Lambda[H]} \left| \sum_{j=0}^p \nu_j\big(f(s+c_j) - f(s)\big) \right|$$

$$\leq Q(s,f) + \max_{\nu \in \Lambda[H]} |d_\nu(s)| + \max_{\nu \in \Lambda[H]} \left| \bigotimes_{j=0}^p (f(s+c_j) - f(s))^{\otimes \nu_j} \right|.$$

This together with (4.7) shows that

$$P(s,f) \leq Q(s,f) + \max_{\nu \in \Lambda[H]} |d_\nu(s)| + \max_{\nu \in \Lambda[H]} \left| \bigotimes_{j=0}^p (f(s+c_j) - f(s))^{\otimes \nu_j} \right|$$

$$- \deg(H)f(s)$$

$$\leq \max_{\mu \in \Lambda[Q]} |b_\mu(s)| + \max_{\nu \in \Lambda[H]} |d_\nu(x)| + \max_{\mu \in \Lambda[Q]} \bigotimes_{j=0}^p \left| (f(s+c_j) - f(s))^{\otimes \mu_j} \right|$$

$$+ \max_{\nu \in \Lambda[H]} \bigotimes_{j=0}^p \left| (f(s+c_j) - f(s))^{\otimes \nu_j} \right| + (\deg(Q) - \deg(H))f(s)$$

$$\leq \max_{\mu \in \Lambda[Q]} |b_\mu(s)| + \max_{\nu \in \Lambda[H]} |d_\nu(x)| + \max_{\mu \in \Lambda[Q]} \bigotimes_{j=0}^p \left| (f(s+c_j) - f(s))^{\otimes \mu_j} \right|$$

$$+ \max_{\nu \in \Lambda[H]} \bigotimes_{j=0}^p \left| (f(s+c_j) - f(s))^{\otimes \nu_j} \right|,$$

since $\{\deg(Q) - \deg(H)\}f(s) \leq 0$. Similarly as to above, we obtain

$$
\begin{aligned}
P(s,f)^+ \leq 2\Big\{ & m\big(r, \max_{\mu \in \Lambda[Q]} b_\mu(s)\big) + m\big(r, \max_{\nu \in \Lambda[H]} d_\nu(s)\big) \\
& + \max_{\mu \in \Lambda[Q]} m\big(r, |(\Delta_{c_j} f)^{\otimes \mu_j}|\big) + \max_{\nu \in \Lambda[H]} m\big(r, |(\Delta_{c_j} f)^{\otimes \nu_j}|\big) \Big\}.
\end{aligned}
\tag{4.8}
$$

Inserting both (4.6) and (4.8) into (4.5), we obtain the desired estimate. □

Remark 4.10. Observe that the estimate for the proximity function of $P(x, f)$ in Theorem 4.9 above may easily be slightly improved by more careful estimates in the proof. However, as we saw in the section on various forms of the lemma on tropical quotients, Theorem 4.9 becomes most useful while considering meromorphic functions of hyper-order $\rho_2 < 1$. As to this case, we apply Theorem 3.27 to obtain the following two corollaries. Note that possible slight improvements to Theorem 4.9 have no effect to the conclusions in these corollaries.

Corollary 4.11. *Let* $H(x, f), P(x, f), Q(x, f)$ *be tropical difference polynomials in* f *and its shifts with small coefficients. If* f *is a tropical meromorphic function of hyper-order* $\rho_2(f) < 1$ *satisfying equation*

$$
H(x, f) \otimes P(x, f) = Q(x, f)
\tag{4.9}
$$

such that $\deg(Q) \leq \deg(H)$ *in* f *and its shifts, then for* δ *with* $0 < \delta < 1 - \rho_2$,

$$
m(r, f) = S_\delta(r, f).
$$

Corollary 4.12. *Let* $P(x, f), Q(x, f)$ *be tropical difference polynomials in* f *and its shifts with small coefficients and* $\alpha \in \mathbb{R}$. *If* f *is a tropical meromorphic function of hyper-order* $\rho_2(f) < 1$ *satisfying equation*

$$
f(x)^{\otimes \alpha} \otimes P(x, f) = Q(x, f)
\tag{4.10}
$$

such that $\deg(Q) \leq \alpha$ *in* f *and its shifts, then for* δ *with* $0 < \delta < 1 - \rho_2$,

$$
m(r, f) = S_\delta(r, f).
$$

Chapter 5

Tropical holomorphic curves

Cartan's generalization of Nevanlinna's second main theorem for holomorphic curves in the finite dimensional projective space is an important extension of value distribution theory, which has surprising applications in the study of meromorphic functions in the complex plane as well. The purpose of this chapter is to introduce a tropical version of Cartan's value distribution theory. Following [67], we will define a tropical analogue of Cartan's characteristic function, and show that it is well defined and reduces exactly to the tropical characteristic function introduced in Chapter 3 in the special case where the tropical holomorphic curve is one-dimensional. The central result of this section is a tropical analogue of Cartan's second theorem. We will show that this result is a natural extension of the tropical second main theorem, Theorem 3.40, by proving that under a certain natural non-degeneracy condition on targets, the tropical Cartan theorem reduces to Theorem 3.40 in the one-dimensional case. The tropical Cartan's second main theorem also implies a second main theorem type inequality, which appears to include a previously unknown ramification type term. We will conclude the section by a discussion on ramification in the tropical meromorphic functions. In order to study the properties of tropical holomorphic curves, we first need to introduce some notions from tropical linear algebra.

5.1 Tropical matrixes and determinants

In this short section we will introduce only the most basic notions related to tropical matrixes and determinants. Let $A = (a_{ij})$ and $B = (b_{ij})$ be $(n+1) \times (n+1)$ matrices, so that the indexes i and j satisfy $i, j \in \{0, \ldots, n\}$. The matrix operations of tropical addition and tropical multiplication are defined by

$$A \oplus B = (a_{ij} \oplus b_{ij})$$

and

$$A \otimes B = \left(\bigoplus_{k=0}^{n} a_{ik} \otimes b_{kj} \right),$$

respectively. We denote by $A^\top$ the transpose of the matrix A defined in the usual way. The matrix A is called **regular** if A contains at least one element different from $0_\circ$ in each row. We adopt the definition of Yoeli [114] and define the tropical determinant $|A|_\circ$ of A by

$$|A|_\circ = \bigoplus a_{0\pi(0)} \otimes a_{1\pi(1)} \otimes \cdots \otimes a_{n\pi(n)},$$

where the tropical summation is taken over all permutations

$$\{\pi(0), \pi(1), \ldots, \pi(n)\}$$

of $\{0, 1, \ldots, n\}$. Due to the fact that there is no subtraction in the tropical arithmetics, the definition of the tropical determinant is exactly the same as the definition of the tropical permanent $\mathrm{per}(A)$ of the matrix A [74, 114]. It follows immediately by the definitions of regularity and the tropical determinant that a matrix A is regular if and only if $|A|_\circ \neq 0_\circ$.

5.2   Tropical Casoratian

Similarly as the Casorati determinant is a difference analogue of the Wronskian determinant, the tropical Casoratian is a natural ultra-discrete version of the Casoratian. The tropical Casorati determinant is in a key role in the formulation of a tropical analogue of Cartan's second main theorem below. In what follows we go through some of the basic properties of the tropical Casoratian applied to a collection of tropical entire functions.

Let $g(x)$ be a tropical entire function, let $n \in \mathbb{N}$, and fix a constant $c \in \mathbb{R} \setminus \{0\}$. Then we denote

$$g(x) \equiv g, \quad g(x+c) \equiv \overline{g}, \quad g(x+2c) \equiv \overline{\overline{g}} \quad \text{and} \quad g(x+nc) \equiv \overline{g}^{[n]}.$$

In other words, we have suppressed the x-dependence of g for the sake of brevity, and, for the same reason we do not state the c-dependence of the shift $x + c$ explicitly from now on. Now the **tropical Casorati determinant** of tropical entire functions $g_0, \ldots, g_n$ is defined by

$$C_\circ(g_0, g_1, \ldots, g_n) = \bigoplus \overline{g}_0^{[\pi(0)]} \otimes \overline{g}_1^{[\pi(1)]} \otimes \cdots \otimes \overline{g}_n^{[\pi(n)]}, \tag{5.1}$$

where the sum is taken over all permutations $\{\pi(0), \ldots, \pi(n)\}$ of $\{0, \ldots, n\}$. The tropical Casorati determinant is also called as the **tropical Casoratian**, and it satisfies the following properties.

Lemma 5.1. *If $g_0, \ldots, g_n$ and h are tropical entire functions, then*

(i) $C_\circ(g_0, g_1, \ldots, g_i, \ldots, g_j, \ldots, g_n) = C_\circ(g_0, g_1, \ldots, g_j, \ldots, g_i, \ldots, g_n)$ *for all $i, j \in \{0, \ldots, n\}$ such that $i \neq j$.*
(ii) $C_\circ(1_\circ, g_1, \ldots, g_n) \geq C_\circ(\overline{g}_1, \ldots, \overline{g}_n)$.
(iii) $C_\circ(0_\circ, g_1, \ldots, g_n) = 0_\circ$.
(iv) $C_\circ(g_0 \otimes h, g_1 \otimes h, \ldots, g_n \otimes h) = h \otimes \overline{h} \otimes \cdots \otimes \overline{h}^{[n]} \otimes C_\circ(g_0, g_1, \ldots, g_n)$.

Proof. The property (i) follows straightforwardly by the definition of the tropical Casoratian. Namely,

$$C_\circ(g_0, g_1, \ldots, g_i, \ldots, g_j, \ldots, g_n)$$
$$= \bigoplus \overline{g}_{\pi(0)}^{[0]} \otimes \overline{g}_{\pi(1)}^{[1]} \otimes \cdots \otimes \overline{g}_{\pi(i)}^{[i]} \otimes \cdots \otimes \overline{g}_{\pi(j)}^{[j]} \otimes \cdots \otimes \overline{g}_{\pi(n)}^{[n]}$$
$$= \bigoplus \overline{g}_{\pi(0)}^{[0]} \otimes \overline{g}_{\pi(1)}^{[1]} \otimes \cdots \otimes \overline{g}_{\pi(j)}^{[j]} \otimes \cdots \otimes \overline{g}_{\pi(i)}^{[i]} \otimes \cdots \otimes \overline{g}_{\pi(n)}^{[n]}$$
$$= C_\circ(g_0, g_1, \ldots, g_j, \ldots, g_i, \ldots, g_n).$$

In order to show that properties (ii) are (iii) are true, we expand the tropical Casoratian $C_\circ(g_0, g_1, \ldots, g_n)$ with respect to the first column vector to obtain

$$C_\circ(g_0, g_1, \ldots, g_n) = g_0 \otimes C_\circ(\overline{g}_1, \ldots, \overline{g}_n) \oplus \cdots \oplus \overline{g}_0^{[n]} \otimes C_\circ(g_1, , \ldots, g_n).$$

From this expression the assertion of property (ii) follows by substituting $g_0 = 1_\circ$. Similarly, property (iii) follows by taking $g_0 = 0_\circ$.

Finally we prove that the property (iv) holds. In order to do this, we first use the fact that tropical determinants are invariant under transposing. This observation implies by (5.1) that

$$C_\circ(g_0, g_1, \ldots, g_n) = \bigoplus \overline{g}_{\pi(0)}^{[0]} \otimes \overline{g}_{\pi(1)}^{[1]} \otimes \cdots \otimes \overline{g}_{\pi(n)}^{[n]}. \tag{5.2}$$

Next, we multiply both sides of equation (5.2) by the tropical entire function $h \otimes \overline{h} \otimes \cdots \otimes \overline{h}^{[n]}$, to obtain

$$h \otimes \overline{h} \otimes \cdots \otimes \overline{h}^{[n]} \otimes C_\circ(g_0, g_1, \ldots, g_n)$$
$$= \bigoplus (\overline{g_{\pi(0)} \otimes h})^{[0]} \otimes (\overline{g_{\pi(1)} \otimes h})^{[1]} \otimes \cdots \otimes (\overline{g_{\pi(n)} \otimes h})^{[n]}$$
$$= \bigoplus (\overline{g_0 \otimes h})^{[\pi(0)]} \otimes (\overline{g_1 \otimes h})^{[\pi(1)]} \otimes \cdots \otimes (\overline{g_n \otimes h})^{[\pi(n)]}$$
$$= C_\circ(g_0 \otimes h, g_1 \otimes h, \ldots, g_n \otimes h).$$

$\square$

The properties (i)-(iv) of Lemma 5.1 are at least to some extent direct analogues of properties of the ordinary Casoratian $C(g_0, g_1, \ldots, g_n)$ applied to the collection $g_0, g_1, \ldots, g_n$ of entire functions. We will next discuss to what extent such analogues exist. If also h is an entire function, then the Casoratian $C(g_0, g_1, \ldots, g_n)$ satisfies the following properties:

(a) $C(g_0, g_1, \ldots, g_i, \ldots, g_j, \ldots, g_n) = -C(g_0, g_1, \ldots, g_j, \ldots, g_i, \ldots, g_n)$ for all $i, j \in \{0, \ldots, n\}$.

(b) $C(1, g_1, \ldots, g_n) = C(\Delta g_1, \ldots, \Delta g_n)$ with $\Delta g = \overline{g} - g$.

(c) $C(0, g_1, \ldots, g_n) = 0$.

(d) $C(hg_0, hg_1, \ldots, hg_n)(x) = \prod_{k=0}^{n} h(x+k) C(g_0, g_1, \ldots, g_n)(x)$.

It is clear that (a) corresponds to (i), (c) to (iii) and (d) to (iv), but the property (b) of the complex plane Casoratian appears to be different to its proposed tropical counterpart, Lemma 5.1 (ii), which is only an inequality. In general the inequality in the opposite direction does not hold, however. For any $x \in \mathbb{R}$, there exists a permutation $\pi_x(j)$ of $\{0, 1, \ldots, n\}$ such that

$$C_\circ(1_\circ, g_1, \ldots, g_n)(x) = \overline{1_\circ}^{[\pi_x(0)]}(x) \otimes \overline{g_1}^{[\pi_x(1)]}(x) \otimes \cdots \otimes \overline{g_n}^{[\pi_x(n)]}.$$

In the special case where $\pi_x(0) = 0$ it follows that $\pi_x(j)$ $(1 \leq j \leq n)$ is a permutation of $\{1, \ldots, n\}$, and so

$$C_\circ(1_\circ, g_1, \ldots, g_n)(x) = \overline{g_1}^{[\pi_x(1)]}(x) \otimes \cdots \otimes \overline{g_n}^{[\pi_x(n)]} \leq C_\circ(\overline{g_1}, \ldots, \overline{g_n})(x).$$

On the other hand, if $\pi(0) = n$, then $\pi_x(j+1)$ $(0 \leq j \leq n-1)$ is a permutation of $\{0, \ldots, n-1\}$ and therefore it follows that

$$C_\circ(1_\circ, g_1, \ldots, g_n)(x) = \overline{g_1}^{[\pi_x(1)]}(x) \otimes \cdots \otimes \overline{g_n}^{[\pi_x(n)]} \leq C_\circ(g_1, \ldots, g_n)(x).$$

If neither of the special cases occur, it is a straightforward matter to select such tropical entire functions $g_0, g_1, \ldots, g_n$ for which the inequality fails to be valid.

5.3   Tropical linear independence

In order to define tropical holomorphic curves and study their properties, we need a notion of linear independence for finite collections of tropical entire functions. The first naive idea of introducing this notion is to simply replace the operations of usual addition and multiplication in the classical definition of linear independence by tropical addition and multiplication, respectively, and call the outcome a definition of tropical linear independence. However, this would lead to considering linear combinations of the form

$$\bigoplus_{\nu=0}^{n} a_\nu \otimes g_\nu \equiv 0_\circ, \quad \text{that is,} \quad \max_{0 \leq \nu \leq n} \{a_\nu + g_\nu\} = -\infty, \qquad (5.3)$$

where $g_0, \ldots, g_n$ are tropical entire functions. The equation (5.3) holds if and only if $a_\nu = 0_\circ$ for all $\nu \in \{0, \ldots, n\}$. Therefore, by this naive notion of tropical linear independence, all finite collections of tropical entire functions are linearly independent, which is clearly an absurd conclusion. Even the two identically same tropical entire functions would be linearly independent! There are several natural ways to go around this problem. But first, we need to define the notion of **tropical linear combination** as follows over $\mathbb{R}_{\max} =: \mathbb{R} \cup \{-\infty\}$.

Definition 5.2. If $g_0, \ldots, g_n$ are tropical entire functions and $a_0, \ldots, a_n \in \mathbb{R}_{\max}$, then

$$f = \bigoplus_{\nu=0}^{n} a_\nu \otimes g_\nu = \bigoplus_{i=0}^{j} a_{k_i} \otimes g_{k_i} \qquad (5.4)$$

is called a tropical linear combination of $g_0, \ldots, g_n$ over $\mathbb{R}_{\max}$, where the index set $\{k_0, \ldots, k_j\} \subset \{0, \ldots, n\}$ is such that $a_{k_i} \in \mathbb{R}$ for all $i \in \{0, \ldots, j\}$, while $a_\nu = 0_\circ$ if $\nu \notin \{k_0, \ldots, k_j\}$.

Tropical linear combination (5.4) may also be written in an inner product form

$$f = (a_0, \ldots, a_n) \otimes (g_0, \ldots, g_n)^\top,$$

by making use of tropical matrix operations defined above. Now we can define the notion of tropical linear independence in a useful way. As mentioned above, there are several ways to do this.

Definition 5.3. It is said that $f_1, \ldots, f_n$ are **weakly linearly independent** over $\mathbb{R}_{\max}$ if there is $i \in \{1, \ldots, n\}$ such that f_i can be expressed as a linear combination of $f_1, \ldots, f_{i-1}, f_{i+1}, \ldots, f_n$.

The notion of tropical linear independence, which fits for our purposes most naturally is due to Gondran and Minoux [28,29]. The following definition is adjusted so that it is applicable for finite collections of tropical meromorphic functions over the max-plus algebra $\mathbb{R}_{\max}$.

Definition 5.4. Tropical meromorphic functions $f_0, \ldots, f_n$ are **linearly dependent** (resp. **independent**) in the **Gondran-Minoux sense** if there exist (resp. there do not exist) two disjoint subsets I and J of $K := \{0, \ldots, n\}$ such that $I \cup J = K$ and

$$\bigoplus_{i \in I} \alpha_i \otimes f_i = \bigoplus_{j \in J} \alpha_j \otimes f_j, \quad \text{that is,} \quad \max_{i \in I}\{\alpha_i + f_i\} = \max_{j \in J}\{\alpha_j + f_j\}, \quad (5.5)$$

where the constants $\alpha_0, \ldots, \alpha_n \in \mathbb{R}_{\max}$ are not all equal to $0_\circ$.

If one of the index sets I or J in equation (5.5) is the empty set, then the corresponding tropical sum vanishes. And by vanishing we mean that the sum is identically equal to $0_\circ = -\infty$. However, this is impossible in the case when $f_0, \ldots, f_n$ in (5.5) are Gondran-Minoux linearly independent. Namely, if say $I = \emptyset$ in (5.5), then $J = K$ and so

$$\alpha_0 \otimes f_0 \oplus \cdots \oplus \alpha_n \otimes f_n = 0_\circ,$$

which is only possible if $\alpha_0 = \cdots = \alpha_n = 0_\circ$. This contradicts the fact that by Definition 5.4 there is at least one non-zero constant amongst $\alpha_0, \ldots, \alpha_n$. As an example of an application of the concept of Gondran-Minoux linear independence to tropical entire functions, we prove a result concerning linear independence of a collection of tropical units. Tropical units are analogues the of exponential function in the complex plane, so the following proposition is a natural analogue of the so-called Borel's lemma (see [96, Theorem A.3.3])

Proposition 5.5. *Let $f_0, \ldots, f_n$ be tropical units such that $f_i \oslash f_j$ are not constants for any distinct i and j. Then $f_0, \ldots, f_n$ are linearly independent in the Gondran and Minoux sense.*

Proof. We assume, conversely to the assertion, that there exists a collection of constants $\alpha_j \in \mathbb{R}_{\max}$, which are not all equal to $0_\circ$, such that

$$\bigoplus_{i \in I} \alpha_i \otimes f_i(t) = \bigoplus_{j \in J} \alpha_j \otimes f_j(t), \quad (5.6)$$

where I and J are disjoint subsets of $K := \{0, \ldots, n\}$ satisfying $I \cup J = K$. Since both the left and the right side of (5.6) are tropical rational functions, there exist a constant $t_0 \in \mathbb{R}$, and indexes $i_0 \in I$ and $j_0 \in J$ such that

$$\alpha_{i_0} \otimes f_{i_0}(t) = \alpha_{j_0} \otimes f_{j_0}(t)$$

for all $t \geq t_0$. From the fact that f_{i_0} and f_{j_0} are both tropical units, we have

$$f_{i_0} \oslash f_{j_0} = \alpha_{j_0} \oslash \alpha_{i_0},$$

which contradicts the assumption that the tropical quotient of the distinct tropical units in the set $\{f_0, \ldots, f_n\}$ is always non-constant. $\quad\square$

In some cases it is possible to represent a tropical linear combination f given by (5.4) using only a subset of the functions $g_0, \ldots, g_n$. There are two situations when this happens. First, if

$$a_{\nu_0} = 0_\circ$$

for at least one index $\nu_0 \in \{0, \ldots, n\}$, and second, if

$$f(x) > a_{\nu_0} \otimes g_{\nu_0}(x)$$

for all $x \in \mathbb{R}$, again for at least one $\nu_0 \in \{0, \ldots, n\}$. In these cases f can be considered to be 'degenerate' in the sense that we do not need to utilize all of the functions $g_0, \ldots, g_n$ in the set $G := \{g_0, \ldots, g_n\}$ to express it. One important example is the case where $g_0, \ldots, g_n$ are Gondran-Minoux linearly dependent. Then any tropical linear combination of $g_0, \ldots, g_n$ can be always written by using only a subset of G, and is therefore degenerate in the above sense. In order to define the notion of degeneracy in exact terms, we introduce the following adaptation of the concepts introduced in [3].

Definition 5.6. Let $G = \{g_0, \ldots, g_n\}(\neq \{0_\circ\})$ be a set of tropical entire functions, linearly independent in the Gondran-Minoux sense, and let

$$\mathcal{L}_G = \text{span}\langle g_0, \ldots, g_n \rangle = \left\{ \bigoplus_{k=0}^{n} a_k \otimes g_k : (a_0, \ldots, a_n) \in \mathbb{R}_{\max}^{n+1} \right\}$$

be their **linear span**. (Note that $\bigoplus_{k=0}^{n} a_k \otimes g_k = 0_\circ$ when each a_k is equal to $0_\circ$ so that $0_\circ \in \mathcal{L}_G$.) The collection G is called as the **spanning basis** of $\mathcal{L}_G$. The shortest length of the representation of $f \in \mathcal{L}_G \setminus \{0_\circ\}$ is defined by

$$\ell(f) = \min \left\{ j \in \{1, \ldots, n+1\} : f = \bigoplus_{i=1}^{j} a_{k_i} \otimes g_{k_i} \right\},$$

where $a_{k_i} \in \mathbb{R}$ with integers $0 \leq k_1 < k_2 < \cdots < k_j \leq n$, and the **dimension** of $\mathcal{L}_G$ is

$$\dim(\mathcal{L}_G) = \max\{\ell(f) : f \in \mathcal{L}_G \setminus \{0_\circ\}\}. \tag{5.7}$$

If $G = \{g_0, \ldots, g_n\}$ is a set of Gondran-Minoux linearly independent tropical entire functions, then the definition of the dimension of the linear span $\mathcal{L}_G$ implies that there exist at least one tropical entire function $f \in \mathcal{L}_G \setminus \{0_\circ\}$ such that

$$f = \bigoplus_{i=1}^{\dim(\mathcal{L}_G)} a_{k_i} \otimes g_{k_i}$$

where $a_{k_i} \in \mathbb{R}$ for all $i \in \{1, \ldots, \dim(\mathcal{L}_G)\}$, and

$$f \neq \bigoplus_{i=1}^{\dim(\mathcal{L}_G)-1} b_{m_i} \otimes g_{m_i}$$

for all collections of constants $b_{m_i} \in \mathbb{R}$ with $i \in \{1, \ldots, \dim(\mathcal{L}_G)-1\}$. If the dimension of the linear span is the same as the number of elements in G, then there is at least one f in $\mathcal{L}_G$ which cannot be represented as a tropical linear combination of only some of the functions $\{g_0, \ldots, g_n\}$. Such tropical linear combinations are referred to from now on as **complete**. The formal definition is as follows.

Definition 5.7. Let $G = \{g_0, \ldots, g_n\}(\neq \{0_\circ\})$ be a set of tropical entire functions, linearly independent in the Gondran-Minoux sense, and let f be a tropical linear combination of $g_0, \ldots, g_n$. If $\ell(f) = n+1$, then f is said to be complete. That is, the coefficients a_k in any expression of f of the form

$$f = \bigoplus_{k=0}^{n} a_k \otimes g_k$$

must satisfy $a_k \in \mathbb{R}$ for all $k \in \{0, \ldots, n\}$ so that $\mathcal{L}_G$ attains its full dimension, $\dim(\mathcal{L}_G) = n+1$.

Definition 5.6 immediately implies that the linear span $\mathcal{L}_G$ satisfies $\dim(\mathcal{L}_G) \geq 1$ for any non-empty finite collection G of tropical entire functions, since $G \subset \mathcal{L}_G$ and $\ell(g) = 1$ when $g \in G \setminus \{0_\circ\}$. The following example illustrates some of the properties of $\dim(\mathcal{L}_G)$.

Example 5.8. In classical linear algebra, if you have a collection of, say, three linearly independent polynomials, the corresponding linear space would always be of dimension three. In this example we will discuss a tropical analogue of this fact by using the notion of Gondran-Minoux linear independence. By choosing

$$g_0(x) = x, \quad g_1(x) \equiv 1_\circ \quad \text{and} \quad g_2(x) = 1_\circ \oslash x,$$

we obtain a collection $G = \{g_0, g_1, g_2\}$ of tropical polynomials which are Gondran-Minoux linearly independent. Now the linear span of this set of tropical polynomials satisfies

$$\text{span}\langle x, 1_\circ, -x \rangle = \{g(x : a, b, c) := \max\{x + a, b, -x + c\} : a, b, c \in \mathbb{R}_{\max}\},$$

which means that $\ell(g(x : a, b, c)) = 3$ if and only if all of the coefficients a, b and c are real (so that none of them is equal to $0_\circ = -\infty$) and satisfy the inequality $b > (a + c)/2$. Therefore the maximum

$$\dim(\mathcal{L}_G) = 3$$

in (5.7) is attained for any collection of coefficients $a, b, c \in \mathbb{R}$ of g such that $b > (a + c)/2$.

Tropical linear changes applied to the function in the spanning basis G may cause a loss of information in a different way to the classical case. Suppose that the collection $\{g_0, g_1, g_2\}$ is replaced by the tropical linear combinations

$$f_0(x) := a_{00} \otimes g_0(x) \oplus a_{01} \otimes g_1(x) = \max\{x + a_{00}, a_{01}\},$$
$$f_1(x) := a_{11} \otimes g_1(x) \oplus a_{12} \otimes g_2(x) = \max\{a_{11}, -x + a_{12}\},$$
$$f_2(x) := a_{20} \otimes g_0(x) \oplus a_{22} \otimes g_2(x) = \max\{x + a_{20}, -x + a_{22}\}$$

which can also be expressed as

$$\begin{pmatrix} f_0(x) \\ f_1(x) \\ f_2(x) \end{pmatrix} = \begin{pmatrix} a_{00} & a_{01} & 0_\circ \\ 0_\circ & a_{11} & a_{12} \\ a_{20} & 0_\circ & a_{22} \end{pmatrix} \otimes \begin{pmatrix} g_0(x) \\ g_1(x) \\ g_2(x) \end{pmatrix}. \tag{5.8}$$

By definition the coefficient matrix is regular in the tropical sense if the tropical determinant of the coefficient matrix does not vanish. This condition is clearly satisfied, for example, if $a_{ij} \neq 0_\circ$ for all $i, j \in \{0, 1, 2\}$ in the coefficient matrix, since then

$$\begin{vmatrix} a_{00} & a_{01} & 0_\circ \\ 0_\circ & a_{11} & a_{12} \\ a_{20} & 0_\circ & a_{22} \end{vmatrix}_\circ = a_{00} \otimes a_{11} \otimes a_{22} \oplus a_{01} \otimes a_{12} \otimes a_{20}$$

$$= \max\{a_{00} + a_{11} + a_{22}, a_{01} + a_{12} + a_{20}\} \in \mathbb{R}.$$

The coefficient matrix does not vanish in the tropical sense also in the case where exactly one of the sets $\{a_{00}, a_{11}, a_{22}\}$ and $\{a_{01}, a_{12}, a_{20}\}$ contains the element $0_\circ$. But this case is degenerate in the sense that at least one of the tropical linear combinations f_j is identically the same as one of the functions g_j. Hence we omit further considerations of this special case.

Let us now consider the matrix equation

$$\begin{pmatrix} f_0(x) \\ f_1(x) \\ f_2(x) \end{pmatrix} = \begin{pmatrix} a_{00} & a_{01} & 0_\circ \\ 0_\circ & a_{11} & a_{12} \\ a_{20} & a_* & a_{22} \end{pmatrix} \otimes \begin{pmatrix} g_0(x) \\ g_1(x) \\ g_2(x) \end{pmatrix} \tag{5.9}$$

obtained from (5.8) by replacing the middle component of the third row in the coefficient matrix. Now

$$f_2(x) = \max\{x + a_{20}, a_*, -x + a_{22}\}, \tag{5.10}$$

and so equation (5.9) is valid under the condition that each component of the third row of the coefficient matrix is different from $0_\circ$, and

$$(a_{20} + a_{22})/2 \geq a_* > 0_\circ.$$

Equation (5.10) gives an alternative representation for f_2, which makes use of all three functions g_0, g_1 and g_2. Despite of this f_2 is not complete in the sense of Definition 5.7, since

$$\ell(f_2) = 2.$$

Similar modification is not possible for either of $f_0(x)$ and $f_1(x)$ due to the forms of the functions $g_2(x)$ and $g_0(x)$.

Even if we impose such conditions on the coefficient matrix of (5.8), which make it regular in the tropical sense, we still cannot represent the function $g_1(x) = 1_\circ$ as a tropical linear combination of the functions $f_0(x)$, $f_1(x)$ and $f_2(x)$, and thus $g_1 \notin \mathrm{span}\langle f_0, f_1, f_2 \rangle$. This is due to the fact that regularity of a tropical matrix does not always imply its invertibility. In this particular case, for example, a straightforward calculation shows that there is no tropical matrix (b_{ij}) $(i, j \in \{0, 1, 2\})$ such that

$$\begin{pmatrix} a_{00} & a_{01} & 0_\circ \\ 0_\circ & a_{11} & a_{12} \\ a_{20} & 0_\circ & a_{22} \end{pmatrix} \otimes \begin{pmatrix} b_{00} & b_{01} & b_{02} \\ b_{10} & b_{11} & b_{12} \\ b_{20} & b_{21} & b_{22} \end{pmatrix} = \begin{pmatrix} 1_\circ & 0_\circ & 0_\circ \\ 0_\circ & 1_\circ & 0_\circ \\ 0_\circ & 0_\circ & 1_\circ \end{pmatrix}.$$

In the previous example we discussed a set of tropical linear combinations $\{f_0, f_1, f_2\}$ which consisted of functions that are not complete over the spanning basis $\{g_0, g_1, g_2\}$. We call such sets of tropical linear combinations **degenerate**. The exact definition is as follows.

Definition 5.9. Let $G = \{g_0, \ldots, g_n\}$ be a set of tropical entire functions, linearly independent in the Gondran-Minoux sense, and let $Q \subset \mathcal{L}_G$ be a collection of tropical linear combinations of G over $\mathbb{R}_{\max}$. The **degree of degeneracy** of Q is defined to be

$$\mathrm{ddg}(Q) = \mathrm{card}\left(\{f \in Q : \ell(f) < n + 1\}\right).$$

If $\mathrm{ddg}(Q) = 0$ then Q is called **non-degenerate**.

This definition implies that the degree of degeneracy of a set of tropical linear combinations over $G = \{g_0, \ldots, g_n\}$ is the number of non-complete elements of the set in question. Moreover, the assumed Gondran-Minoux linear independence of G in the definition enables the construction of a set Q of $n + 1$ complete linear combinations over G. One can say that $\mathrm{card}(Q)$

attains its 'full dimension', $n+1$, as that of the spanned space L_G by G, if each element of Q is complete. In this way the number of complete elements of Q is the 'actual dimension' of the subspace spanned by Q, and so $\mathrm{ddg}(Q)$ is the 'codimension' of the subspace spanned by Q.

5.4   Tropical holomorphic curves

We will now generalize some of the key concepts of tropical value distribution theory to the **tropical projective space** $\mathbb{TP}^n$. The space $\mathbb{TP}^n$ is defined in the following way. Let the equivalence relation $\sim$ be defined so that

$$(a_0, a_1, \ldots, a_n) \sim (b_0, b_1, \ldots, b_n)$$

if and only if

$$(a_0, a_1, \ldots, a_n) = \lambda \otimes (b_0, b_1, \ldots, b_n) := (\lambda \otimes b_0, \lambda \otimes b_1, \ldots, \lambda \otimes b_n)$$

for some $\lambda \in \mathbb{R}$. We denote by $[a_0 : a_1 : \cdots : a_n]$ the equivalence class of $(a_0, a_1, \ldots, a_n)$. When $a_0 \in \mathbb{R}$, we may take

$$(a_1 \oslash a_0, \ldots, a_n \oslash a_0) \in \mathbb{R}_{\max}^n$$

as a representative element of $[a_0 : a_1 : \cdots : a_n]$. The space $\mathbb{TP}^n$ is now defined as a quotient space of $\mathbb{R}_{\max}^{n+1} \setminus \{\mathbf{0}_\circ\}$ by the equivalence relation $\sim$, where $\mathbf{0}_\circ = (0_\circ, \ldots, 0_\circ)$ is the zero element of $\mathbb{R}_{\max}^{n+1}$.

We consider as an example the one-dimensional tropical projective space $\mathbb{TP}^1$. It turns out that the space $\mathbb{TP}^1$ is the same as the **completed** max-plus semiring

$$\mathbb{R}_{\max} \cup \{+\infty\} = \mathbb{R} \cup \{\pm\infty\}.$$

This can be seen by making use of a map

$$h : \mathbb{TP}^1 \to \mathbb{R}_{\max} \cup \{+\infty\},$$

which maps the element $[1_\circ : a] \in \mathbb{TP}^1$ onto

$$a \oslash 1_\circ = a \in \mathbb{R}_{\max} \cup \{+\infty\}$$

in the case when $a \in \mathbb{R}_{\max}$, and the element $[0_\circ : a] \in \mathbb{TP}^1$ onto

$$a \oslash 0_\circ = +\infty \in \mathbb{R}_{\max} \cup \{+\infty\}$$

when $a \in \mathbb{R}$.

With the definition of the tropical projective space at hand, we can now define the notion of a tropical holomorphic curve.

Definition 5.10. Let $[a_0 : \cdots : a_n] \in \mathbb{TP}^n$ be the equivalence class of $(a_0, \cdots, a_n) \in \mathbb{R}^{n+1}_{\max} \setminus \{\mathbf{0}_\circ\}$, and let

$$f = [g_0 : \cdots : g_n] : \mathbb{R} \to \mathbb{TP}^n$$

be a tropical holomorphic map where $g_0, \ldots, g_n$ are tropical entire functions and do not have any roots which are common to all of them. Denote

$$\mathbf{f} = (g_0, \ldots, g_n) : \mathbb{R} \to \mathbb{R}^{n+1}.$$

Then the map $\mathbf{f}$ is called a **reduced representation** of the **tropical holomorphic curve** f in $\mathbb{TP}^n$.

We will show later on in this chapter that tropical holomorphic curves are natural extensions of tropical meromorphic functions in the sense that the family of 1-dimensional tropical holomorphic curves can be identified with the family of tropical meromorphic functions.

In the next definition we will introduce a tropical analogue of the Cartan characteristic function for tropical holomorphic curves.

Definition 5.11. If $f : \mathbb{R} \to \mathbb{TP}^n$ is a tropical holomorphic curve with a reduced representation $\mathbf{f} = (g_0, \ldots, g_n)$, then

$$T_{\mathbf{f}}(r) = \frac{1}{2} \left(F(r) + F(-r) \right) - F(0), \qquad F(x) = \max\{g_0(x), \ldots, g_n(x)\},$$

is said to be the **tropical Cartan characteristic** function of f.

In the definition of the tropical Cartan characteristic we have picked a reduced representation $\mathbf{f}$ of the considered tropical holomorphic curve, and defined the characteristic in terms of $\mathbf{f}$. In general such a logic may produce a characteristic function which depends on the selected reduced representation of the considered holomorphic curve. Clearly such a characteristic would not give a consistent description of the growth of the tropical holomorphic curve, for instance. We will show next that the characteristic introduced in Definition 5.11 is devoid of such problems.

Proposition 5.12. *The tropical Cartan characteristic function $T_{\mathbf{f}}(r)$ is independent of the reduced representation of the tropical holomorphic curve f.*

Proof. Let $f : \mathbb{R} \to \mathbb{TP}^n$ be a tropical holomorphic curve with two different reduced representations,

$$\mathbf{f} := (g_0, g_1, \ldots, g_n)$$

and

$$\widetilde{\mathbf{f}} := (\tilde{g}_0, \tilde{g}_1, \ldots, \tilde{g}_n).$$

Let $x \in \mathbb{R}$ be selected arbitrarily. By the definition of projective coordinates it follows that there exists a $\lambda = \lambda(x) \in \mathbb{R}$, which depends in general on x, such that

$$\tilde{g}_j(x) = g_j(x) + \lambda(x)$$

for all $0 \le j \le n$. From the properties of a reduced representation it follows that there are no common roots for the tropical entire functions in either of the collections $g_0(x), \ldots, g_n(x)$ and $\tilde{g}_0(x), \ldots, \tilde{g}_n(x)$. This fact implies that

$$\lambda(x) \equiv \tilde{g}_j(x) - g_j(x) \tag{5.11}$$

is a tropical meromorphic function such that

$$\omega_\lambda(x) \equiv \omega_{\tilde{g}_j}(x) - \omega_{g_j}(x)$$

for all $j \in \{0, \ldots, n\}$.

Suppose now that $x_0 \in \mathbb{R}$ is a pole of $\lambda(x)$. Then it follows by (5.11), and the fact that g_j and $\tilde{g}_j$ are tropical entire for all $j \in \{0, \ldots, n\}$, that x_0 is a common root of $g_0(x), \ldots, g_n(x)$. This is a contradiction to the definition of a tropical holomorphic curve $\mathbf{f}$. Similarly, if $x_0 \in \mathbb{R}$ is a root of $\lambda(x)$, then it is also a common root of $\tilde{g}_0(x), \ldots, \tilde{g}_n(x)$, which contradicts the fact that these functions define $\widetilde{\mathbf{f}}$. The only remaining possibility is that $\lambda(x)$ is a tropical unit, and so of the form

$$\lambda(x) = \alpha x + \beta$$

for all $x \in \mathbb{R}$, where the constants $\alpha \in \mathbb{R}$ and $\beta \in \mathbb{R}$ are independent of $r = |x|$. Thus by (5.11), we have

$$\tilde{g}_j(x) = g_j(x) + \alpha x + \beta,$$

which implies that

$$\tilde{F}(x) = \max\{\tilde{g}_0(x), \ldots, \tilde{g}_n(x)\} = F(x) + \alpha x + \beta$$

and

$$\tilde{F}(0) = F(0) + \beta.$$

Finally, this implies that

$$\frac{1}{2}\left(\tilde{F}(r) + \tilde{F}(-r)\right) - \tilde{F}(0) = \frac{1}{2}\left(F(r) + F(-r)\right) - F(0),$$

which is the same as $T_{\mathbf{f}}(r) = T_{\widetilde{\mathbf{f}}}(r)$. $\qquad\square$

The previous result implies that the tropical Cartan characteristic function of a tropical holomorphic curve f does not depend on the reduced representation $\mathbf{f}$ chosen to represent it. Therefore it is unnecessary to refer to the reduced representation $\mathbf{f}$ in the notation $T_{\mathbf{f}}(r)$ of the characteristic function, and so we will simply use the notation $T_f(r)$ from now on.

We will next demonstrate that, analogously to the classical characteristics, the tropical Cartan characteristic is a non-negative function of r. For this purpose we need the following lemma, which will also be applied in the proof of the tropical analogue of the second main theorem for holomorphic curves towards the end of this chapter.

Lemma 5.13. *Let g be a tropical entire function. Then*

$$g(r) + g(-r) \geq 2g(0)$$

for all $r \geq 0$.

Proof. The assertion is trivial if $r = 0$, so we suppose that $r > 0$. Since g is piecewise linear function, there exists $\alpha, \beta \in \mathbb{R}$ and an interval $[r_1, r_2] \subset \mathbb{R}$ containing the origin such that $r_1 < r_2$ and

$$g(x) = \alpha x + \beta \tag{5.12}$$

for all $x \in [r_1, r_2]$. Define the tropical unit u as

$$u(x) := \alpha x + \beta \tag{5.13}$$

for all $x \in \mathbb{R}$. Then, since g is entire its graph is convex, and so

$$g(x) \geq u(x) \tag{5.14}$$

for all $x \in \mathbb{R}$. Therefore, in particular, by substituting $x = \pm r$ into (5.14) it follows that

$$g(r) + g(-r) \geq u(r) + u(-r) = 2\beta. \tag{5.15}$$

Finally, since we have chosen the interval $[r_1, r_2]$ in such a way that it contains the origin, it follows by combining (5.12) and (5.13) that

$$g(0) = u(0) = \beta. \tag{5.16}$$

Thus the assertion follows by combining (5.15) and (5.16). $\square$

Now the fact that tropical Cartan characteristic is always non-negative follows as a simple corollary of Lemma 5.13.

Proposition 5.14. *Let f be a tropical holomorphic curve. Then $T_f(r) \geq 0$ for all $r \geq 0$.*

Proof. By definition f is of the form $f = [g_0 : \cdots : g_n]$, where $g_0, \ldots, g_n$ are tropical entire functions without common roots. Hence,

$$T_f(r) = \frac{1}{2}\left(F(r) + F(-r)\right) - F(0),$$

where

$$F(x) = \max\{g_0(x), \ldots, g_n(x)\}$$

is in fact a tropical entire function as a tropical linear combination of tropical entire functions $g_0(x), \ldots, g_n(x)$. Thus by Lemma 5.13, we have $F(r) + F(-r) \geq 2F(0)$ for all $r \geq 0$ from which the assertion follows. $\square$

We will next show that in the special case where the tropical projective space is one-dimensional the corresponding Cartan characteristic function coincides, up to a constant, with the tropical Nevanlinna characteristic introduced in Chapter 1 when the one-dimensional tropical holomorphic curve

$$f = [g : h] : \mathbb{R} \to \mathbb{TP}^1$$

is perceived as a tropical meromorphic function expressed as a quotient of the tropical entire functions g and h. But first, we need to prove that all tropical meromorphic functions admit such a representation $f = g \oslash h$. The following lemma due to Tsai treats the case of tropical rational functions. Recall that a real piecewise linear continuous function is tropical rational if it has only finitely many roots and poles.

Lemma 5.15 ([106, Theorem 7.3]). *If $f : \mathbb{R} \to \mathbb{R}$ is tropical meromorphic then it is tropical rational if and only if it can be written in the form*

$$f(t) = \frac{a_0 \oplus a_1 \otimes t^{\otimes l_1} \oplus \cdots \oplus a_p \otimes t^{\otimes l_p}}{b_0 \oplus b_1 \otimes t^{\otimes s_1} \oplus \cdots \oplus b_q \otimes t^{\otimes s_q}} \oslash, \tag{5.17}$$

where $p, q \in \mathbb{Z}_{\geq 0}$, $0 < l_1 < \cdots < l_p$, $0 < s_1 < \cdots < s_q$, and the coefficients $a_0, \ldots, a_p$ and $b_0, \ldots, b_q$ are real constants.

Proof. The proof given here is from [67] and it follows a different method to the original one given by Tsai [106].

Suppose first that f is given by the formula (5.17). The right side of (5.17) defines a piecewise linear continuous function consistently to the assumption that f is tropical meromorphic. Moreover, $\omega_f(x) = 0$ for all $x \in \mathbb{R}$ except possibly for those points where the maximum in the numerator or the denominator of the tropical fraction (5.17) is attained simultaneously at more than one tropical term. Since there are at most finitely many such points, it follows that f has finitely many roots and poles and is thus a tropical rational function.

Assume now that f is a tropical rational function. By the definition, f has finitely many roots, say $\alpha_1, \ldots, \alpha_n$, and finitely many poles, say $\beta_1, \ldots, \beta_m$. Such an f can be written in the form

$$f(t) = \frac{A_0 \oplus A_1 \otimes t^{\otimes L_1} \oplus \cdots \oplus A_n \otimes t^{\otimes L_n}}{B_0 \oplus B_1 \otimes t^{\otimes S_1} \oplus \cdots \oplus B_m \otimes t^{\otimes S_m}} \oslash \tag{5.18}$$

where $A_0 \in \mathbb{R}$ and $B_0 \in \mathbb{R}$ can be selected arbitrarily, and

$$A_i = A_0 - \alpha_1 \omega_f(\alpha_1) - \alpha_2 \omega_f(\alpha_2) - \cdots - \alpha_i \omega_f(\alpha_i),$$
$$L_i = \omega_f(\alpha_1) + \cdots + \omega_f(\alpha_i),$$
$$B_j = B_0 - \beta_1 |\omega_f(\beta_1)| - \beta_2 |\omega_f(\beta_2)| - \cdots - \beta_j |\omega_f(\beta_j)|,$$
$$S_j = |\omega_f(\beta_1)| + \cdots + |\omega_f(\beta_j)|,$$

for $i \in \{1, \ldots, n\}$ and $j \in \{1, \ldots, m\}$. This means that $0 < L_1 < \cdots < L_n$ and $0 < S_1 < \cdots < S_n$. $\qquad\square$

Before applying the previous lemma to find a representation for tropical meromorphic functions in terms of one-dimensional tropical holomorphic curves, we briefly discuss the uniqueness of the representation (5.18) of a tropical rational function. If f is any tropical rational function, which has roots $\alpha_1, \ldots, \alpha_n$ and poles $\beta_1, \ldots, \beta_m$, then it may seem by a quick look that the representation (5.18) for $f(t)$ would be unique up to two arbitrary constants $A_0 \in \mathbb{R}$ and $B_0 \in \mathbb{R}$. However, this is not the case. The function $f(t)$ in (5.18) can be multiplied with an arbitrary tropical unit without any effect on the location of the roots and poles or their multiplicity. And finally, there may be terms in (5.18) which do not contribute anything to the maximum. For instance, the numerator

$$f_1(t) := A_0 \oplus A_1 \otimes t^{\otimes L_1} \oplus \cdots \oplus A_n \otimes t^{\otimes L_n}. \tag{5.19}$$

of $f(t)$ in (5.18) forms a convex hull. If the graph of a monic $A_{n+1} \otimes t^{\otimes L_{n+1}}$ lies completely below the bordering polygonal line defined by $f_1(t)$, then the tropical addition of $A_{n+1} \otimes t^{\otimes L_{n+1}}$ to $f_1(t)$ has no effect on the form of f_1.

The shortest possible expression for $f_1(t)$ can be obtained by taking

$$g_j = t^{\otimes L_j} \quad \text{for all} \quad j \in \{0, \ldots, N\},$$

and by selecting a spanning basis $G \subset \{g_0, \ldots, g_N\}$ so that f_1 becomes a complete tropical linear combination of G over $\mathbb{R}_{\max}$ in the sense of Definition 5.7. Another way to describe unique representations of tropical rational functions has been given by Tsai [106, section 3] in his discussion on maximally represented polynomials.

All meromorphic functions in the complex plane can be represented as a quotient of two entire functions, which do not have any common roots. Now we are ready to show that a completely analogous result holds also for tropical meromorphic functions.

Proposition 5.16. *For any tropical meromorphic function f there exist tropical entire functions g and h such that $f = h \oslash g$, and g and h do not have any common roots.*

Proof. Let $n \in \mathbb{N}$ and denote $I_n = [-n, n]$. Let the function

$$f_n : I_n \to \mathbb{R}$$

be defined in such a way that $f_n(t) = f(t)$ for all $t \in I_n$. By definition, roots and poles of a tropical meromorphic function have no finite limit points, and so f_n has finitely many of them. This implies that there exists a tropical rational function R_n such that $R_n(t) = f_n(t)$ for all $t \in I_n$ and R_n does not have any roots or poles outside of I_n. Lemma 5.15 implies that R_n can be expressed in the form

$$R_n(t) = P_n(t) \oslash Q_n(t),$$

where $P_n(t)$ and $Q_n(t)$ are tropical polynomials without common roots. By definition P_n and Q_n satisfy

$$P_{n+1}(t) \equiv P_n(t) \quad \text{and} \quad Q_{n+1}(t) \equiv Q_n(t)$$

for all $t \in I_n$, and therefore

$$h(t) := \lim_{k \to \infty} P_k(t) = P_n(t) \quad \text{and} \quad g(t) := \lim_{k \to \infty} Q_k(t) = Q_n(t)$$

for all $t \in I_n$. The assertion now follows by defining

$$h = \lim_{n \to \infty} P_n \quad \text{and} \quad g = \lim_{n \to \infty} Q_n,$$

where the limits converge locally uniformly. $\qquad \square$

Proposition 5.16 gives us the means to define tropical meromorphic functions in terms of one-dimensional tropical holomorphic curves. We will next show that corresponding characteristic functions are the same, up to an additive constant.

Proposition 5.17. *If $f = h \oslash g$ is a tropical meromorphic function such that $f(0) = 1_\circ = 0$, then*

$$T_f(r) = T(r, f),$$

where $\mathbf{f} = (g, h)$ is a reduced representation of the holomorphic curve f : $\mathbb{R} \to \mathbb{TP}^1$ given by $f = [g : h]$.

Proof. The characteristic function $T_f(r)$ is defined as

$$T_f(r) = \frac{1}{2}\left(F(r) + F(-r)\right) - F(0),$$

where

$$F(x) = \max\{g(x), h(x)\}.$$

By making use of the notation

$$(h - g)^+(x) = \max\{(h - g)(x), 0\},$$

it therefore follows that

$$F(x) = (h - g)^+(x) + g(x) \tag{5.20}$$

for all $x \in \mathbb{R}$. By applying (5.20) with $x = \pm r$, we have

$$T_f(r) = m(r, h - g) + \frac{1}{2}g(r) + \frac{1}{2}g(-r) - (h - g)^+(0) - g(0). \tag{5.21}$$

By splitting the function g into a difference of two non-negative functions as

$$g = g^+ - (-g)^+,$$

it follows that

$$g(r) + g(-r) = g^+(r) + g^+(-r) - (-g)^+(r) - (-g)^+(-r)$$

for all $r \geq 0$, and so

$$\frac{1}{2}g(r) + \frac{1}{2}g(-r) = m(r, g) - m(r, 1_\circ \oslash g). \tag{5.22}$$

In addition, the tropical Jensen formula (1.10) yields

$$m(r, g) - m(r, 1_\circ \oslash g) = N(r, 1_\circ \oslash g) - N(r, g) + g(0), \tag{5.23}$$

and so, by combining (5.21) with (5.22) and (5.23), we finally have

$$T_f(r) = m(r, h \oslash g) - N(r, g) + N(r, 1_\circ \oslash g) - (h - g)^+(0). \tag{5.24}$$

Since g and h are tropical entire without common roots, it follows that $N(r, g) \equiv 0$ and

$$N(r, 1_\circ \oslash g) = N(r, h \oslash g).$$

Thus (5.24) becomes

$$T_f(r) = T(r, h \oslash g) - (h \oslash g)^+(0) = T(r, f) - f^+(0),$$

which yields the assertion. $\qquad\square$

Remark 5.18. Without the normalization $f(0) = 1_\circ$, the statement of Proposition 5.17 becomes

$$T_f(r) = T(r, f) - f^+(0).$$

Since Proposition 5.17 shows that the characteristic functions $T_f(r)$ and $T(r, f)$ are equal up to a constant if the tropical meromorphic function f is interpreted as a one-dimensional tropical holomorphic curve $f : \mathbb{R} \to \mathbb{TP}^1$, it follows that the order and hyper-order, for instance, remain unchanged if we replace $T(r, f)$ with $T_f(r)$ in the definitions. If we have a tropical holomorphic curve $f : \mathbb{R} \to \mathbb{TP}^n$ in the dimension $n \geq 2$ we can still obtain a useful inequality in terms of tropical linear combinations of the coordinates of f. Before introducing this inequality we need to introduce a counting function for the common roots of two tropical entire functions.

Definition 5.19. Let h_1 and h_2 be tropical entire functions. We define

$$N_{\min}(r, 0, h_1, h_2) = \min\{N(r, 1_\circ \oslash h_1), N(r, 1_\circ \oslash h_2)\}.$$

Lemma 5.20. *Let* $g = [g_0 : \cdots : g_n] : \mathbb{R} \to \mathbb{TP}^n$ *be a tropical holomorphic curve, and let*

$$\hat{f} = \bigoplus_{\nu=0}^{n} \hat{a}_\nu \otimes g_\nu \tag{5.25}$$

and

$$\tilde{f} = \bigoplus_{\nu=0}^{n} \tilde{a}_\nu \otimes g_\nu \tag{5.26}$$

be tropical linear combinations of the $n + 1$ *tropical entire functions* $g_0, \ldots, g_n$. *Then*

$$T\left(r, \hat{f} \oslash \tilde{f}\right) + N_{\min}\left(r, 0, \hat{f}, \tilde{f}\right) \leq T_g(r) + \max\{g_0(0), \ldots, g_n(0)\}$$
$$+ C(\tilde{a}, \hat{a}) - \tilde{f}(0),$$

where

$$C(\tilde{a}, \hat{a}) = \max_{j,k \in \{0,\ldots,n\}} \{\hat{a}_j, \tilde{a}_k\}. \tag{5.27}$$

Proof. Since $\tilde{f}$ and $\hat{f}$ are both tropical linear combinations of tropical entire functions, they themselves are tropical entire as well. We define u to be a tropical entire function which has roots exactly at the points where $\tilde{f}$

and $\hat{f}$ have common roots, taking multiplicities into account. This means that u satisfies

$$\omega_u(x) = \min\{\omega_{\hat{f}}(x), \omega_{\tilde{f}}(x)\} \tag{5.28}$$

for all $x \in \mathbb{R}$. In addition we normalize u in such a way that $u(0) = \tilde{f}(0)$. Define

$$\hat{w} := \hat{f} \oslash u \quad \text{and} \quad \tilde{w} := \tilde{f} \oslash u. \tag{5.29}$$

Then

$$\omega_{\hat{w}} = \omega_{\hat{f}} - \omega_u \geq 0$$

and

$$\omega_{\tilde{w}} = \omega_{\tilde{f}} - \omega_u \geq 0,$$

and thus $\hat{w}$ and $\tilde{w}$ are tropical entire functions satisfying

$$\tilde{w}(0) = \tilde{f}(0) - u(0) = 1_\circ.$$

Suppose that there exists a point $x_0 \in \mathbb{R}$ at which both $\hat{w}$ and $\tilde{w}$ have a root, which means that $\omega_{\hat{w}}(x_0) > 0$ and $\omega_{\tilde{w}}(x_0) > 0$. But then by (5.29)

$$\min\{\omega_{\hat{f}}(x_0), \omega_{\tilde{f}}(x_0)\} > \omega_u(x_0),$$

which contradicts (5.28). Thus the tropical entire functions $\hat{w}$, $\tilde{w}$ and u satisfy $\hat{f} = u \otimes \hat{w}$ and $\tilde{f} = u \otimes \tilde{w}$, and moreover $\hat{w}$ and $\tilde{w}$ do not have any common roots.

Proposition 5.17 yields

$$T\left(r, \hat{f} \oslash \tilde{f}\right) = T\left(r, \hat{w} \oslash \tilde{w}\right) = \frac{1}{2}\left(G(r) + G(-r)\right) - G(0), \tag{5.30}$$

where $G(x) = \max\{\hat{w}(x), \tilde{w}(x)\}$. Note that $G(0)$ is non-negative by the definition of G. By defining

$$F(x) = \max\{g_0(x), \ldots, g_n(x)\}$$

we obtain directly from (5.26) and (5.25) that

$$\max\{\hat{f}(x), \tilde{f}(x)\} \leq F(x) + C(\tilde{a}, \hat{a}) \tag{5.31}$$

for all $x \in \mathbb{R}$, where $C(\tilde{a}, \hat{a})$ is as in (5.27). Now, by combining Definition 5.11 with equations (5.30) and (5.31), it follows that

$$T\left(r, \hat{f} \oslash \tilde{f}\right) \leq T_g(r) + F(0) - \frac{1}{2}\left(u(r) + u(-r)\right) + C(\tilde{a}, \hat{a}). \tag{5.32}$$

Using the tropical Jensen formula (1.10), we have

$$\frac{1}{2}\left(u(r) + u(-r)\right) = m(r, u) - m(r, 1_\circ \oslash u) = N(r, 1_\circ \oslash u) - N(r, u) + u(0),$$

where $N(r, u) = 0$ for all r since u is tropical entire. Thus (5.32) can be written as

$$T\left(r, \hat{f} \oslash \tilde{f}\right) \leq T_g(r) - N(r, 1_\circ \oslash u) + C(\tilde{a}, \hat{a})$$

$$+ \max\{g_0(0), \ldots, g_n(0)\} - \tilde{f}(0),$$

which yields the assertion. $\qquad\square$

5.5 Second main theorem for tropical holomorphic curves

Cartan [15] introduced a natural generalization of Nevanlinna's second main theorem for holomorphic curves in $\mathbb{P}^n(\mathbb{C})$ by showing that if $f : \mathbb{C} \to \mathbb{P}^n(\mathbb{C})$ is a linearly non-degenerate holomorphic curve and $H_0, \ldots, H_q$ hyperplanes of $\mathbb{P}^n(\mathbb{C})$ in general position, then

$$(q - n)T_f(r) \le \sum_{j=0}^{q} N_f(r, H_j) - N_W(r, 0) + O(\log^+(rT_f(r)))$$

as $r \to \infty$ outside of a set of finite linear measure. Here $T_f(r)$ now denotes the usual **Cartan characteristic**

$$T_f(r) = \frac{1}{2\pi} \int_0^{2\pi} U(re^{i\theta})d\theta - U(0), \qquad U(z) = \max_{0 \le j \le n} \log |g_j(z)|,$$

and $N_W(r, 0)$ is a ramification term defined in terms of the Wronskian determinant of the entire component functions of f. See Appendix A for a short description of Cartan's value distribution theory in $\mathbb{C}$. In what follows we introduce a second main theorem for tropical holomorphic curves, which is a tropical analogue of Cartan's second main theorem. Similarly as Cartan's result implies Nevanlinna's second main theorem as a direct corollary in the case when the considered holomorphic curve is of dimension one, its tropical analogue implies, under a natural condition ensuring that the target values are in a sense genuine, a tropical second main theorem due to Laine and Tohge. We will prove this fact at the end of this section after proving the actual tropical analogue of Cartan's result.

Cartan also conjectured that if instead of being completely linearly non-degenerate, the image of a holomorphic curve to $\mathbb{P}^n(\mathbb{C})$ spans a linear subspace of codimension s, then the inequality

$$(q - n - s)T_f(r) \le \sum_{j=0}^{q} N_f(r, H_j) - \frac{n+1}{n+1-s}N_W(r, 0) + O(\log^+(rT_f(r)))$$

holds nearly everywhere. Cartan's conjecture was proved by Nochka [82] in 1983, almost half a century after its original formulation. The following theorem is a tropical analogue of Nochka's generalization of Cartan's result.

Theorem 5.21. *Let q and n be positive integers with $q > n$, and let $\varepsilon > 0$. Given $n + 1$ tropical entire functions $g_0, \ldots, g_n$ without common roots, and linearly independent in Gondran-Minoux sense, let the $q + 1$ tropical linear combinations $f_0, \ldots, f_q$ of the g_j over the semi-ring $\mathbb{R}_{\max}$ be defined by*

$$f_k(x) = a_{0k} \otimes g_0(x) \oplus a_{1k} \otimes g_1(x) \otimes \cdots \otimes a_{nk} \otimes g_n(x), \quad 0 \le k \le q.$$

Let $\lambda = ddg(\{f_{n+1}, \ldots, f_q\})$ and

$$L = \frac{f_0 \otimes f_1 \otimes \cdots \otimes f_q}{C_\circ(f_0, f_1, \cdots, f_n)} \oslash. \tag{5.33}$$

If the tropical holomorphic curve g of $\mathbb{R}$ into $\mathbb{TP}^n$ with reduced representation $\mathbf{g} = (g_0, \ldots, g_n)$ is of hyper-order

$$\rho_2 := \rho_2(\mathbf{g}) < 1, \tag{5.34}$$

then

$$(q - n - \lambda)T_g(r) \leq N\big(r, 1_\circ \oslash L\big) - N(r, L) + o\left(\frac{T_g(r)}{r^{1-\rho_2-\varepsilon}}\right), \tag{5.35}$$

where r approaches infinity outside an exceptional set of finite logarithmic measure.

Proof. In order to get a better hold on the function L introduced in (5.33) we first introduce an auxiliary function

$$\tilde{L} = \frac{f_0 \otimes \overline{f}_1 \otimes \cdots \otimes \overline{f}_n^{[n]} \otimes f_{n+1} \otimes \cdots \otimes f_q}{C_\circ(f_0, f_1, \ldots, f_n)} \oslash,$$

which will turn out to be in sense more compatible with the properties of the tropical Casoratian than L. Next we will show that replacing L by $\tilde{L}$ in the right side of inequality (5.35) will only affect the error term of the inequality. The first step is to ensure that the conditions needed to apply Theorem 3.15 are satisfied for the functions f_j. Since $g_0, \ldots, g_n$ have no common roots, for each root x_0 of f_j there will always be at least one function g_k, with $k = 0, \ldots, n$, such that g_k does not have a root at x_0. Hence by Lemma 5.20, we have

$$N(r, 1_\circ \oslash f_j) \leq \sum_{k=0}^{n} N(r, g_k \oslash f_j)$$

$$\leq \sum_{k=0}^{n} T(r, g_k \oslash f_j) \tag{5.36}$$

$$\leq (n+1)T_g(r) + O(1),$$

which implies that

$$N(r, 1_\circ \oslash f_j) \leq (n+2)T_g(r) \tag{5.37}$$

for all r large enough. Thus by the assumption $\rho_2(g) < 1$ it follows that

$$\eta_j := \limsup_{r \to \infty} \frac{\log\log N(r, 1_\circ \oslash f_j)}{\log r} < 1$$

for all $j = 1, \ldots, n$, and so Theorem 3.15 is applicable for $f_1, \ldots, f_n$. Hence,

$$N\left(r, 1_\circ \oslash \overline{f}_j^{[j]}\right) \leq N\left(r + j, 1_\circ \oslash f_j\right)$$

$$= N\left(r, 1_\circ \oslash f_j\right) + o\left(\frac{N\left(r, 1_\circ \oslash f_j\right)}{r^{1-\eta_j-\varepsilon}}\right),$$

where $j = 1, \ldots, n$ and r tends to infinity outside of an exceptional set of finite logarithmic measure. Similarly,

$$N\left(r, 1_\circ \oslash f_j\right) \leq N\left(r + j, 1_\circ \oslash \overline{f}_j^{[j]}\right)$$

$$= N\left(r, 1_\circ \oslash \overline{f}_j^{[j]}\right) + o\left(\frac{N\left(r, 1_\circ \oslash f_j\right)}{r^{1-\eta_j-\varepsilon}}\right),$$

and so in fact, using in addition (5.37) to estimate the error term, we have an asymptotic equality

$$N\left(r, 1_\circ \oslash \overline{f}_j^{[j]}\right) = N\left(r, 1_\circ \oslash f_j\right) + o\left(\frac{T_g(r)}{r^{1-\rho_2-\varepsilon}}\right) \tag{5.38}$$

for all $j = 1, \ldots, n$, where r tends to infinity outside of an exceptional set of finite logarithmic measure. Therefore,

$$N\left(r, 1_\circ \oslash \tilde{L}\right) - N(r, \tilde{L})$$

$$= N\left(r, \frac{1_\circ}{f_0 \otimes \overline{f}_1 \otimes \cdots \otimes \overline{f}_n^{[n]} \otimes f_{n+1} \otimes \cdots \otimes f_q} \oslash\right)$$

$$- N\left(r, \frac{1_\circ}{C_\circ(f_0, \ldots, f_n)} \oslash\right)$$

$$= \sum_{j=0}^{n} N\left(r, 1_\circ \oslash \overline{f}_j^{[j]}\right) + N\left(r, \frac{1_\circ}{f_{n+1} \otimes \cdots \otimes f_q} \oslash\right)$$

$$- N\left(r, \frac{1_\circ}{C_\circ(f_0, \ldots, f_n)} \oslash\right)$$

$$= \sum_{j=0}^{n} N\left(r, 1_\circ \oslash f_j\right) + N\left(r, \frac{1_\circ}{f_{n+1} \otimes \cdots \otimes f_q} \oslash\right)$$

$$- N\left(r, \frac{1_\circ}{C_\circ(f_0, \ldots, f_n)} \oslash\right) + o\left(\frac{T_g(r)}{r^{1-\rho_2-\varepsilon}}\right)$$

$$= N\left(r, \frac{1_\circ}{f_0 \otimes \cdots \otimes f_q} \oslash\right) - N\left(r, \frac{1_\circ}{C_\circ(f_0, \ldots, f_n)} \oslash\right) + o\left(\frac{T_g(r)}{r^{1-\rho_2-\varepsilon}}\right)$$

$$= N\left(r, 1_\circ \oslash L\right) - N(r, L) + o\left(\frac{T_g(r)}{r^{1-\rho_2-\varepsilon}}\right)$$

$$\tag{5.39}$$

for all r outside of an exceptional set of finite logarithmic measure. Hence, in order to prove that the assertion of the theorem is valid it is sufficient to show that (5.35) holds with L replaced by $\tilde{L}$.

Lemma 5.1 implies that

$$C_\circ(f_0, f_1, \ldots, f_n) = f_0 \otimes \overline{f}_0 \otimes \cdots \otimes \overline{f}_0^{[n]} \otimes C_\circ(1_\circ, f_1 \oslash f_0, \ldots, f_n \oslash f_0).$$

Therefore the auxiliary function $\tilde{L}$ satisfies the equation

$$v = \tilde{L} \otimes K, \tag{5.40}$$

where

$$v = f_{n+1} \otimes \cdots \otimes f_q$$

and

$$K = C_\circ(1_\circ, f_1 \oslash f_0, \ldots, f_n \oslash f_0) \otimes (\overline{f}_0 \oslash \overline{f}_1) \otimes \cdots \otimes (\overline{f}_0^{[n]} \oslash \overline{f}_n^{[n]}). \tag{5.41}$$

Hence, by (5.40) and the tropical Jensen formula (1.10), it follows that

$$\begin{aligned}
\frac{1}{2}v(r) + \frac{1}{2}v(-r) &= \frac{1}{2}\tilde{L}(r) + \frac{1}{2}\tilde{L}(-r) + \frac{1}{2}K(r) + \frac{1}{2}K(-r) \\
&= \frac{1}{2}\tilde{L}^+(r) - \frac{1}{2}(-\tilde{L})^+(r) + \frac{1}{2}\tilde{L}^+(-r) - \frac{1}{2}(-\tilde{L})^+(-r) \\
&\quad + \frac{1}{2}K^+(r) - \frac{1}{2}(-K)^+(r) + \frac{1}{2}K^+(-r) - \frac{1}{2}(-K)^+(-r) \\
&= m(r, \tilde{L}) - m(r, 1_\circ \oslash \tilde{L}) + m(r, K) - m(r, 1_\circ \oslash K) \\
&= N(r, 1_\circ \oslash \tilde{L}) - N(r, \tilde{L}) + m(r, K) - m(r, 1_\circ \oslash K) \\
&\quad + \tilde{L}(0).
\end{aligned} \tag{5.42}$$

From (5.39) and (5.41) it follows that the right-hand side of (5.42) is otherwise compatible with the right-hand side of (5.35), except that the constant term $\tilde{L}(0)$ depends on $\tilde{L}$ rather than L. The dependence on the auxiliary function $\tilde{L}$ in the error term can be removed by observing that

$$\tilde{L} = L \otimes \frac{\overline{f}_1 \otimes \cdots \otimes \overline{f}_n^{[n]}}{f_1 \otimes \cdots \otimes f_n} \oslash,$$

which implies, together with equation (5.42), that

$$\begin{aligned}
\frac{1}{2}v(r) + \frac{1}{2}v(-r) &\leq N(r, 1_\circ \oslash \tilde{L}) - N(r, \tilde{L}) + m(r, K) - m(r, 1_\circ \oslash K) \\
&\quad + L(0) + \sum_{j=1}^{n} \left(\max_{0 \leq i \leq n}\{a_{ij} + g_i(j)\} - \max_{0 \leq i \leq n}\{a_{ij} + g_i(0)\} \right).
\end{aligned} \tag{5.43}$$

Moreover, the function K in (5.41) satisfies

$$K = C_\circ(1_\circ, f_1 \oslash f_0, \ldots, f_n \oslash f_0) \oslash \left((\overline{f}_1 \oslash \overline{f}_0) \otimes \cdots \otimes (\overline{f}_n^{[n]} \oslash \overline{f}_0^{[n]}) \right),$$

which means that K is formed completely out of tropical sums and products of the form

$$(\overline{f}_j^{[l]} \oslash \overline{f}_0^{[l]}) \oslash (\overline{f}_j^{[m]} \oslash \overline{f}_0^{[m]})$$

where $l, m \in \{0, 1, \ldots, n\}$ and $j \in \{1, \ldots, q\}$. Theorem 3.27 therefore implies that

$$m(r, K) \leq \sum_{j=1}^{q} \sum_{k=0}^{q} o\left(\frac{T(r, f_j \oslash f_0)}{r^{1-\rho_2-\varepsilon}} \right) \tag{5.44}$$

as r approaches infinity outside of an exceptional set of finite logarithmic measure. The use of Lemma 5.20 to simplify the error term in (5.44) yields

$$m(r, K) = o\left(\frac{T_g(r)}{r^{1-\rho_2-\varepsilon}} \right) \tag{5.45}$$

for all r outside of a set of finite logarithmic measure. Hence, we have by (5.39), (5.43) and (5.45) that

$$\frac{1}{2}v(r) + \frac{1}{2}v(-r) \leq N(r, 1_\circ \oslash L) - N(r, L) + L(0)$$
$$+ \sum_{j=1}^{n} \left(\max_{0 \leq i \leq n} \{a_{ij} + g_i(j)\} - \max_{0 \leq i \leq n} \{a_{ij} + g_i(0)\} \right)$$
$$+ o\left(\frac{T_g(r)}{r^{1-\rho_2-\varepsilon}} \right)$$
$$\tag{5.46}$$

giving us, up to a constant, the right-hand side of (5.35).

The next step is to find a lower bound for the function $\frac{1}{2}v(r) + \frac{1}{2}v(-r)$. In order to do this, we write the functions f_ν, $(n+1 \leq \nu \leq q)$, in the form

$$f_\nu(x) = \max_{j \in I_\nu}\{a_{j\nu} + g_j(x)\}, \quad a_{j\nu} \in \mathbb{R}, \tag{5.47}$$

where the index sets $I_\nu \subset \{0, 1, \ldots, n\}$ are of cardinality $\text{card}(I_\nu)$. Now, by Lemma 5.13 and (5.47), it follows that

$$\frac{1}{2}f_\nu(r) + \frac{1}{2}f_\nu(-r) \geq \max_{j \in I_\nu}\{a_{j\nu} + g_j(0)\}$$

for all $\nu \in \{n+1, \ldots, q\}$, and so

$$
\frac{1}{2}v(r) + \frac{1}{2}v(-r) = \frac{1}{2} \sum_{\nu=n+1}^{q} (f_\nu(r) + f_\nu(-r))
$$

$$
= \frac{1}{2} \sum_{\substack{\nu=n+1 \\ \mathrm{card}(I_\nu)=n+1}}^{q} (f_\nu(r) + f_\nu(-r))
$$

$$
+ \frac{1}{2} \sum_{\substack{\nu=n+1 \\ \mathrm{card}(I_\nu)<n+1}}^{q} (f_\nu(r) + f_\nu(-r))
$$

$$
\geq \frac{1}{2} \sum_{\substack{\nu=n+1 \\ \mathrm{card}(I_\nu)=n+1}}^{q} (f_\nu(r) + f_\nu(-r))
$$

$$
+ \sum_{\substack{\nu=n+1 \\ \mathrm{card}(I_\nu)<n+1}}^{q} \max_{j \in I_\nu} \{a_{j\nu} + g_j(0)\}
$$

$$
\geq (q-n-\lambda)\left(T_g(r) + \max_{j \in \{0,\ldots,n\}} \{g_j(0)\}\right)
$$

$$
+ \sum_{\substack{\nu=n+1 \\ \mathrm{card}(I_\nu)=n+1}}^{q} \min_{j \in I_\nu} \{a_{j\nu}\} + \sum_{\substack{\nu=n+1 \\ \mathrm{card}(I_\nu)<n+1}}^{q} \max_{j \in I_\nu} \{a_{j\nu} + g_j(0)\}.
$$

$$
\tag{5.48}
$$

We have hence obtained the desired lower bound for the function $\frac{1}{2}v(r) + \frac{1}{2}v(-r)$. The assertion of the theorem follows by combining the upper bound (5.46) with the lower bound (5.48), and incorporating the constant terms into the error term $o(T_g(r)r^{-1+\rho_2+\varepsilon})$. □

5.6   Ramification

One of the differences between the second main theorem of Nevanlinna theory, and its tropical version is the lack of obvious tropical analogue of ramification. The tropical version of Cartan's second main theorem, Theorem 5.21 above, offers some clues on how to define a tropical analogue of ramification. We will discuss this phenomenon in this section by a comparison of the complex analytical and tropical theories.

Let $g_0, g_1, \ldots, g_n$ be a collection of entire functions in the complex plane such that they do not have any zeros that are common to all of them. Then the classical Cartan characteristic function of $g = (g_0, g_1, \ldots, g_n)$ is defined by

$$T_f(r) = \frac{1}{2\pi} \int_0^{2\pi} U(re^{i\theta})d\theta - U(0), \qquad U(z) = \max_{0 \le j \le n} \log |g_j(z)|.$$

The number

$$\lambda := \dim \left\{ (c_0, c_1, \ldots, c_n) \in \mathbb{C}^{n+1} \ \Big| \ \sum_{j=0}^n c_j g_j \equiv 0 \right\}$$

plays an important role in Cartan's and Nochka's second main theorems. The constant λ satisfies $0 \le \lambda \le n - 1$, and

$$\dim \left\{ (c_0, c_1, \ldots, c_n) \in \mathbb{C}^{n+1} \ \Big| \ \sum_{j=0}^n c_j f_j \equiv 0 \right\} = \lambda$$

for any collection $f_0, f_1, \ldots, f_n$ of $n+1$ linear combinations of $g_0, g_1, \ldots, g_n$ over $\mathbb{C}$ in general position. The system g is said to be **degenerated** if the constant λ is strictly positive. Now the Cartan-Nochka second main theorem states that for any $q + 1$ such linear combinations $f_0, f_1, \ldots, f_n$, we have the inequality

$$(q - n - \lambda)T_f(r) \le \sum_{j=0}^q N_{n-\lambda}(r, 1/f_j) + O\big(\log r T_f(r)\big), \qquad (5.49)$$

where r tends to infinity outside a set of finite linear measure. The integrated truncated counting function $N_{n-\lambda}(r, 1/f_j)$ takes into account the zeros of f_j of order m exactly $\min\{m, n - \lambda\}$ times in the counting. Each counting function $N_{n-\lambda}(r, 1/f_j)$ on the right side of inequality (5.49) is obtained by combining together the contribution from the corresponding non-truncated counting function $N(r, 1/f_j)$ with the negative contribution from the ramification term, which is expressed in terms of the Wronskian determinant of the basis function $g_0, g_1, \ldots, g_n$. The tropical analogue, Theorem 5.21, of Cartan's theorem has an analogue of sorts for the ramification term expressed in terms of the Casoratian of the tropical linear combinations of the tropical entire functions in a spanning basis. Despite of this, finding a way to define consistent truncation for tropical linear combinations of tropical entire functions appears to be difficult. Namely, Theorem 5.21 implies that

$$(q - n - \lambda)T_g(r) \le \sum_{j=0}^q N(r, 1_\circ \oslash f_j) - N(r, 1_\circ \oslash C_\circ(f_0, \ldots, f_n)) + o\left(\frac{T_g(r)}{r^{1-\rho_2-\varepsilon}}\right),$$

$$(5.50)$$

which is an inequality of a similar type than Cartan's second main theorem, and the term

$$N(r, 1_\circ \oslash C_\circ(f_0, \dots, f_n)) \tag{5.51}$$

plays the role of the ramification term. There are immediate differences to the classical case, however. First, the counting function (5.51) is dependent on the tropical linear combinations $f_0, \dots, f_n$ of $g_0, \dots, g_n$, rather than the functions $g_0, \dots, g_n$ themselves, as is the case in the classical Cartan second main theorem. The reason for this stems from the properties of tropical matrixes, which in general cannot be inverted even if they are regular. In the classical Cartan second main theorem the Wronskian of the linear combinations f_j can be expressed in terms of the Wronskian of the base functions g_i under a natural generality assumption on the f_j's. Non-invertibility of tropical matrixes means that this phenomenon does not have a tropical analogue. Second, the interpretation of the term (5.51) proves to be very difficult, even if we restrict our considerations to the special case where the functions in the spanning basis and the linear combinations are identically the same. This special case is highly non-trivial as it is the same case which implies the (one-dimensional) tropical second main theorem by Laine and Tohge [69] as is seen below. We illustrate this with the following example.

Example 5.22. Let f_0 and f_1 be any tropical entire functions defined so that in the interval $[-1, 2]$ they satisfy

$$f_0(x) = \begin{cases} 1 & (x \le 0) \\ x + 1 & (x \ge 0) \end{cases}, \quad \text{and} \quad f_1(x) = \begin{cases} -(x - 1) & (x \le 1) \\ 2(x - 1) & (x \ge 1) \end{cases}. \tag{5.52}$$

One can for example choose f_0 and f_1 to be tropical polynomials, which are defined by the formulas (5.52) for all $x \in \mathbb{R}$. Now, looking at the Casoratian of these functions with $c = 1$, we have

$$C_\circ(f_0, f_1) = \max\{f_0(x + 1) + f_1(x), f_0(x) + f_1(x + 1)\}$$
$$= \begin{cases} -x + 2 & (x \le -1) \\ 3 & (-1 \le x \le 0) \\ 3 & (0 \le x \le 2/3) \\ 3x + 1 & (2/3 \le x \le 1) \\ 3x + 1 & (1 \le x). \end{cases}$$

which means that $C_\circ(f_0, f_1)$ has a root of multiplicity 1 at the point $x = -1$, and another root of multiplicity 3 at $x = 2/3$. The entire functions f_0 and f_1 have roots at the points $x = 0$ and $x = 1$, but at these points the

Casoratian of f_0 and f_1 is regular. By associating the roots of f_0 and f_1 at $x = 0, 1$ with the nearby roots of their Casoratian, we can obtain a reduction in the multiplicity of the roots of f_0 and f_1 in a similar fashion as in the complex analytic case. Namely, the contribution from the four points $x = -1, 0, 1$ and $x = 2/3$ to the combination of the non-integrated counting functions

$$n(r, 1_\circ \oslash f_0) + n(r, 1_\circ \oslash f_1) - n(r, 1_\circ \oslash C_\circ(f_0, f_1))$$

is $-1(= 0 - 1)$, $+1 = (1 - 0)$, $3 = (3 - 0)$ and $-3 = (0 - 3)$, respectively. However, it is not clear whether this reduction follows a consistent principle such as the truncation rule in the second main theorem by Cartan and Nochka.

In the previous example the sum of multiplicities of the roots of f_0 and f_1 is exactly the same as the sum of multiplicities of the roots of the tropical Casoratian $C_\circ(f_0, f_1)$. Similar examples can be constructed in a straightforward way in which the same phenomenon presents itself. However, the question of whether or not this happens generally is so far open even in the case $n = 1$.

Let us consider the one-dimensional case a while longer. According to the classical second main theorem in the complex plane,

$$(q - 1)T(r, f) \leq \overline{N}(r, f) + \sum_{k=1}^{q} \overline{N}\big(r, 1/(f - a_k)\big) + O\big(\log rT(r, f)\big)$$

holds for all r outside of a set of finite linear measure, provided that the target values $a_1, \ldots, a_q \in \mathbb{C}$ are distinct. Here the counting function $\overline{N}(r, f)$ is a measure of the poles of the meromorphic function f with their multiplicities ignored. From the properties of the counting function $\overline{N}(r, f)$ it follows immediately that

$$\overline{N}(r, f) \equiv \overline{N}(r, f^{(j)})$$

for all $j \in \mathbb{N}$. This property is unique for the pole counting function $\overline{N}(r, f)$, and it clearly does not remain valid for any other targets of f. An analogous property can be obtained for tropical meromorphic functions ψ of hyperorder strictly less than one. Namely, then we have

$$\overline{N}\big(r, \psi(x)\big) = \overline{N}\big(r, \psi(x + j)\big) + o\left(\frac{T(r, \psi)}{r^{1 - \rho_2 - \varepsilon}}\right)$$

again for all $j \in \mathbb{N}$, where the asymptotic identity holds for any $\varepsilon > 0$, and for all $r \to \infty$ outside of a set of finite logarithmic measure. This analogue is so far the only known tropical property which has to do with truncation.

5.7   Second main theorem as an application of the one-dimensional case

In this subsection we will show that Theorem 5.21 essentially implies a one-dimensional tropical second main theorem due to Laine and Tohge. Their result is stated as Theorem 3.44 in Chapter 3. We obtain the following Corollary 5.23 as a consequence of Theorem 5.21.

Corollary 5.23. *If f is a non-constant tropical meromorphic function of hyper-order $\rho_2 < 1$, if $\varepsilon > 0$, and if $q \geq 1$ distinct values $a_1, \ldots, a_q \in \mathbb{R}$ satisfy*

$$\max\{a_1, \ldots, a_q\} < \inf\{f(\alpha) : \omega_f(\alpha) < 0\}, \tag{5.53}$$

and

$$f \not\equiv f \oplus a_k \tag{5.54}$$

for all $k = 1, \ldots, q$, then

$$qT(r, f) \leq \sum_{k=1}^{q} N(r, 1_\circ \oslash (f \oplus a_k)) + o\left(\frac{T(r, f)}{r^{1-\rho_2-\varepsilon}}\right) \tag{5.55}$$

for all r outside of an exceptional set of finite logarithmic measure.

As we mentioned above, the statements of Theorem 3.44 and Corollary 5.23 are essentially the same. Before proving the corollary, we will explain what exactly we mean by this. As one can see by comparing the assertions of Corollary 5.23 and Theorem 3.44, they are identical, and the slight difference between these results comes from the assumptions. The assumption (3.11) of Theorem 3.44 is the same as the corresponding assumption (5.53) of Corollary 5.23, but the assumptions (5.54) and (3.12) are different, and neither one of them implies the other. The failure of (5.54) implies the existence of a value a_k such that $a_k \leq f(x)$ for all $x \in \mathbb{R}$, which means that the quantity

$$\inf\{f(\alpha) : \omega_f(\alpha) > 0\}$$

on the left-hand side of (3.12) is not smaller than a_k. Therefore, the assumption (3.12) is satisfied, when each a_j in Corollary 5.23 is not $0_\circ$, unless (5.54) holds. The purpose of the condition $f \not\equiv f \oplus a_k$ is to ensure that the targets a_k are real actual targets, meaning that none of the tropical sums $f \oplus a_k$ is identically the same as f.

Proof of Corollary 5.23. The first step is to apply Proposition 5.16 to represent the tropical meromorphic function f in the form

$$f = g_1 \oslash g_0,$$

where g_0 and g_1 are tropical entire functions without common roots. We will apply Theorem 5.21 with $n = 1$, $c = 1$, and g_0 and g_1 as the functions in the spanning basis. One of the conditions that needs to be satisfied so that we can do this is that g_0 and g_1 are linearly independent in the Gondran-Minoux sense. But in this special case this follows from the assumption f is non-constant. Now we define $g = [g_0 : g_1]$, $f_0 = g_0$, $f_1 = g_1$, and

$$f_k = (a_{k-1} \otimes g_0) \oplus (1_\circ \otimes g_1) \tag{5.56}$$

for all $k \in \{2, \ldots, q+1\}$. With these choices we have

$$
\begin{aligned}
C_\circ(f_0, f_1) &= (f_0(x) \otimes f_1(x+1)) \oplus (f_0(x+1) \otimes f_1(x)) \\
&= (f_1(x+1) \oslash f_1(x)) \oplus (f_0(x+1) \oslash f_0(x)) \otimes f_0(x) \otimes f_1(x),
\end{aligned}
$$

and so

$$L(x) = f_2(x) \otimes \cdots \otimes f_{q+1}(x) \oslash \left((f_1(x+1) \oslash f_1(x)) \oplus (f_0(x+1) \oslash f_0(x))\right).$$

By defining

$$D(x) := (f_1(x+1) \oslash f_1(x)) \oplus (f_0(x+1) \oslash f_0(x))$$

it follows that

$$N(r, 1_\circ \oslash L) \le \sum_{k=2}^{q+1} N(r, 1_\circ \oslash f_k) + N(r, D)$$

and

$$N(r, L) = N(r, 1_\circ \oslash D).$$

Therefore, by the tropical Jensen formula,

$$
\begin{aligned}
N(r, 1_\circ \oslash L) - N(r, L) &\le \sum_{k=2}^{q+1} N(r, 1_\circ \oslash f_k) + N(r, D) - N(r, 1_\circ \oslash D) \\
&= \sum_{k=2}^{q+1} N(r, 1_\circ \oslash f_k) - m(r, D) + m(r, 1_\circ \oslash D) \\
&\quad + D(0),
\end{aligned}
\tag{5.57}
$$

where the constant $D(0) = \max\{g_1(1) - g_1(0), g_0(1) - g_0(0)\}$ depends only on the function f. Since

$$1_\circ \oslash D(\pm r) = \min\{f_1(\pm r) - f_1(\pm r + 1), f_0(\pm r) - f_0(\pm r + 1)\}$$
$$\leq f_0(\pm r) \oslash f_0(\pm r + 1),$$

it follows by (5.38) that

$$m(r, 1_\circ \oslash D) - m(r, D)$$

$$= \frac{1}{2}(1_\circ \oslash D)(r) + \frac{1}{2}(1_\circ \oslash D)(-r)$$

$$\leq \frac{1}{2}(f_0(r) \oslash f_0(r+1)) + \frac{1}{2}(f_0(-r) \oslash f_0(-r+1))$$

$$= m(r, f_0(x) \oslash f_0(x+1)) - m(r, f_0(x+1) \oslash f_0(x))$$

$$= N(r, f_0(x+1) \oslash f_0(x)) - N(r, f_0(x) \oslash f_0(x+1)) + O(1)$$

$$= N(r, 1_\circ \oslash f_0(x)) - N(r, 1_\circ \oslash f_0(x+1)) + O(1) = o\left(\frac{T_g(r)}{r^{1-\rho_2-\varepsilon}}\right),$$
$$(5.58)$$

where r tends to infinity outside of an exceptional set of finite logarithmic measure. Using (5.57) and (5.58), we have

$$N(r, 1_\circ \oslash L) - N(r, L) \leq \sum_{k=2}^{q+1} N\left(r, 1_\circ \oslash f_k\right) + o\left(\frac{T_g(r)}{r^{1-\rho_2-\varepsilon}}\right), \quad (5.59)$$

where again r tends to infinity outside of an exceptional set of finite logarithmic measure.

The assumption (5.53) implies that the roots of g_0 are exactly the poles of $a_k \oplus f$ for all $k = 1, \ldots, q$, and the multiplicities are the same as well. Hence, equation (5.56) yields

$$N\left(r, 1_\circ \oslash f_k\right) = N(r, 1_\circ \oslash (f \oplus a_{k-1})) \qquad (5.60)$$

for all $k = 2, \ldots, q+1$. In addition, by the assumptions (5.53) and (5.54) it follows that

$$f \not\equiv f \oplus a_k \not\equiv a_k \qquad \text{for all } k \in \{1, \ldots, q\},$$

which implies

$$g_1 \oslash g_0 \not\equiv f_{k+1} \oslash g_0 \not\equiv a_k \qquad \text{for all } k \in \{1, \ldots, q\}. \quad (5.61)$$

Taking into account the assumption that f is non-constant, the condition (5.61) implies that $\{f_2, \ldots, f_{q+1}\}$ is non-degenerate, and thus Theorem 5.21 applies with $\lambda = 0$ and $n = 1$. By (5.59) we therefore have

$$qT_g(r) \leq \sum_{k=2}^{q+1} N\left(r, 1_\circ \oslash f_k\right) + o\left(\frac{T_g(r)}{r^{1-\rho_2-\varepsilon}}\right), \quad (5.62)$$

where r tends to infinity outside of an exceptional set of finite logarithmic measure. We conclude by applying Proposition 5.17 with $f = g_0 \oslash g_1$ and $g = [g_0 : g_1]$, and by using (5.60), that equation (5.62) takes the desired form

$$qT(r, f) \leq \sum_{k=1}^{q} N\big(r, 1_\circ \oslash (f \oplus a_k)\big) + o\left(\frac{T(r, f)}{r^{1-\rho_2-\varepsilon}}\right),$$

where the exceptional set for r is again of finite logarithmic measure. $\quad\square$

Chapter 6

Representations of tropical periodic functions

In this chapter, our key result is to giving a representation of tropical meromorphic, periodic functions in terms of certain elementary tropical meromorphic functions that have, essentially, a sawtooth waveform, with their width, height and inclination depending on two parameters. Before proceeding to this representation, we first make some simple observations concerning periodic functions in the tropical setting. Without restricting generality, we assume in this chapter that tropical meromorphic functions f under consideration here are of period 1, unless otherwise specified. Then, trivially, f satisfies the elementary equation $y(x + 1) \oslash y(x) = 0$, that is, $y(x + 1) - y(x) = 0$.

6.1 Representations of tropical periodic functions

We first observe that tropical periodic functions have, roughly speaking, equally many roots and poles. Recalling the notion of multiplicity (of roots and poles), and denoting by spt ω_f the support of the multiplicity function ω_f, we state

Theorem 6.1. *Given a tropical meromorphic 1-periodic function f, it is true that*

$$\sum_{c \in (spt\ \omega_f) \cap [0,1)} \omega_f(c) = 0,$$

i.e.

$$\sum_{a \in [0,1),\, \omega_f(a) > 0} \tau_f(a) = \sum_{b \in [0,1),\, \omega_f(b) < 0} \tau_f(b).$$

Proof. Denote by $0 = c_0 < c_1 < \cdots < c_K < c_{K+1} = 1$ all elements in (spt ω_f) $\cap [0, 1)$. Computing the slopes of f, which are well-determined outside of spt ω_f, we readily obtain

$$\sum_{c \in (\text{spt } \omega_f) \cap (0,1)} \omega_f(c) = \sum_{j=1}^{K} \omega_f(c_j) = \sum_{k=1}^{K} \left(f' \left(\frac{c_{j+1} + c_j}{2} \right) - f' \left(\frac{c_j + c_{j-1}}{2} \right) \right)$$

$$= f' \left(\frac{1 + c_K}{2} \right) - f' \left(\frac{c_1}{2} \right) = f' \left(\frac{c_K - 1}{2} \right) - f' \left(\frac{c_1}{2} \right) = -\omega_f(0),$$

completing the proof. $\qquad\qquad\square$

Remark 6.2. Observe that whenever $\sum_{c \in (\text{spt } \omega_f) \cap [0,1)} |\omega_f(c)| = 0$, then f must be constant in $[0, 1)$, hence in $\mathbb{R}$ by 1-periodicity.

Remark 6.3. Theorem 6.1 has a certain similarity to the behavior of elliptic functions in the complex plane, as well as to the argument principle in classical complex analysis.

Corollary 6.4. *Given a tropical meromorphic periodic function f, then*

$$\sum_{c \in (\text{spt } \omega_f) \cap [x, x+n)} \omega_f(c) = 0$$

for each $x \in \mathbb{R}$ and each $n \in \mathbb{N}$.

In order to obtain the anticipated representation of tropical periodic functions, the following simple proposition appears to be useful:

Proposition 6.5. *Suppose f, g are tropical meromorphic functions. Then $\omega_f = \omega_g$ on a closed interval $I \subset \mathbb{R}$ if and only if $f - g (= f \oslash g)$ is linear on I.*

Proof. Suppose first that $\omega_f = \omega_g$ on a closed interval $I \subset \mathbb{R}$, and define $h := f \oslash g$. Of course, h is tropical meromorphic and $\omega_h = 0$ on I. Then the right and left derivatives of h satisfy $h'_+(x) = h'_-(x)$ for each $x \in I$. Therefore, h' exists and is constant on I, meaning that h must be linear on I.

On the other hand, if h is linear, then $\omega_h = \omega_f - \omega_g = 0$. $\qquad\square$

Next we need to consider certain special tropical meromorphic 1-periodic functions defined as

$$\pi^{(a,b)}(x) := \frac{1}{a+b} \min\{a(x-[x]), b(1-(x-[x]))\},$$

where $a > 0, b > 0$ are arbitrary real parameters. Observe that

$$\pi^{(a,b)}(0) = \frac{1}{a+b} \min\{0,b\}$$

and, in fact, $\pi^{(a,b)}(n) = 0$ for each $n \in \mathbb{Z}$. As one can immediately see, the graph of $\pi^{(a,b)}$ has a sawtooth waveform of width 1, with its height and inclination depending on the parameters (a,b). More precisely, $\pi^{(a,b)}(x) \geq 0$ for each $x \in \mathbb{R}$, and the (local) maxima are attained at the points $a(x - [x]) = b(1-(x-[x]))$, i.e., at $x = \frac{b}{a+b} + n, n \in \mathbb{Z}$. Therefore, the height of the graph equals to

$$\pi^{(a,b)}\left(\frac{b}{a+b} + n\right) = \frac{ab}{(a+b)^2},$$

see the adjacent Figures 6.1-6.3 depicting the graphs of $\pi^{(1,1)}, \pi^{(1,6)}$ and $\pi^{(8,2)}$. For simplicity, we call functions of type $\pi^{(a,b)}$ sawtooth functions in what follows.

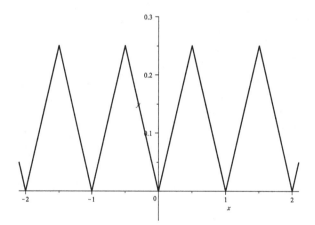

Fig. 6.1 $\pi^{(1,1)}$

Moreover, it is immediate to see that $\omega_{\pi^{(a,b)}}(n) = 1$ and $\omega_{\pi^{(a,b)}}(\frac{b}{a+b} + n) = -1$ whenever $n \in \mathbb{Z}$, while $\omega_{\pi^{(a,b)}}(x) = 0$ for each $x \in \mathbb{R} \setminus \mathbb{Z}$ which is not of the form $\frac{b}{a+b} + \mathbb{Z}$. Hence, $\pi^{(a,b)}$ is a tropical meromorphic 1-periodic

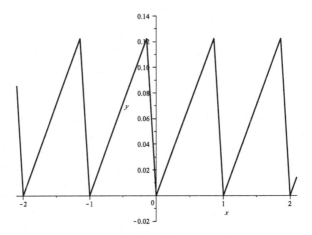

Fig. 6.2 $\pi^{(1,6)}$

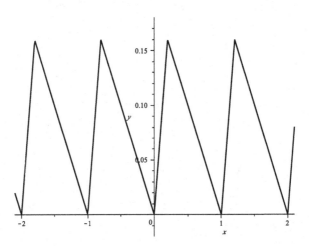

Fig. 6.3 $\pi^{(8,2)}$

function that has roots of multiplicity 1 exactly at the points of $\mathbb{Z}$ and poles of multiplicity one exactly at the points of $\frac{b}{a+b} + \mathbb{Z}$.

To obtain the representation of tropical periodic functions in terms of the sawtooth functions $\pi^{(a,b)}$, let f now be an arbitrary tropical meromorphic 1-periodic, non-linear function, and let $(\text{spt } \omega_f) \cap [0,1) = \{c_0, c_1, \ldots, c_K\}$

be the set of its roots and poles in the period interval $[0, 1)$. We may now fix (essentially unique) non-negative parameters a_k, b_k such that $c_k = \frac{b_k}{a_k + b_k}$ for each $k = 1, \ldots, K$. Define then

$$\hat{f}(x) := -\sum_{k=1}^{K} \omega_f(c_k)\pi^{(a_k, b_k)}(x) + f(0).$$

Now we immediately see, by properties of the functions $\pi^{(a_k, b_k)}$ that $\hat{f}(0) = f(0)$ and

$$\hat{f}(c_1) = \hat{f}\left(\frac{b_1}{a_1 + b_1}\right) = -\omega_f(c_1)\pi^{(a_1, b_1)}\left(\frac{b_1}{a_1 + b_1}\right) + f(0),$$

hence

$$\hat{f}(c_1) = -\omega_f(c_1)\frac{a_1 b_1}{(a_1 + b_1)^2} + f(0) = c_1(c_1 - 1)\omega_f(c_1) + f(0). \tag{6.1}$$

To compute the roots and poles of $\hat{f}$, The following simple observation appears to be useful:

Proposition 6.6. *If g is a finite linear combination of tropical meromorphic functions, say $g = \sum A_k g_k$, then $\omega_{\sum A_k g_k}(x) = \sum A_k \omega_{g_k}(x)$ for each $x \in \mathbb{R}$.*

Proof. This immediately follows by the definition of ω and the linearity of derivative. $\qquad\square$

By Proposition 6.6, we see, for each $j = 1, \ldots, K$, that

$$\omega_{\hat{f}}(c_j) = \sum_{k=1}^{K}(-\omega_f(c_k))\omega_{\pi^{(a_k, b_k)}}(c_j) = -\omega_f(c_j)\omega_{\pi^{(a_j, b_j)}}(c_j) = \omega_f(c_j),$$

since $\omega_{\pi^{(a_j, b_j)}}(c_j) = -1$ and $\omega_{\pi^{(a_k, b_k)}}(c_j) = 0$ for $k \neq j$. Moreover, by Theorem 6.1,

$$\omega_{\hat{f}}(0) = \sum_{k=1}^{K}(-\omega_f(c_k))\omega_{\pi^{(a_k, b_k)}}(0) = \sum_{k=1}^{K}(-\omega_f(c_k)) = \omega_f(0).$$

Since we now have that $\omega_f(x) = \omega_{\hat{f}}(x)$ holds at each $x \in \mathbb{R}$, we conclude by Proposition 6.5 that $f(x) = \hat{f}(x)$ for all $x \in \mathbb{R}$. Therefore, we have proved the following representation theorem:

Theorem 6.7. *Non-constant tropical meromorphic 1-periodic functions f have the representation*

$$f(x) = -\sum_{j=1}^{K} \omega_j(c_j)\pi^{(a_j, b_j)}(x) + f(0),$$

where $\{c_1, \ldots, c_K\} = spt\, \omega_f \cap [0, 1)$ *and* $\pi^{(a_j, b_j)}$ *are sawtooth functions such that* $c_j = b_j/(a_j + b_j)$ *for* $j = 1, \ldots, K$.

The following two examples are intended to illuminate the previous reasoning:

Example 6.8. Consider the 1-periodic function $f_1(x)$ defined by connecting the points $f_1(0) = f_1(1/2) = f_1(1) = 0, f_1(2/3) = 1/18$ with a piecewise linear continuous curve. Then, using the preceding notations, we have $c_1 = 1/2, c_2 = 2/3, \omega_{f_1}(c_1) = 1/3$ and $\omega_{f_1}(c_2) = -1/2$. Applying Theorem 6.7, we see that

$$f_1(x) = -\frac{1}{3}\pi^{(1,1)}(x) + \frac{1}{2}\pi^{(1,2)}(x)$$

$$= -\frac{1}{6}\min\{-x + [x], -1 + x - [x]\} + \frac{1}{6}\min\{x - [x], 2(1 - x + [x])\}$$

and, in fact,

$$f(x) = \left\{ \begin{array}{ll} 0 & (0 \leq x - [x] \leq 1/2), \\ \frac{1}{3}(x - [x]) - \frac{1}{6} & (1/2 \leq (x - [x]) \leq 2/3), \\ -\frac{1}{6}(x - [x]) + \frac{1}{6} & (2/3 \leq (x - [x]) \leq 1). \end{array} \right.$$

See the adjacent Figure 6.4. Moreover, we have $\omega_{f_1}(0) + \omega_{f_1}(c_1) + \omega_{f_1}(c_2) = 0$ by Theorem 6.1, and as one can see in Figure 6.4 as well.

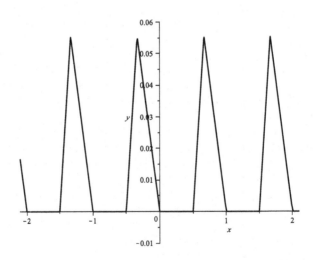

Fig. 6.4 $-\frac{1}{3}\pi^{(1,1)} + \frac{1}{2}\pi^{(1,2)}$

Example 6.9. Now consider the 1-periodic function $f_2(x)$ defined by the points $f_2(0) = f_2(1) = 1/2$, $f_2(1/2) = 3/4$, $f_2(3/4) = 1/4$ in a similar way as f_1 in Example 6.8. Then we have $\omega_{f_2}(1/2) = -5/2$ and $\omega_{f_2}(3/4) = 3$. Therefore, f_2 can be represented as

$$f_2(x) = \frac{5}{2}\pi^{1,1}(x) - 3\pi^{1,3}(x) + \frac{1}{2},$$

see Figure 6.5.

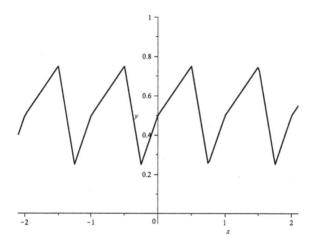

Fig. 6.5 $\frac{5}{2}\pi^{(1,1)} - 3\pi^{(1,3)} + \frac{1}{2}$

Theorem 6.10. *All non-constant tropical 1-periodic meromorphic functions f satisfy $T(r, f) \asymp \kappa r^2$ for some $\kappa > 0$, hence are of order 2. In particular, this holds for all sawtooth functions $\pi^{(a,b)}$.*

Proof. First note that

$$m(r, \pi^{(a,b)}) \le |\pi^{(a,b)}(r)| + |\pi^{(a,b)}(-r)| \le \frac{2ab}{(a+b)^2}.$$

By Theorem 6.5, $m(r, f) = O(1)$. Therefore, it is sufficient to consider the counting function $N(r, f)$ only. Denote now by $n_K(f)$ the number of poles of f in the multiple period interval $[-K, K)$ for $K \in \mathbb{N}$. Then, clearly, $n_K(f) = Kn_1(f) =: KC$ by 1-periodicity. Moreover, $n(1, f) + \tau_f(-1) = n(1, f) + n(0, f) = n_1(f) = C$ as the pole at $r = -1$ contributes to $n_1(f)$ but not to $n(1, f)$, and $\tau_f(-1) = n(0, f)$ by periodicity. Also, $n(K, f) + n(0, f) = Kn_1(f)$ for each $K \in \mathbb{N}$. Hence, for $r \geq 1$, we get

$$N(r, f) = \frac{1}{2} \int_0^r n(t, f)dt \leq \frac{1}{2}n([r] + 1, f)r \leq ([r] + 1)Cr \leq O(r^2),$$

and so $T(r, f) = O(r^2)$.

If it now happens that $N(r, f) = O(r^{2-\varepsilon})$, then we have

$$n(r, f) \leq \frac{2}{r}N(2r, f) \leq \frac{2}{r}C(2r)^{2-\varepsilon} = \tilde{C}r^{1-\varepsilon}$$

for some $C, \tilde{C} > 0$. Therefore,

$$\tilde{C}r^{1-\varepsilon} \geq n(r, f) \geq n([r], f) = [r]n_1(f) - n(0, f) \geq (r - 1)n_1(f) - n(0, f),$$

and so

$$n_1(f) \leq \frac{\tilde{C}r^{1-\varepsilon} + n(0, f)}{r - 1} \to 0$$

as $r \to \infty$. This means that f has no poles in the period interval $[0, 1)$. Similarly, by Theorem 6.1, there are no roots in $[0, 1)$ as well, and therefore, f must be constant, completing the proof. $\qquad\square$

In some of the later reasoning, we need to consider tropical c-periodic meromorphic functions such that their period $c \neq 0, 1$. In particular, we need to consider their order:

Theorem 6.11. *All non-constant tropical c-periodic meromorphic functions f satisfy $T(r, f) \asymp \kappa r^2$ for some $\kappa > 0$, and are, therefore, or order 2.*

Proof. By periodicity, it is clear that f is bounded, and therefore, $m(r, f) = O(1)$. An easy modification of the preceding proof to compute $N(r, f)$ now applies. Denote, this time, by $n_K(f)$ the number of poles of f in the interval $[-Kc, Kc)$ for $K \in \mathbb{N}$. Then $n_K(f) = Kn_1(f) =: KC$, say, by periodicity, and $n(c, f) + \tau_f(-c) = n(c, f) + n(0, f) = n_1(f) = C$ as the (possible) pole at $r = -c$ contributes to $n_1(f)$ but not to $n(c, f)$, and $\tau_f(-c) = n(0, f)$ by periodicity. Therefore, $n(cK, f) + n(0, f) = KC$ for each $K \in \mathbb{N}$. Hence, assuming that $r \geq c$, say, we have $Kc \leq r < (K + 1)c$ for some $K \in \mathbb{N}$, and therefore,

$$N(r, f) = \frac{1}{2} \int_0^r n(t, f)dt \leq \frac{1}{2}(K + 1)Cr \leq \frac{1}{2}(r + c)Cr \leq Cr^2,$$

and so $T(r, f) = O(r^2)$.

If it happens that $N(r, f) = O(r^{2-\varepsilon})$, a slight modification of corresponding part of the proof of Theorem 6.10 applies to show that f has no poles in the period interval $[0, c)$. Again by Theorem 6.1, there are no roots as well, and f has to be a constant, a contradiction. $\qquad\square$

Another way of proving Theorem 6.11 is as follows. Say that f is a periodic tropical meromorphic function of period $c > 0$. Then $F(x) := f(cx)$ satisfies

$$F(x + 1) = f(cx + c) = f(cx) = F(x)$$

for all $x \in \mathbb{R}$, and so F is a periodic tropical meromorphic function of period 1. Since F is of order 2 by Theorem 6.10, and

$$T(r, F) = T_F(r) = T_f(|c|r) = T(|c|r, f)$$

by Proposition 5.17, it follows that f is of order 2 as well.

6.2 Ultra-discrete theta functions

In his paper [81], A. Nobe has ultra-discretized the elliptic theta function $\vartheta_{00}(z, \tau)$ and obtained a piecewise quadratic function denoted by $\Theta_0(x)$, that is,

$$\Theta_0(x) = -x^2 + \max_{n=-\infty}^{+\infty} \left\{ 2nx - n^2 \right\}. \qquad (6.2)$$

Such an agreement as $2nx - n^2 \to -\infty$ $(n \to \pm\infty)$ for each fixed $x \in \mathbb{R}$ assures us the convergence of this infinite max-plus series. Then the function $\Theta_0(x) + x^2$ is, of course, a tropical entire function in our setting. The theta function $\vartheta_{00}(z, \tau)$ is given by

$$\vartheta_{00}(z, \tau) = \sum_{n=-\infty}^{+\infty} e^{\pi i n^2 \tau} e^{2\pi i n z},$$

where $\tau \in \mathbb{H} = \{ z \in \mathbb{C} : \operatorname{Im} z > 0 \}$ and $z \in \mathbb{C}$. By using Jacobi's transformation formula,

$$\vartheta_{00}\left(\frac{z}{\tau}, -\frac{1}{\tau} \right) = e^{-\pi i/4} \tau^{1/2} e^{\pi i z^2/\tau} \vartheta_{00}(z, \tau),$$

the expression is reduced into

$$\vartheta_{00}(z, \tau) = e^{\pi i/4} \tau^{-1/2} e^{-\pi i z^2/\tau} \sum_{n=-\infty}^{+\infty} e^{-\pi i n^2/\tau} e^{2\pi i n z/\tau}$$

where $-1/\tau \in \mathbb{H}$ and $z/\tau \in \mathbb{C}$ ($\tau \neq 0$). Let $\tau = i\pi\varepsilon(\in \mathbb{H})$, where ε is a positive number. For $x = \mathrm{Re}\, z$, $z \in \mathbb{C}$, we have

$$\vartheta_{00}(x, i\pi\varepsilon) = e^{-k\pi i}(\pi\varepsilon)^{-1/2}e^{-x^2/\varepsilon}\sum_{n=-\infty}^{+\infty}e^{(2nx-n^2)/\varepsilon}$$

for some integer k. As its ultra-discrete limit

$$\Theta_0(x) = \lim_{\varepsilon \to +0}\varepsilon\log\vartheta_{00}(x, i\pi\varepsilon),$$

Nobe's ultra-discrete elliptic theta function (6.2) is finally obtained. In fact, for any $N \in \mathbb{N}$, the function

$$\varepsilon\log\sum_{n=-N}^{N}\exp\big[(2nx-n^2)/\varepsilon\big]$$

uniformly converges to

$$g_N(x) = \max_{n=-N}^{N}\{2nx-n^2\}$$

in the limit $\varepsilon \to +0$ and moreover this $g_N(x)$ uniformly converges to

$$\max_{n=-\infty}^{\infty}\{2nx-n^2\}$$

on $|x| < N+\frac{1}{2}$ in the limit $N \to \infty$. Here $\log w$ stands for the principal value of the natural logarithm of a complex number w. The representation (6.2) immediately shows that $\Theta_0(x)$ is 1-periodic, since

$$-(x+1)^2 + \max_{n=-\infty}^{+\infty}\big\{2n(x+1)-n^2\big\} = -x^2 + \max_{n=-\infty}^{+\infty}\big\{2(n-1)x-(n-1)^2\big\}.$$

Thus we have actually

$$\Theta_0(x+m) = \Theta_0(x) \quad \text{for all } m \in \mathbb{Z}.$$

This means that if $\Theta_0(x)$ were a tropical entire function, then it would be of order 2 as before by applying the tropical Nevanlinna theory. Although this is not the case, we will give further explanation in what sense the growth of $\Theta_0(x)$ can be regarded to be of order 2 at the end of this section. Before the observation, let us now put

$$f(x) := \Theta_0\left(\frac{1}{2}x\right) + \frac{1}{4}x^2 = \max_{n=-\infty}^{+\infty}\{nx-n^2\}$$

and $g(x) := \max_{n\in\mathbb{N}}\{nx-n^2\} = \bigoplus_{n=1}^{\infty}(x^{\otimes n}\oslash n^2)$ so that

$$f(x) = g(x) \oplus 1_\circ \oplus g(-x), \quad x \in \mathbb{R}.$$

In Nobe's paper [81], Jacobi's elliptic functions $\operatorname{sn} u$, $\operatorname{cn} u$ and $\operatorname{dn} u$ have been also ultra-discretized into piecewise linear and 1-periodic functions through the $\Theta_0(x)$. For example, the ultra-discretization of $\operatorname{dn} u$ is denoted by $-J(x)$ and takes the local maximum at $x \in \mathbb{Z}$ and the local minimum at $x \in \mathbb{Z} + \frac{1}{2}$. This function is indeed given by

$$
\begin{aligned}
-J(x) &= \Theta_0\left(\frac{1}{2}\right) + \Theta_0(x) - \Theta_0\left(x + \frac{1}{2}\right) \\
&= f(1) - \frac{1}{4} + f(2x) - x^2 - f(2x+1) + x^2 + x + \frac{1}{4} \\
&= f(2x) - f(2x+1) + x + f(1) \\
&= \{g(2x) \oplus 1_\circ \oplus g(-2x)\} \oslash \{g(2x+1) \oplus 1_\circ \oplus g(-2x-1)\} \otimes x.
\end{aligned}
$$

Note that

$$
f(1) = \max_{n=-\infty}^{+\infty} \{n - n^2\} = 0.
$$

Hence, when $x > 0$, this tropical meromorphic function is obtained by an ultra-discrete analogue of the 'logarithmic shift' (or the ultra-discrete Cole-Hopf transformation) of the tropical entire function $g(2x) \oplus 1_\circ$ associated with the tropical unit x, that is,

$$
\begin{aligned}
\{g(2x) \oplus 1_\circ\} &\oslash \{g(2x+1) \oplus 1_\circ\} \otimes x \\
&= \max_{n \in \mathbb{N}}\{2nx - n^2\} \oslash \max_{n \in \mathbb{N}}\{2nx - n^2 + n\} \otimes x.
\end{aligned}
$$

Note that both $f(x)$ and $g(x)$ are not periodic any more, while $-J(x)$ is a 1-periodic function. Therefore, the order of growth of the function $-J(x)$ is again exactly equal to 2 and it can be also observed by studying the distribution of roots of poles of $-J(x)$. On the other hand, Nevanlinna theory on tropical meromorphic functions introduced in Chapter 3 tells us that the order of growth of this function should be equal to that of the tropical entire function $g(2x)$ due to the expression above. Recall that the growth is measured by its Nevanlinna characteristic function $T(r, -J)$, but it is mainly controlled by the counting function of the poles of $-J$, that is, the roots of the entire function $g(2x+1) \oplus 1_\circ$. In fact, the growth of $|J(x)|$ behaves as the minority then, since this tropical meromorphic function is obtained as an ultra-discrete logarithmic shift of a tropical entire function. As an application of the result in Chapter 2 we confirm the tropical entire function $g(x)$ is of order

$$
\limsup_{n \to \infty} \frac{\log n^2}{\log(n^2/n)} = 2
$$

and of type

$$c(2) \limsup_{n \to \infty} \frac{n^2}{(n^2/n)^2} = \frac{1}{4},$$

respectively, so that for any given $\epsilon > 0$, it behaves like

$$(1/4 - \epsilon)x^2 \le |g(x)| \le (1/4 + \epsilon)x^2$$

as $x \to +\infty$ in a large set of $\mathbb{R}_{\ge 0}$.

Remark 6.12. Now we observe the function x^2 which is, of course, neither tropical entire nor tropical meromorphic function in our setting, This does not, however, have any singularity which provides its pole or root for a tropical function, so that one might also as well to regard it as a tropical unit, or a **genus** in function theory, having the same order

$$\limsup_{r \to \infty} \frac{\log M(r, x^2)}{\log r} = \limsup_{r \to \infty} \frac{\log r^2}{\log r} = 2$$

and the same type

$$c(2) \limsup_{r \to \infty} \frac{M(r, x^2)}{r^2} = \frac{1}{4} \limsup_{r \to \infty} \frac{r^2}{r^2} = \frac{1}{4}$$

to those of the tropical entire function $f(x)$.

Combining the above results, we may think that the function $\Theta_0(x) = f(2x) \oslash x^2$ behaves essentially like a tropical meromorphic function of order 2.

Chapter 7

Applications to ultra-discrete equations

In this chapter we look into methods of solving ultra-discrete equations in terms of tropical meromorphic functions. We present general solutions of certain relatively simple classes of ultra-discrete equations, and classes of special rational and hypergeometric solutions of ultra-discrete Painlevé equations. In addition, we describe how to apply tropical Nevanlinna theory to study the growth of tropical meromorphic solutions of ultra-discrete equations.

7.1 First-order ultra-discrete equations

We start this chapter by considering simple first-order equations of type
$$y(x + 1) = y(x) + c, \qquad c \in \mathbb{R}, \tag{7.1}$$
which may equivalently be written as
$$y(x + 1) \oslash y(x) = c.$$
Equation (7.1) may now immediately be solved completely. In fact, a special solution $f_0(x) := cx$ is at once verified. Let then f be an arbitrary tropical meromorphic solution to equation (7.1). Then the difference $f(x) \oslash x^{\otimes c} = f(x) - cx$ is clearly tropical meromorphic and 1-periodic. Therefore, all tropical meromorphic solutions to (7.1) have a unique representation as the sum of the linear polynomial cx and a tropical tropical meromorphic 1-periodic function. Recalling Theorem 6.7 in Chapter 6 we obtain

Theorem 7.1. *All non-constant tropical meromorphic solutions f to (7.1) have a representation*
$$f(x) = -\sum_{j=1}^{K} \omega_j(c_j) \pi^{(a_j, b_j)}(x) + cx + f(0),$$

where $\{c_1, \ldots, c_K\} = spt\, \omega_f \cap [0, 1)$ and $\pi^{(a_j, b_j)}$ are sawtooth functions such that $c_j = b_j/(a_j + b_j)$ for $j = 1, \ldots, K$.

Remark 7.2. For a later need, we add here the following observation: Considering equation

$$y(x + 1) = y(x) + \Pi_1(x), \tag{7.2}$$

where Π_1 is a tropical 1-periodic meromorphic function, then we may formally apply the preceding to conclude that all solutions f of (7.2) are of the form

$$f(x) = \Xi_1(x) + x\Pi_1(x),$$

where Ξ_1 is 1-periodic as well. Insisting that f is tropical meromorphic, it follows that both $\Xi_1(x)$ and $x\Pi_1(x)$ should be tropical meromorphic. But then, $x\Pi_1(x)$ has to be locally linear, meaning that it has to be locally of type cx. By continuity, we conclude that Π_1 is constant. Therefore, (7.2) reduces back to (7.1).

We next proceed to considering ultra-discrete equations of type

$$y(x + 1) = y(x)^{\otimes c} = cy(x), \qquad c \in \mathbb{R}. \tag{7.3}$$

Clearly, any solution to (7.3) is constant, provided $c = 0$. Therefore, we assume now that $c \neq 0$. Halburd and Southall showed in [46], Lemma 4.1, that equation $y(x + 1) = ny(x)$, $n \in \mathbb{N}$, permits non-constant tropical meromorphic solutions with integer slopes if and only if $n = \pm 1$. In the case of real slopes, the situation becomes less trivial. We start with

Theorem 7.3. *Suppose $c = \pm 1$ in $y(x+1) = cy(x)$. If $c = 1$, resp. $c = -1$, then all non-constant tropical meromorphic functions are 1-periodic, resp. 2-periodic and anti-1-periodic, and of order 2.*

Proof. If $c = 1$, then all non-constant solutions to $y(x + 1) = y(x)$ are trivially 1-periodic. In the case $c = -1$, i.e. the case $y(x + 1) = -y(x)$, let f be an arbitrary non-constant tropical meromorphic solution. Clearly, f is anti-1-periodic. Since $f(x + 2) = -f(x + 1) = f(x)$, f is 2-periodic. As for the claim concerning the order of f, it is sufficient to refer to Theorem 6.11 in Chapter 6. $\qquad\square$

As for the general case $c \neq 0, \pm 1$, we prove the following representation result:

Theorem 7.4. *Let f be a non-constant tropical meromorphic solution to (7.3), where $c \neq 0, \pm 1$. Then f can be represented as*

$$f(x) = \sigma(c) \sum_{j=1}^{q} \omega_f(x_j) e_c(x - x_j), \tag{7.4}$$

where $\{x_1, \ldots, x_q\} = \mathrm{spt}\, \omega_f \cap [0, 1)$ and where $\sigma(c) = c/(c-1)$ for $|c| > 1$, and $\sigma(c) = c/(1-c)$ for $0 < |c| < 1$.

Proof. First observe that spt $\omega_f \cap [0, 1)$ is a finite set. Secondly, spt $\omega_f \cap [0, 1)$ must be non-empty. Indeed, if spt $\omega_f \cap [0, 1) = \emptyset$, then f is non-constant and linear, and a contradiction follows at once: Since $f(x) = \alpha x + \beta$ for some real coefficients α, β, then $\alpha x + \alpha + \beta = c\alpha x + c\beta$, implying that $c = 1$, $\alpha = 0$ and $f = \beta$. But now, equation (7.3) implies that f is constant, a contradiction.

Define now a tropical meromorphic function as follows:

$$g(x) = \sigma(c) \sum_{j=1}^{q} \omega_f(x_j) e_c(x - x_j),$$

where $\sigma(c)$ is as determined in the claim. By elementary properties of tropical exponential functions, we see that g is a solution to (7.3):

$$g(x + 1) = \sigma(c) \sum_{j=1}^{q} \omega_f(x_j) e_c(x + 1 - x_j)$$

$$= c \left(\sigma(c) \sum_{j=1}^{q} \omega_f(x_j) e_c(x - x_j) \right) = cg(x).$$

Consider now the interval $[0, 1)$ for a while. If now $x \notin \{x_1, \ldots, x_q\}$, then we trivially have $\omega_f(x) = 0$. Recalling Proposition 6.6 in Chapter 6, we infer that

$$\omega_g(x) = \omega_f(x_j) \sigma(c) \omega_{e_c}(x - x_j) = 0,$$

since $\omega_{e_c}(x - x_j) = 0$. Recalling next Proposition 2.7, we know that $\omega_{e_c}(0) = c/(c-1)$, whenever $|c| > 1$, and $\omega_{e_c}(0) = c/(1-c)$, if $0 < |c| < 1$. Therefore,

$$\omega_g(x_j) = \omega_f(x_j) \sigma(c) \omega_{e_c}(0) = \omega_f(x_j)$$

for all $j = 1, \ldots, q$. By the definition of tropical exponential functions e_c, we may now extend, by induction, this reasoning to $[n, n+1)$ for each $n \in \mathbb{Z}$ to conclude that $\omega_g(x) = \omega_f(x)$ for all $x \in \mathbb{R}$. By Proposition 6.5 in Chapter 6, we observe that $f - g$ is a linear function in $\mathbb{R}$, hence $f(x) = g(x) + Ax + B$ for some real constants A, B. Therefore,

$$0 = f(x+1) - cf(x) - (1 - c)Ax + A + (1 - c)B,$$

resulting in $A = B = 0$. This completes the proof. $\qquad\square$

Remark 7.5. By [84], p. 295–300, and [65], Theorem 3.5, the solution space to (7.3) is one-dimensional with respect to 1-periodic coefficients, and so, given a solution f and a 1-periodic locally linear function π, then πf formally solves (7.3). Observe, however, that by the representation (7.4), f is locally non-constant. Therefore, πf is not tropical meromorphic except for the case that π is locally constant, reducing to a constant by the continuity of tropical meromorphic functions.

Corollary 7.6. *A non-constant tropical meromorphic solution f of equation (7.3), $c \neq 0$, is of hyper-order $\rho_2(f) < 1$ if and only if $c = \pm 1$.*

7.2 Second-order ultra-discrete equations

This section is devoted to considering equations of type

$$y(x + 1) + y(x - 1) = cy(x), \qquad c \in \mathbb{R}, \tag{7.5}$$

equivalently written in tropical notation as

$$y(x + 1) \otimes y(x - 1) = y(x)^{\otimes c}.$$

For an earlier consideration of the special case

$$y(x + 1) + y(x - 1) = cy(x), \qquad c \in \mathbb{Z},$$

in the restricted setting of integer slopes, see [46]. Here Halburd and Southall proved, see [46, Lemma 4.2], that tropical meromorphic solutions of finite order with integer slopes exist if and only if $c = 0, \pm 1, \pm 2$. In our present setting of real slopes, a much more detailed consideration is needed. Actually, we proceed to showing in this chapter, that tropical meromorphic solutions exist for all $c \in \mathbb{R}$. More precisely, if $c = 0$, then all solutions to (7.5) are 2-periodic, and therefore of order 2 by Theorem 6.11 in Chapter 6. Assume, from now on, that $c \neq 0$. If $|c| \leq 2$, then all tropical meromorphic solutions are of finite order, and in fact of order $\rho \leq 2$, while if $|c| > 2$, then all tropical meromorphic solutions must be of hyper-order $\rho_2 = 1$. We divide our consideration in this section in several parts, according to the value of c. In all parts, key role is played by the quadratic $\lambda^2 - c\lambda + 1 = 0$, and its roots a, b that satisfy, of course, $a + b = c$ and $ab = 1$. An immediate consequence, repeatedly applied in this section, is that (7.5) may be written in the form

$$y(x + 1) - ay(x) = b(y(x) - ay(x - 1)). \tag{7.6}$$

Theorem 7.7. *All non-constant tropical meromorphic solutions of*

$$y(x + 1) + y(x - 1) = 2y(x) \tag{7.7}$$

are represented as linear combinations of the linear identity function $L(x) = x$ and of 1-periodic tropical meromorphic functions. Therefore, real multiples of $L(x)$ are of order $\rho = 1$, while all other non-constant tropical meromorphic solutions are of order $\rho = 2$.

Proof. In this case, $a = b = 1$, and (7.6) takes the form

$$y(x + 1) - y(x) = y(x) - y(x - 1).$$

This special equation is now easily solved completely for tropical meromorphic solutions. In fact, all 1-periodic tropical meromorphic functions clearly solve this equation. Another possibility for a tropical meromorphic solution f is that $y(x) - y(x - 1)$ is 1-periodic, hence $y(x) - y(x - 1) = \Pi_1(x)$, say. But Remark after Theorem 7.1 tells that Π_1 reduces to a constant, and by Theorem 7.1, f must be a linear combination of $L(x)$ and a 1-periodic tropical meromorphic function. □

Theorem 7.8. *All non-constant tropical meromorphic solutions of*

$$y(x + 1) + y(x - 1) = -2y(x) \tag{7.8}$$

are linear combinations of 2-periodic functions that are anti-1-periodic. Moreover, each such solution has a representation

$$y(x) = \sum_{j=1}^{K} (-1)^{[x - c_j]} \pi_{y,j}(x),$$

where $\pi_{y,j}$ is 1-periodic, and $c_1, \ldots, c_K$ are the slope discontinuities of y in the interval $[0, 2)$.

Proof. First observe that equation (7.8) cannot have constant solutions except for $y = 0$. Now, (7.6) takes the form

$$y(x + 1) + y(x) = -(y(x) + y(x - 1)). \tag{7.9}$$

Denote $g(x) := y(x) + y(x - 1)$. Then, by (7.9), g is anti-1-periodic, i.e. $g(x + 1) = g(x)$, and 2-periodic. By Theorem 6.7 in Chapter 6, we easily obtain a representation for $g(x)$. To this end, define $\xi := x/2$ and $h(\xi) := g(x)$. Clearly, h is 1-periodic and tropical meromorphic in the variable ξ. Moreover, $h(\xi + 1/2) = -h(\xi)$. Let $d_1, \ldots, d_K$ denote the slope discontinuities of g in $[0, 2)$. Then, $d_1/2, \ldots, d_K/2$ are the slope discontinuities of h in $[0, 1)$. By Theorem 6.7 in Chapter 6,

$$h(\xi) = \sum_{j=1}^{K} \pi_{h,j}(\xi) + h(0),$$

where each $\pi_{h,j}(\xi)$, $j = 1, \ldots, K$, is a suitable 1-periodic function. Defining now $\pi_{g,j}(x) := \pi_{h,j}(x/2) = \pi_{h,j}(\xi)$, $j = 1, \ldots, K$, we obtain for g the representation

$$g(x) = \sum_{j=0}^{K} \pi_{g,j}(x),$$

where $\pi_{g,0}(x) \equiv 0$. Therefore, we now have to solve

$$y(x) + y(x - 1) = \sum_{j=0}^{K} \pi_{g,j}(x).$$

Hence, all solutions y are now obtained as the sum of a solution of the homogeneous equation

$$y(x) + y(x - 1) = 0$$

and a special solution y_j of

$$y(x) + y(x - 1) = \pi_{g,j}(x), \qquad j = 1, \ldots, K. \tag{7.10}$$

But now, by Remark to Theorem 7.1 above, each $\pi_{g,j}(x)$ reduces to a constant, hence vanishing identically, and we have $g(x) \equiv g(0)$ by the continuity of g. But by anti-1-periodicity of g, $g(0) = 0$, and so $g = 0$. Therefore, to solve equation (7.8), it remains to solve the homogeneous equation $y(x) + y(x - 1) = 0$. This immediately completes the first claim that y is 2-periodic and anti-1-periodic.

Denoting now by $c_1, \ldots, c_K$, say, the slope discontinuities of y in $[0, 2)$, we may repeat the reasoning above to obtain the representation

$$y(x) = \sum_{j=1}^{K} \chi_{y,j}(x),$$

where the 2-periodic components $\chi_{y,j}(x)$ depend on the slope values $\omega_{y,j}$ at c_j, as determined in Theorem 6.7 in Chapter 6. By the anti-1-periodicity property, it is easy to verify that this representation may be written in the form

$$y(x) = \sum_{j=1}^{K} (-1)^{[x-c_j]} \pi_{y,j}(x),$$

where each $\pi_{y,j}(x)$ is 1-periodic. Indeed, if $c_j < 1$, then $y(0) = -\pi_{y,j}(0)$, $y(2) = -\pi_{y,j}(2) = -\pi_{y,j}(0)$ and $y(1) = \pi_{y,j}(1) = \pi_{y,j}(0)$, and similarly in the case that $c_j \geq 1$. $\qquad\square$

Theorem 7.9. *Suppose $|c| > 2$ in (7.5). Then all tropical meromorphic solutions to (7.5) may be represented as a finite sum of certain tropical exponential functions as follows:*

$$y(x) = \sum_{j=1}^{p} \alpha_j e_a(x - a_j) + \sum_{j=1}^{q} \beta_j e_b(x - b_j), \qquad (7.11)$$

where a, b are the (real, distinct, and $\neq 0, \pm 1$) roots of $\lambda^2 - c\lambda + 1 = 0$, and a_j, resp. b_j, are the slope discontinuities of y in the interval $[0, 1)$, resp. $[0, 2)$. Some, or even all, of the coefficients β_j may vanish.

Proof. Looking at (7.6), and denoting $g(x) := y(x) - ay(x-1)$, we observe that g satisfies the equation

$$g(x+1) - bg(x) = 0. \qquad (7.12)$$

By Theorem 7.4, g has a representation

$$g(x) = \sum_{j=1}^{q} \beta_j e_b(x - b_j), \qquad (7.13)$$

where $b_1, \ldots, b_q$ are the slope discontinuities of g in the interval $[0, 1)$. From $g'(x) = y'(x) - ay'(x-1)$ we conclude that $b_1, \ldots, b_q$ are determined by the slope discontinuities of y in the interval $[0, 2)$. However, it may happen that such slope discontinuities of y at b_0, say, and $b_0 + 1$ cancel each other when computing g. To avoid complications in notations, we assume that

$b_1, \ldots, b_q$ contain such points as well, setting $\beta_j = 0$ for these cases. This includes the possibility that all of the coefficients $\beta_j = 0$. Therefore, we now have to solve

$$y(x+1) - ay(x) = b \sum_{j=1}^{q} \beta_j e_b(x - b_j). \qquad (7.14)$$

By linearity, all solutions are now obtained as linear combinations of solutions of the corresponding homogeneous equation

$$y(x+1) - ay(x) = 0 \qquad (7.15)$$

and at least one of the special solutions to

$$y_j(x+1) - ay_j(x) = b\beta_j e_b(x - b_j) \qquad (7.16)$$

for each $j = 1, \ldots, q$. Now, solutions to (7.15) may be represented in the form

$$y_H(x) = \sum_{j=1}^{p} \alpha_j e_a(x - a_j), \qquad (7.17)$$

where $a_1, \ldots, a_p$ are the slope discontinuities of y_H in the interval $[0, 1)$. Given j, to determine a special solution to (7.16), we only need to substitute $y_j(x) = B_j e_b(x - b_j)$ into (7.16) to determine the constant coefficient $B_j = b\beta_j/(b - a)$. Combining these considerations, we obtain the asserted representation

$$y(x) = \sum_{j=1}^{p} \alpha_j e_a(x - a_j) + \sum_{j=1}^{q} \frac{b\beta_j}{b - a} e_b(x - b_j).$$

$\square$

Remark 7.10. By [84], p. 295–300, again, the solution space (7.3), whenever $|c| > 2$, is two-dimensional, and similarly as in our Remark to Theorem 7.4, the 1-periodic multipliers reduce to constants as the basic solutions in this theorem are never locally constants as tropical exponentials.

Theorem 7.11. *All non-constant tropical meromorphic solutions f to (7.5) in the case of $|c| > 2$ are of hyper-order $\rho_2(f) = 1$.*

Proof. Given f, consider the representation (7.11) in Theorem 7.9. If it happens that $g(x) = y(x) - ay(x - 1)$ vanishes, then $y(x) = \sum_{j=1}^{p} \alpha_j e_a(x - a_j)$ by (7.11). Since each tropical exponential is of hyper-order $\rho_2 = 1$, we first see that $\rho_2(y) \leq 1$. On the other hand, since the slope discontinuities a_j are distinct, we may fix one of the $e_a(x - a_1), \ldots, e_a(x - a_p)$, say $e_a(x - a_s)$.

Since $e_a(x - a_s)$ has no poles and has zeros exactly at $a_s + k$, $k \in \mathbb{Z}$, of multiplicity $(1 - 1/a)a^k$, see Proposition 1.23, we may apply the Jensen formula to conclude that

$$T(\gamma r, f) \geq T(\gamma r, 1 \oslash f) + f(0) \geq N(\gamma r, 1 \oslash f) + f(0)$$

$$\geq \frac{\gamma - 1}{2} n(r, 1 \oslash f) + f(0) \geq \left(1 - \frac{1}{a}\right) \frac{\gamma - 1}{2} \sum_{k=1}^{[r]} a^k + f(0)$$

$$\geq \left(1 - \frac{1}{a}\right) \frac{\gamma - 1}{2} a^{[r]} + f(0) \geq \frac{a - 1}{a^2} \cdot \frac{\gamma - 1}{2} a^r + f(0).$$

It is now an easy exercise to see that $\rho_2(f) \geq 1$.

If $g(x) = y(x) - y(x - 1)$ does not vanish, then we may apply the representation $g(x) = \sum_{j=1}^{q} \beta_j e_b(x - b_j)$ in (7.13) and the reasoning above to conclude that $\rho_2(g) \geq 1$. But then $g(x) = y(x) - ay(x - 1)$ implies that $\rho_2(y) \geq 1$ as well. The converse inequality again follows from (7.11), and we are done. $\qquad\square$

We next proceed to considering equation (7.5) in the remaining case that $0 < |c| < 2$. Then, clearly, the roots $\lambda_{\pm}$ of $\lambda^2 - c\lambda + 1 = 0$ are complex conjugates and they satisfy $\lambda_+ \lambda_- = 1$. Therefore, we may write $\lambda_+ = re^{i\theta}$ and $\lambda_- = \frac{1}{r}e^{-i\theta}$, where $r \geq 1$ and $\theta \in [0, 2\pi)$. However, c is real, hence $\operatorname{Im} c = \operatorname{Im}(\lambda_+ + \lambda_-) = (r - \frac{1}{r})\sin\theta = 0$. This means that we must have either $r = 1$ or $\theta = 0$ or $\theta = \pi$. But if now $\theta = 0$ or $\theta = \pi$, then $|c| = 2$, resulting in $r = 1$. Therefore, we have $c = \lambda_+ + \lambda_- = e^{i\theta} + e^{-i\theta} = 2\cos\theta$, implying that (7.5) reduces now into

$$y(x + 1) + y(x - 1) = (2\cos\theta)y(x). \tag{7.18}$$

In order to construct solutions to (7.18), first observe that

$$Y(x) := e^{i\theta[x]}\left(x - [x] + \frac{1}{e^{i\theta} - 1}\right)$$

is a formal solution, verified by a routine computation. This immediately means that $y_1(x) := \operatorname{Re} Y(x)$ and $y_2(x) := \operatorname{Im} Y(x)$ are (formal) solutions to (7.18) as well. By a routine computation again, we see that

$$y_1(x) = (\cos(\theta[x]))(x - [x]) + \frac{(\cos(\theta[x]))(\cos\theta - 1) + (\sin(\theta[x]))\sin\theta}{2(1 - \cos\theta)} \tag{7.19}$$

and

$$y_2(x) = (\sin(\theta[x]))(x - [x]) + \frac{(\sin(\theta[x]))(\cos\theta - 1) + (\cos(\theta[x]))\sin\theta}{2(1 - \cos\theta)}. \tag{7.20}$$

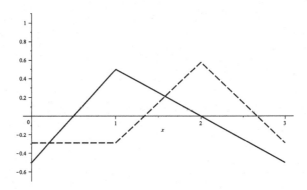

Fig. 7.1   Graph of y_1 and y_2.

For the graphs of these solutions, see Fig. 7.1.

These functions are locally linear, and they are tropical meromorphic, as they are continuous at every $x \in \mathbb{R}$, including all integers $m \in \mathbb{Z}$. We leave these verifications (that are nothing but elementary computations) as exercises for the reader.

So, we have arrived at the following

Theorem 7.12. *Equation (7.5) admits tropical meromorphic solutions for all $c \in \mathbb{R}$.*

Remark 7.13. Observe that the solutions (7.19) and (7.20) to equation $y(x+1) + y(x-1) = (2\cos\theta)y(x)$ are periodic as soon as $\theta = 2\pi/r$ for a rational r. On the other hand, if θ is an irrational multiple of 2π, then these two solutions are non-periodic.

It is clear that the solutions (7.19) and (7.20) to equation $y(x+1) + y(x-1) = -y(x)$ are linearly independent in the Gondran-Minoux sense, see Definition 5.4. Looking at Theorem 7.9, this raises the problem whether all tropical meromorphic solutions to $y(x+1) + y(x-1) = (2\cos\theta)y(x)$ would be suitable linear combinations of (7.19) and (7.20). To illustrate this situation, first observe that all tropical meromorphic solutions to

$$y(x+1) + y(x-1) + y(x) = 0 \tag{7.21}$$

are 3-periodic. This readily follows by combining (7.21) with $y(x+2) + y(x) + y(x+1) = 0$, which results in $y(x+2) = y(x-1)$. Clearly, given an arbitrary such solution y and an arbitrary 1-periodic tropical meromorphic function $\phi(x)$, then $\phi(x)y(x)$ is a formal solution. Contrary to the case of tropical exponentials in Theorem 7.9 that are never locally constant and so such a product $\phi(x)$ has to be constant to ensure that $\phi(x)y(x)$ remains to be tropical meromorphic. In the case of (7.19) and (7.20) solving equation $y(x+1) + y(x-1) = -y(x)$, the conclusion is the same that a 1-periodic multiplier $\phi(x)$ must be a constant. Since (7.19) is not locally constant, this is trivial, while for (7.20), the argument follows from the fact that (7.20) is locally non-constant in the interval $[1,3]$, hence $\phi(x)$ has to be constant there, and by 1-periodicity, it is constant everywhere. On the other hand, a suitable 3-periodic tropical multiplier $\phi(x)$ to (7.20), say, can easily be found so that $\phi(x)y_2(x)$ is tropical meromorphic. Indeed, define $\phi(x)$ as $\phi(x) := \frac{3}{2} - 2|x - \frac{1}{2}|$ for $0 \le x \le 1$ and as the constant $\phi(x) := \frac{1}{2}$ for $1 \le x \le 3$. Then $\phi(x)y_2(x)$ is tropical meromorphic, see Fig. 7.2.

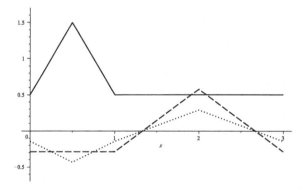

Fig. 7.2   Graph of y_2, ϕ and ϕy_2.

However, $\phi(x)y_2(x)$ does not solve equation $y(x+1) + y(x-1) + y(x) = 0$. In fact, taking $x = 3/2$, we obtain that

$$\phi(5/2)y(5/2) + \phi(1/2)y(1/2) + \phi(3/2)y(3/2) = -1/(2\sqrt{3}) \ne 0.$$

It remains open, whether an arbitrary tropical meromorphic solution to $y(x+1) + y(x-1) = (2\cos\theta)y(x)$ could be represented as a linear combination (with constant coefficients) of (7.19) and (7.20) and their shifts, say. As a possible example, see the solution $y(x)$ on p. 919 in [69].

We now close this part by proving

Theorem 7.14. *Let $y(x)$ be a non-trivial tropical meromorphic solution* to $y(x + 1) + y(x - 1) = 2(\cos\theta)y(x)$ *with* $\theta \in (0, \pi)$. *Then* y *is of order* $\rho(y) = 2$.

Proof. By a remark above, if $2\pi\theta$ is a rational number, then y is periodic, and therefore of order $\rho(y) = 2$ by Theorem 6.11. In the remaining non-periodic case, the solutions of the quadratic $\lambda^2 - (2\cos\theta)\lambda + 1 = 0$ are $\cos\theta \pm i\sin\theta$, and therefore we may write equation (7.6) as

$$y(x+1)-(\cos\theta+i\sin\theta)y(x) = (\cos\theta-i\sin\theta)(y(x)-(\cos\theta+i\sin\theta)y(x-1)).$$

Considering now $g(x) = s(x) + it(x) := y(x) - (\cos\theta + i\sin\theta)y(x - 1)$, where $s(x) = y(x) - (\cos\theta)y(x - 1)$ and $t(x) = (\sin\theta)y(x - 1)$, it is a trivial computation to see that $|g(x)|$ is 1-periodic, since $g(x+1) = (\cos\theta - i\sin\theta)g(x)$. Moreover,

$$\begin{pmatrix} s(x + 1) \\ t(x + 1) \end{pmatrix} = \begin{pmatrix} \cos\theta & \sin\theta \\ -\sin\theta & \cos\theta \end{pmatrix} \begin{pmatrix} s(x) \\ t(x) \end{pmatrix}. \tag{7.22}$$

This means geometrically that the shift $(s(x), t(x)) \to (s(x+1), t(x+1))$ of the vector $(s(x), t(x))$ describes a rotation in the plane. Since the image of $(s([0, 1]), t([0, 1]))$ is bounded by continuity, the vector $(s(x), t(x))$ remains bounded over the whole real axis $\mathbb{R}$. Moreover,

$$\begin{pmatrix} y(x) \\ y(x - 1) \end{pmatrix} = \begin{pmatrix} 1 - \cos\theta \\ 0 - \sin\theta \end{pmatrix}^{-1} \begin{pmatrix} s(x) \\ t(x) \end{pmatrix},$$

and we conclude that $y(x)$ is a bounded (non-periodic) tropical mero-morphic function, hence its proximity function is bounded as well. In order to see what happens for the counting function, we may formally differentiate $g(x) = y(x) - (\cos\theta + i\sin\theta)y(x - 1)$ to obtain $g'(x) = y'(x) - (\cos\theta + i\sin\theta)y'(x - 1)$. Then it is clear all possible discontinu-ities of $g'(x)$ are determined by the slope discontinuities of $y(x)$ in the interval $[-1, 1]$. Since $y(x)$ is continuous, the multiplicities of poles of $y(x)$ remain bounded in the interval $[-1, 1]$, and by the 1-periodicity of $g(x)$, this pattern repeats for all intervals of length 2, starting from $x = 1$ and preceding $x = -1$. Therefore, $N(r, y) = \kappa r^2$ for some positive κ. Since the proximity function is bounded, the assertion immediately follows. $\square$

7.3 What is the general solution to ultra-discrete equations?

So far in this chapter we have considered the existence of tropical meromorphic solutions for natural but relatively special classes of ultra-discrete
equations (7.3) and (7.5), and we have given explicit representations of general solutions of the equations. In this section we consider the question of
finding the general solution to broader classes of ultra-discrete equations.
We illustrate the difficulty of answering this question by making elementary
observations of the relation between a simple discrete Riccati equation and
its ultra-discrete analogue. The discrete Riccati equation

$$x_{n+1} = a + b - \frac{ab}{x_n}, \quad a, b \in \mathbb{C} \ (a \neq b) \tag{7.23}$$

has the general solution

$$x_n = \frac{a^n + cb^n}{a^{n-1} + cb^{n-1}}, \quad n \in \mathbb{Z}_{\geq 0}, \quad (c \in \mathbb{C}). \tag{7.24}$$

By writing equation (7.23) as

$$x_{n+1}x_n + ab = (a + b)x_n \tag{7.25}$$

and transforming (7.25) using an ultra-discrete limit (see Section B.5 in
Appendix B) into an ultra-discrete equation, it follows that, for $A, B \in \mathbb{R}_{\max}$,

$$X_{x+1} \otimes X_n \oplus A \otimes B = (A \oplus B) \otimes X_n. \tag{7.26}$$

Equation (7.26) can be written in the conventional algebra as

$$\max\{X_{n+1} + X_n, A + B\} = X_n + \max\{A, B\},$$

which is equivalent to the ultra-discretization of (7.23), that is,

$$\max\{X_{n+1}, -X_n + A + B\} = \max\{A, B\}. \tag{7.27}$$

We also transform the solution (7.24) into the sequence

$$X_n = \left(A^{\otimes n} \oplus C \otimes B^{\otimes n}\right) \oslash \left(A^{\otimes (n-1)} \oplus C \otimes B^{\otimes (n-1)}\right), \quad n \in \mathbb{Z}_{\geq 0},$$

that is again in conventional algebra,

$$X_n = \max\{nA, nB+C\} - \max\{(n-1)A, (n-1)B+C\}, \quad n \in \mathbb{Z}_{\geq 0} \tag{7.28}$$

for $C \in \mathbb{R}_{\max}$. Then we can confirm that the ultra-discrete Riccati equation
(7.27) is solved by (7.28). However, our interest lies in the existence of

tropical meromorphic solutions, so that we are to show that the ultra-discrete equation

$$\max\{f(x+1), -f(x)+\alpha+\beta\} = \max\{\alpha, \beta\}, \quad \alpha, \beta \in \mathbb{R}_{\max} \ (\alpha \neq \beta) \quad (7.29)$$

has a tropical meromorphic solution, indeed a tropical rational solution

$$f(x) = \max\{\alpha x, \beta x + \gamma\} - \max\{\alpha x - \alpha, \beta x - \beta + \gamma\} \qquad (7.30)$$

for $\gamma \in \mathbb{R}_{\max}$. It is straightforward to see that this is the case, since the left-hand side of (7.29) is then

$$
\begin{aligned}
&= \max\{\max\{\alpha x + \alpha, \beta x + \beta + \gamma\}, \max\{\alpha x + \beta, \beta x + \alpha + \gamma\}\} \\
&\quad - \max\{\alpha x, \beta x + \gamma\} \\
&= \max\{\alpha x + \max\{\alpha, \beta\}, \beta x + \max\{\alpha, \beta\} + \gamma\} \\
&\quad - \max\{\alpha x, \beta x + \gamma\} \\
&= \max\{\alpha, \beta\}.
\end{aligned}
$$

Now we ask whether equation (7.29) permits the function (7.30) as its general solution. This could be confirmed in such a way as an ultra-discrete analogue of the way to solve discrete equation (7.23) by means of Cole-Hopf transformations. The discrete Cole-Hopf transformation, $x_n \to u_n$, is then given by $x_n = \frac{u_n}{u_{n-1}}$ and its ultra-discrete analogue in our setting will be $f(x) = g(x) \oslash g(x-1) = g(x) - g(x-1)$ for a tropical entire function $g(x)$ satisfying

$$\max\{g(x+1), g(x-1) + \alpha + \beta\} = g(x) + \max\{\alpha, \beta\}, \qquad (7.31)$$

which is also written as

$$g(x+1) \oplus \alpha \otimes \beta \otimes g(x-1) = (\alpha \oplus \beta) \otimes g(x).$$

This is the ultra-discrete analogue of the homogeneous linear discrete equation

$$u_{n+1} + abu_{n-1} = (a+b)u_n \quad \text{or} \quad u_{n+1} - (a+b)u_n + abu_{n-1} = 0,$$

whose general solution is

$$u_n = c_1 a^n + c_2 b^n, \quad c_1, c_2 \in \mathbb{R}.$$

Translating this u_n in max-plus algebra, we have the solution of the form in tropical linear combination

$$g(x) = \gamma_1 \otimes \alpha^{\otimes x} \oplus \gamma_2 \otimes \beta^{\otimes x} = \max\{\alpha x + \gamma_1, \beta x + \gamma_2\}.$$

Here $\gamma_j = \gamma_j(x)$ $(j = 1, 2)$ are not only elements in $\mathbb{R}_{\max}$ but also in the family of 1-periodic tropical meromorphic functions. In fact, we could verify this fact as follows: Let us first write (7.31) in a 'homogeneous' form

$$\max\left\{g(x+1) - \big(g(x) + \max\{\alpha, \beta\}\big), -g(x) + \big(g(x-1) + \min\{\alpha, \beta\}\big)\right\} = 0.$$
(7.32)

For this identity to hold it is required that at each point $x \in \mathbb{R}$ we have either $g(x + 1) = g(x) + \max\{\alpha, \beta\}$ or $g(x) = g(x - 1) + \min\{\alpha, \beta\}$. Then we assume that the latter holds on the whole $\mathbb{R}$, so that we see that $g(x)$ is given as $\min\{\alpha, \beta\}x + \gamma_2(x)$ and satisfies (7.32) since

$$\max\left\{\min\{\alpha, \beta\} - \max\{\alpha, \beta\}, 0\right\} = 0.$$

On the other hand, if we assume the former holds on the whole $\mathbb{R}$, then it follows that $g(x) = \max\{\alpha, \beta\}x + \gamma_1(x)$ which satisfies (7.32) by

$$\max\left\{0, -\max\{\alpha, \beta\} + \min\{\alpha, \beta\}\right\} = 0.$$

Further, such a tropical linear combination of two special solutions $\alpha^{\otimes x}$ and $\beta^{\otimes x}$,

$$\max\left\{\max\{\alpha, \beta\}x + \gamma_1(x), \min\{\alpha, \beta\}x + \gamma_2(x)\right\}$$

over the field of 1-periodic tropical meromorphic functions on $\mathbb{R}$ is also a solution to (7.32). In fact,

$$g(x) = \begin{cases} \max\{\alpha, \beta\}x + \gamma_1(x), & \text{when } x \geq \frac{\gamma_2(x) - \gamma_1(x)}{\max\{\alpha, \beta\} - \min\{\alpha, \beta\}}, \\ \min\{\alpha, \beta\}x + \gamma_2(x), & \text{when } x < \frac{\gamma_2(x) - \gamma_1(x)}{\max\{\alpha, \beta\} - \min\{\alpha, \beta\}}. \end{cases}$$

When $x \geq \frac{\gamma_2(x) - \gamma_1(x)}{\max\{\alpha, \beta\} - \min\{\alpha, \beta\}}$,

$$g(x + 1) = \max\{\alpha, \beta\}x + \max\{\alpha, \beta\} + \gamma_1(x) = g(x) + \max\{\alpha, \beta\}x.$$

If $x - 1 < \frac{\gamma_2(x) - \gamma_1(x)}{\max\{\alpha, \beta\} - \min\{\alpha, \beta\}}$,

$$g(x - 1) = \min\{\alpha, \beta\}x - \min\{\alpha, \beta\} + \gamma_2(x),$$

so that the left-hand side of equation (7.32) becomes

$$\max\left\{0, -\Big[\big(\max\{\alpha, \beta\} - \min\{\alpha, \beta\}\big)x - \big(\gamma_2(x) - \gamma_1(x)\big)\Big]\right\} = 0.$$

Similarly, when

$$x < \frac{\gamma_2(x) - \gamma_1(x)}{\max\{\alpha, \beta\} - \min\{\alpha, \beta\}} \quad \text{and} \quad x + 1 \geq \frac{\gamma_2(x) - \gamma_1(x)}{\max\{\alpha, \beta\} - \min\{\alpha, \beta\}},$$

we see that the left-hand side of (7.32) is

$$\max\left\{(\max\{\alpha,\beta\}-\min\{\alpha,\beta\})x-(\gamma_2(x)-\gamma_1(x)),0\right\}=0.$$

Hence we have shown that the tropical meromorphic function

$$g(x)=\{\max\{\alpha,\beta\}x+\gamma_1(x),\min\{\alpha,\beta\}+\gamma_2(x)\} \qquad (7.33)$$

is the solution to the second-order ultra-discrete equation (7.32), which possesses two 'integration constants' $\gamma_1(x)$ and $\gamma_2(x)$. (In fact, when both of the γ_j are constants, the function (7.33) reduces to a tropical rational function of degree $\max\{\alpha,\beta\}$.) However, it is not quite clear whether (7.33) describes a general solution of (7.32), since we could choose an arbitrarily *dense* decomposition of $\mathbb{R}$ into infinitely many intervals on which only one of the equations

$$g(x+1)=g(x)+\max\{\alpha,\beta\} \quad \text{and} \quad g(x)=g(x-1)+\min\{\alpha,\beta\}$$

holds. For example, we could have no concrete method to find a solution $g(x)$ which satisfies equation (7.32) by

$$\begin{cases} g(x+1)=g(x)+\max\{\alpha,\beta\} \text{ on } [a_n,a_{n+1}) \text{ if } |n| \text{ is odd}, \\ g(x)=g(x-1)+\min\{\alpha,\beta\} \text{ on } [a_n,a_{n+1}) \text{ if } |n| \text{ is even}, \end{cases}$$

for a strictly increasing sequence $\{a_n\}_{n=-\infty}^{+\infty}$ satisfying

$$-\infty \leftarrow \cdots < a_n < a_{n-1} < \cdots \to +\infty \quad \text{and} \quad \sup_{n\in\mathbb{Z}}(a_{n+1}-a_n)<\frac{1}{10}.$$

Then, given a point x, we would face to a 'parity' problem and could not determine whether either of $g(x+1)-g(x)$ and $g(x)-g(x-1)$ attains either of α and β at all. Even though, it seems quite natural to imagine that such a solution must however be far from a piecewise linear and continuous function on $\mathbb{R}$. Hence we may be able to concentrate on tropical meromorphic solutions which have the discrete or difference counterparts by a formal transformation as in the above sense.

7.4   Slow growth criterion as a detector of ultra-discrete Painlevé equations

We have seen in Sections 7.1 and 7.2 that the general solutions of certain types of ultra-discrete equations are either of order 2, or of hyper-order 1, depending on the parameter values of the equation. Looking at such a big gap in the growth of solutions corresponding to a slight change in the parameters, in the light of observations made by Arnol'd [5] and Veselov [107] who noticed that low growth (or complexity) of a mapping is a sign of its integrability, it seems justified to assume that there is something very special about the equations with slowly growing solutions. Halburd and Southall [46] suggested that the existence of sufficiently many finite-order tropical meromorphic solutions of an ultra-discrete equation is a necessary condition for the equation to be of Painlevé type. Before continuing with the details of their approach, we give a brief description of what are ultra-discrete Painlevé equations.

Ultra-discrete Painlevé equations can be obtained from discrete Painlevé equations by using an ultra-discretization method introduced by Tokihiro et al. [105]. The basic idea is to introduce a new variable X through $x = e^{X/\varepsilon}$, where x is the dependent variable of the discrete Painlevé equation being transformed, and take the limit $\varepsilon \to 0$ appropriately. As observed by G. Grammaticos, Y. Ohta, A. Ramani, D. Takahashi, and K. M. Tamizhmani in [31], in order to construct the ultra-discrete analogue of the Painlevé equations using the ultra-discretization method, we must start with the discrete form that allows the ultra-discrete limit to be taken. This means that it is a good idea to start with discrete Painlevé equations of multiplicative type in the sense by Grammaticos, Nijhoff and Ramani in [30]. These equations have consistency in the stringent requirement that x should be positive, which enables the change of variables $x = e^{X/\varepsilon}$ to be performed. Examples of such discrete first, second and third Painlevé equations of multiplicative type have been obtained by Ramani and Grammaticos [92] as follows:

$$x_{n+1} x_n^{\sigma} x_{n-1} = \alpha \lambda^n x_n + 1, \quad \sigma = 0, 1, 2, \qquad (\text{d-}P_I^{\sigma})$$

$$x_{n+1} x_{n-1} = \frac{\lambda^n (1 + \alpha \lambda^n / x_n)}{1 + x_n}, \qquad (\text{d-}P_{II}\text{-i})$$

$$x_{n+1} x_{n-1} / x_n = \frac{1 + \alpha \lambda^n / x_n}{1 + \beta x_n \lambda^n}, \qquad (\text{d-}P_{II}\text{-ii})$$

$$x_{n+1} x_{n-1} / x_n^2 = \frac{(1 + \alpha \lambda^x / x_n)(1 + \beta \lambda^n / x_n)}{(1 + \gamma x_n \lambda^n)(1 + \delta x_n \lambda^n)}, \qquad (\text{d-}P_{III})$$

where $\sigma = 0, 1, 2$ and α, β, δ, γ and λ are constants. The ultra-discretization procedure, when applied to equations $(\text{d-}P_I^{\sigma})$–$(\text{d-}P_{III})$, yields

$$\sigma y(x) + y(x + 1) + y(x - 1) = \max\{y(x) + x, 0\}, \qquad (\text{u-}P_I)$$

$$y(x + 1) + y(x - 1) = \max\{0, x - y(x)\} - \max\{0, y(x) + x - a\}, \qquad (\text{u-}P_{II}\text{-i})$$

$$y(x+1) + y(x-1) - y(x) = \max\{0, x - y(x)\} - \max\{0, y(x) + x - a\},$$
$$\text{(u-}P_{II}\text{-ii)}$$

$$y(x+1) + y(x-1) - 2y(x) = \max\{0, x - y(x)\} - \max\{0, y(x) + x - a\}$$
$$+ \max\{0, x - y(x) + b\} - \max\{0, y(x) + x + b\}, \qquad \text{(u-}P_{III}\text{)}$$

respectively, where $\sigma = 0, 1, 2$, $\lambda(> 0)$, α, β, $a(> 0)$ and $b(> 0)$ are all constants. In order to deduce the ultra-discrete analogue of these d-P's, we may also transform formally by translation of $(\times, /, +, {}^n)$ with $(\otimes, \oslash, \oplus, {}^{\otimes n})$ instead of the limiting procedure. Recently, Ormerod and Yamada [87] have found ultra-discrete versions of each Painlevé equation of multiplicative type corresponding to affine Weyl groups A_1 to E_8.

A more comprehensive description of how to obtain ultra-discrete Painlevé equations, and how they are related to continuous and discrete Painlevé equations, is given in Appendix B below.

Returning now back to finite-order tropical meromorphic solutions of ultra-discrete equations, Halburd and Southall considered the class

$$y(x+1) \otimes y(x-1) = R(x, y(x)), \qquad (7.34)$$

where R is tropical rational function in x and $y(x)$. They proved the existence of infinitely tropical meromorphic solutions of (7.34) in the following way. First fix the values $y(0)$ and $y(1)$ as real numbers of your choice, and then define

$$y(2) := R(1, y(1)) - y(0). \qquad (7.35)$$

By connecting the points $y(0)$, $y(1)$ and $y(2)$ with any continuous piecewise linear curve, and using the equation (7.34) to extend this curve from the interval $[0, 2]$ to the whole real line, we obtain a tropical meromorphic solution of (7.34). This solution conforms with our choice of the piecewise linear curve on the interval $[0, 2)$, and it satisfies (7.35). Halburd and Southall then went on to look at special cases of (7.34) assuming the existence of finite-order tropical meromorphic solutions, and proved, for example, that

$$y(x+1) \otimes y(x-1) = y(x)^{\otimes n}$$

has non-trivial finite-order tropical meromorphic solutions with integer slopes if and only in $n \in \{0, \pm 1, \pm 2\}$. In addition, they proved that if $y(x)$ is a tropical meromorphic solution of

$$y(x+1) + 3y(x) + y(x-1) = \max\{y(x) + K, 0\}, \qquad K > 0, \qquad (7.36)$$

such that $y(0) > 0$ and $y(1) < -\max\{y(0), K\}$, then $y(x)$ is of infinite order. Both of these results support the idea that the existence of sufficiently many finite-order tropical meromorphic solutions corresponds to ultra-discrete Painlevé equations.

The following corollary of Theorem 4.3 gives a powerful tool for ruling out the possibility of existence tropical entire solutions of hyper-order less than one for very general classes of ultra-discrete equations.

Corollary 7.15. *If*

$$Q(x, y(x)) = P(x, y(x)), \qquad (7.37)$$

where Q and P are tropical difference polynomials with small tropical meromorphic coefficients, has a tropical entire solution of hyper-order less than one, then

$$\deg_y(P(x, y(x))) = \deg_y(Q(x, y(x))).$$

Proof. Theorems 4.3 and 3.27 imply that

$$m(r, P(x, y(x))) = \deg_y(P(x, y(x)))m(r, y(x)) + S_\delta(r, y)$$

and

$$m(r, Q(x, y(x))) = \deg_y(Q(x, y(x)))m(r, y(x)) + S_\delta(r, y).$$

Since y is tropical entire, we have $m(r, y(x)) \equiv T(r, y(x))$. Therefore,

$$\deg_y(P(x, y(x)))T(r, y(x)) = \deg_y(P(x, y(x)))T(r, y(x)) + S_\delta(r, y)$$

from which the assertion follows. □

7.5 Tropical rational solutions to ultra-discrete Painlevé equations

In this section we follow the observations by D. Takahashi, T. Tokihiro, G. Grammaticos, Y. Ohta and A. Ramani in [103], where they constructed particular solutions to the ultra-discrete Painlevé equations u-P_I, u-P_{II} and u-P_{III}. In fact, examples of rational solutions for the second and third discrete Painlevé equations are also derived by Takahashi et al. in [103]. Special-functions solutions of discrete Painlevé equations are also obtained as solutions to an auxiliary equation of Riccati form, such as

$$x_{n+1} = -\frac{\alpha_n x_n + \beta_n}{\gamma_n x_n + \delta_n},$$

and there is also a discrete analogue of the Airy equation for the discrete second Painlevé equation and a q-difference form of Bessel's equation for the discrete third Painlevé equation, respectively. There are two main approaches to obtain special solutions to discrete Painlevé equations. One is the method based on the *auto-Bäcklund and Schlesinger transforms* and the other is that to deduce *Casorati-type solutions*.

In the paper [103] by Takahashi et al. the special solutions of u-P_{II} and u-P_{III} are given as the ultra-discrete analogue of the Casorati determinant rational solutions of the corresponding discrete Painlevé equations. Then the key question is what the 'τ function' in the ultra-discrete world? In fact, it is well known that the continuous and discrete Painlevé equations can be transformed into bilinear forms involving the so-called τ functions which are entire functions associated with the singularities of those equations. Essentially, we hope to express the possible meromorphic solutions as the logarithmic derivative of this entire function. For the continuous Painlevé equations, the τ functions of rational solutions are expressed by a Wronskian determinant with simple polynomial components, while for the discrete Painlevé equations, these are done by a Casorati determinant as the discrete analogue of the Wronskian. The strategy of finding the special solutions is to transform the u-P_{II} and u-P_{III} into some ultra-discrete bilinear forms in terms of the τ functions. This has been done through the variable transformation

$$W(x) = \tau(x-1) - \tau(x-2) = \tau(x-1) \oslash \tau(x-2)$$

or

$$W(x) = \tau(x-1) - \tau(x-3) = \tau(x-1) \oslash \tau(x-3),$$

and the desired ultra-discrete bilinear equation is

$$\tau(x)+\max\big\{\tau(x-2), \tau(x-1)+x-a\big\} = \tau(x-3)+\max\big\{\tau(x-1), \tau(x-2)+x\big\}$$

or

$$\tau(x)+\max\big\{\tau(x-3), \tau(x-1)+x-a\big\} = \tau(x-4)+\max\big\{\tau(x-1), \tau(x-3)+x\big\},$$

for u-P_{II} of the first or second kind, written as u-P_{II}-i or u-P_{II}-ii, respectively. We will not go into the details concerning special solutions of u-P_{III}, but note that it can be decomposed into two equations and one of them is the u-P_{II} of the first kind and the same variable transformation as in u-P_{II}-i implies two good ultra-discrete bilinear equations to be solved.

The equation u-P_{II}-i admits rational solutions for $a = 4m$ with a non-negative integer m. In fact, the τ function for the tropical rational solution can be constructed by

$$\tau(x) = \sum_{j=0}^{m-1} \max\{0, x - 3j\} = \bigotimes_{j=0}^{m-1} \{1 \oplus x \oslash (3j)\}$$

as in its max-plus product form, which is also expressed as in its max-plus summation form

$$\tau(x) = \max_{0 \le j \le m} \left\{ jx - \frac{3}{2}j(j-1) \right\} = \bigoplus_{0 \le j \le m} \left\{ x^{\otimes j} \oslash \left(\frac{3}{2}j(j-1) \right) \right\}.$$

On the other hand, for the tropical rational solution to the equation u-P_{II}-ii for $a = 6m$ with a non-negative integer m, the corresponding τ function is

$$\tau(x) = \sum_{j=0}^{m-1} \max\{0, x - 4j\} = \bigotimes_{j=0}^{m-1} \{1 \oplus x \oslash (4j)\}$$

as in its max-plus product form, which is also expressed as in its max-plus summation form

$$\tau(x) = \max_{0 \le j \le m} \{jx - 2j(j-1)\} = \bigoplus_{0 \le j \le m} \left\{ x^{\otimes j} \oslash (2j(j-1)) \right\}.$$

In each case, these two expressions for the $\tau(x)$ look similar to the Casorati determinant and the expansion of the determinant, respectively.

Concerning the first Painlevé equations, which do not however have any particular solutions though, they can be approached more directly using concrete expressions by means of Jacobian elliptic functions and therefore by means of elliptic theta functions.

7.6 Ultra-discrete hypergeometric solutions to ultra-discrete Painlevé equations

In this section we look at special solutions of ultra-discrete Painlevé equations expressed in terms of the ultra-discrete hypergeometric function. In his paper [86], C. M. Ormerod studies the ultra-discretization of a q-difference analogue of the Painlevé equation q-P_{III} of the form

$$y(qz)y(z/q) = \frac{a_3 a_4 \big(y(z) + a_1 z\big)\big(y(z) + a_2 z\big)}{\big(y(z) + a_3\big)\big(y(z) + a_4\big)}$$

and proves the following result:

Theorem 7.16 ([86, Theorem 1]). *The ultra-discrete Painlevé equation* u-P_{III} *given by*

$$w(x + Q) + w(x - Q) = A_3 + A_4 + \max\{w, A_1 + x\} + \max\{w, A_2 + x\} \\ - \max\{w, A_3\} - \max\{w, A_4\},$$

where Q is a fixed positive rational number, with parameters specified by

$$A_1 = -(r + 1)Q, \quad A_2 = Q + A_1 + A_3 - A_4 = -rQ, \quad A_3 = 0, \quad A_4 = 0,$$

where $w(0) = 0$ and $r > 1$, is solved by the function

$$w(x) = \max_{k \in \mathbb{N}}\{-krQ - k^2Q + kx\} - \max_{k \in \mathbb{N}}\{-krQ - (k + 1)kQ + kx\}$$

for all $r > 1$.

Especially, he derives these transcendental solutions by considering a framework in which the ultra-discretization process arises as a restriction of a non-archimedean valuation over a field. Note that such a solution is the ultra-discrete logarithmic shift $f(x) \oslash f(x-Q)$ of the tropical entire function

$$f(x) := \max_{k \in \mathbb{N}}\{-krQ - k^2Q + kx\}$$

(or the ultra-discrete Cole-Hopf transformation of $f(x - Q)$), whose order and type are obtained by

$$\limsup_{k \to \infty} \frac{\log(krQ + k^2Q)}{\log\{(krQ + k^2Q)/k\}} = \limsup_{k \to \infty} \frac{\log k^2Q(1 + r/k)}{\log kQ(1 + r/k)} = 2$$

and

$$c(2)\limsup_{k \to \infty} \frac{krQ + k^2Q}{\{(krQ + k^2Q)/k\}^2} = \frac{1}{4}\limsup_{k \to \infty} \frac{k^2Q(1 + r/k)}{k^2Q^2(1 + r/k)^2} = \frac{1}{4Q},$$

respectively. In fact, the above solution is an ultra-discrete hypergeometric solution which solves simultaneously the ultra-discrete Riccati equations

$$\begin{cases} w(x + Q) = \max\{x - rQ - w(x), 0\} - \max\{-w(x), 0\}, \\ w(x - Q) = \max\{x - (r + 1)Q - w(x), 0\} - \max\{-w(x), 0\}. \end{cases}$$

This transcendental solution is given by means of a basic ultra-discrete hypergeometric function

$$_2\Phi_1\left(\begin{matrix} A, \ B \\ C \end{matrix}; \ Q, \ x\right) = \max_{k \in \mathbb{N}}\{kx + c_k\}$$

with

$$c_k = k \left[\max\left\{ A, -\frac{k-1}{2}Q \right\} + \max\left\{ B, -\frac{k-1}{2}Q \right\} \right.$$
$$\left. - \max\left\{ C, -\frac{k-1}{2}Q \right\} - \max\left\{ Q, -\frac{k-1}{2}Q \right\} \right],$$

which converges when $x < C + Q - A - B$. The above $f(x)$ is therefore equal to

$$_2\Phi_1 \left(\begin{matrix} -\infty, & -\infty \\ & Q \end{matrix}; \ Q, \ (x - rQ) + Q \right)$$
$$= \max_{k\in\mathbb{N}} \left\{ k(x - rQ) + kQ + k\left(-\frac{k-1}{2}Q \right) + k\left(-\frac{k-1}{2}Q \right) - kQ - kQ \right\}$$
$$= \max_{k\in\mathbb{N}} \left\{ kx - krQ - k^2Q \right\}$$

for $Q > 0$. Readers are encouraged to learn Ormerod's discussion with a non-archimedean valuation in his paper [86] and also to confirm the fundamental notation of q-hypergeometric series, for example, in [27] or [53], if necessary. Note also that Murata [79] has introduced a method of finding exact solutions of ultra-discrete Painlevé equations, which include ultra-discrete hypergeometric solutions.

As mentioned above, we may apply the tropical Nevanlinna theory and estimate the growth order of the tropical meromorphic function $w(x)$ directly. Then we see that it is the same as that of the entire function $f(x)$. First let us consider Theorem 7.16 in view of the correspondence between q-series and ultra-discrete series. The solution to u-P_{III} is obtained by the ultra-discrete logarithmic shift $f(x) \oslash f(x \oslash Q) = f(x) - f(x - Q)$ of the ultra-discrete hypergeometric series

$$_2\Phi_1 \left(\begin{matrix} -\infty, & -\infty \\ & Q \end{matrix}; \ Q, \ (x - rQ) + Q \right) = \bigoplus_{k=0}^{\infty} (-Q)^{\otimes(rk+k^2)} \otimes x^{\otimes k}.$$

On the other hand, the above-mentioned q-P_{III} admits a special solution in terms of Heine's series or q-hypergeometric series

$$_2\phi_1 \left(\begin{matrix} 0, & 0 \\ & q \end{matrix}; \ q, \ zq^{-r} \right) = \sum_{k=0}^{\infty} \frac{q^{-rk}}{(q:q)_k^2} z^k. \tag{7.38}$$

By writing this function as $J(z)$, a special solution $w(z)$ to the q-P_{III} can be expressed as the difference analogue of logarithmic derivatives of the form

$$\{ J(qz) - J(z) \}/J(z) = J(qz)/J(z) - 1.$$

In fact, with the concrete parameters the q-P_{III} is of the form

$$w(qz)w(z/q) = -\frac{\big(w(z) + q^{-r}z\big)\big(w(z) - q^{-r-1}z\big)}{\big(w(z) + 1\big)\big(w(z) - 1\big)},$$

since the $w(z)$ solves also the q-difference Riccati equation

$$w(qz) = \frac{w(z) + q^{-r}z}{w(z) + 1}$$

and therefore also

$$w(z/q) = -\frac{w(z) - q^{-r-1}z}{w(z) - 1}.$$

We may confirm this fact directly by using the identity

$$J(qz) - 2J(z) = (q^{-r-1}z - 1)J(z/q).$$

Unfortunately, we are now put into an unfavorable situation with the series (7.38). This series converges only at $z = 0$, since $q = t^{-Q}$ is between 0 and 1 for $Q > 0$ so that we have

$$\frac{q^{-rk}}{(q;q)_k^2} \geq \left(\frac{1}{q}\right)^{rk} \to +\infty \quad (k \to +\infty).$$

But this trouble is resolved, if one takes the match of $0 = t^{-\infty}$ for $-\infty$ into account and is convinced to replace $q = t^{-Q}$ by its reciprocal $1/q = t^Q$ in the $J(z)$. Therefore we define the entire function $y(z)$ as

$$_2\phi_1\left(\begin{matrix} 0 \, , \, 0 \\ 1/q \end{matrix}; \, 1/q, \, zq^r\right) = \sum_{k=0}^{\infty} \frac{q^{rk+k(k+1)}}{(q:q)_k^2} z^k. \tag{7.39}$$

Then the meromorphic function $\omega(z) = \frac{y(z/q)}{y(z)} - 1$ solves the q-P_{III} of the form

$$\omega(qz)\omega(z/q) = -\frac{\big(\omega(z) + q^r z\big)\big(\omega(z) - q^{r+1}z\big)}{\big(\omega(z) + 1\big)\big(\omega(z) - 1\big)}.$$

Here we note that $f(z) = y(z/q)$ is expressible in the q-hypergeometric series

$$f(z) = \sum_{k=0}^{\infty} \frac{q^{rk+k^2}}{(q;q)_k^2} z^k, \tag{7.40}$$

which has a good correspondence with the tropical entire function

$$f(x) = \bigoplus_{k=0}^{\infty} (-Q)^{\otimes(rk+k^2)} \otimes x^{\otimes k}. \tag{7.41}$$

Since we can estimate the coefficients as

$$\frac{q^{rk+k^2}}{(1-q)^{2k}} \geq \frac{q^{rk+k^2}}{(q;q)_k^2} \geq q^{rk+k^2} \exp\left(2q\frac{1-q^k}{1-q}\right),$$

and thus

$$-\log\left(\frac{q^{rk+k^2}}{(q;q)_k^2}\right) = (rk+k^2)\log(1/q) + O(k)$$

and

$$-\frac{1}{k}\log\left(\frac{q^{rk+k^2}}{(q;q)_k^2}\right) = (r+k)\log(1/q) + O(1),$$

we see that the q-order and q-type of our entire function $f(z)$ in (7.40) are respectively

$$\limsup_{k\to\infty}\frac{2k}{k} = 2 \quad \text{and} \quad c(2)\limsup_{k\to\infty}\frac{k}{(r+k)\log(1/q)} = \frac{1}{4\log(1/q)},$$

which agree with the order 2 and type $1/(4Q)$ of the tropical entire function $f(x)$ in (7.41).

It has turned out now that there are at least two difficulties or uncertainties in our formal translation as mentioned above, when we attempt to restore it with the data in the max-plus series on $\mathbb{R}$ which had been produced by the ultra-discretization process to an entire function on $\mathbb{C}$ given by a q-series. One is how to treat the q-shifted factorials $(q;q)_n$ in the coefficients of the q-series as well as its ultra-discrete analogue $[-Q;-Q]_n$ in the coefficients of the tropical series. The other is about the location where the once missing number $1 = t^{-0}$ as well as -1 in our tropical series should be revived in the corresponding q-series as its 'ancestor' or in q-difference equations to be satisfied by the function whose expansion is the q-series. This will be studied again in Appendix B.

7.7   An ultra-discrete operator

In this section we study an ultra-discrete operator, denoted $u\delta_Q$, for $Q > 0$ that we might call **max-plus differential** operator here. In fact, we define

$$u\delta_Q f(x) := \max\{f(x+Q), f(x)\} - x - Q,$$

and the function $f(x)$ given formally by a max-plus series expansion of the form (7.42) below. Note that this definition

$$u\delta_Q f(x) = \frac{f(Q\otimes x) \oplus f(x)}{(Q\otimes x)\oplus x}\oslash$$

is a 'symbol-for-symbol' translation of the definition $\delta_q f(z)$ of the q-derivative operator observed in Section C.2.2 of Appendix C below, where analogous treatment for the complex q-difference operator has been given. Now by a simple computation it follows that this operator actually possesses a 'linearity' in the sense of the max-plus algebra, that is, it satisfies both

$$u\delta_Q\big(c \otimes f(x)\big) = c \otimes u\delta\big(f(x)\big) \quad (c \in \mathbb{R})$$

and

$$u\delta_Q\big(f(x) \oplus g(x)\big) = \big(u\delta_Q f(x)\big) \oplus \big(u\delta_Q g(x)\big).$$

As ultra-discrete counterparts of $[n]_q = \sum_{j=0}^{n-1} q^j$ and $[n]_q! = \prod_{k=1}^{n}[k]_q$ with $[0]_q = 1$ and $[0]_q! = 1$, here we introduce for $Q > 0$ the notations

$$\langle n \rangle_Q = \bigoplus_{j=0}^{n-1} Q^{\otimes j} = (n-1)Q \quad \text{and}$$

$$\langle n \rangle_Q! = \bigotimes_{k=1}^{n} \langle k \rangle_Q = \sum_{k=1}^{n}(k-1)Q = \frac{n(n-1)}{2}Q,$$

for $n \in \mathbb{N}$. We define $\langle 0 \rangle_Q = \langle 0 \rangle_Q! = 1_\circ = 0$ in addition.

Consider a function $f(x)$ given formally by

$$f(x) = \bigoplus_{n=0}^{\infty} a_n \otimes x^{\otimes n} \oslash \langle n \rangle_Q!, \quad a_n \in \mathbb{C} \quad (n \in \mathbb{Z}_{\geq 0}). \tag{7.42}$$

For $n \geq 1$ we have

$$u\delta_Q\big(x^{\otimes n}\big) = \max\{n(x+Q), nx\} - x - Q$$
$$= (n-1)x + (n-1)Q = \langle n \rangle_Q \otimes x^{\otimes(n-1)},$$

while for the sake of convenience let us define $u\delta_Q\big(x^{\otimes 0}\big) = u\delta_Q(1_\circ) = 0_\circ = -\infty$ and $u\delta_Q(0_\circ) = 0_\circ$, respectively. Then we obtain $a_n = u\delta_Q^n f(0_\circ)$ in (7.42) for each $n \in \mathbb{Z}_{\geq 0}$ under the assumption of the possibility to operate $u\delta_Q$ termwise, that is,

$$u\delta_Q\big(f(x)\big) = \bigoplus_{n=0}^{\infty} a_n \otimes u\delta_Q\big(x^{\otimes n}\big) \oslash \langle n \rangle_Q!$$

$$= \bigoplus_{n=1}^{\infty} a_n \otimes x^{\otimes(n-1)} \oslash \langle n-1 \rangle_Q!$$

$$= \bigoplus_{n=0}^{\infty} a_{n+1} \otimes x^{\otimes n} \oslash \langle n \rangle_Q!.$$

In fact we have

$$f(0_\circ) = a_0, \quad \text{and} \quad u\delta_q f(0_\circ) = a_1$$

and the general term $a_n = u\delta_q^n f(0_\circ)$ is obtained by

$$u\delta_Q^n(f(x)) = \bigoplus_{k=n}^{\infty} a_k \otimes u\delta_Q^n(x^{\otimes k}) \oslash \langle k \rangle_Q!$$

$$= \bigoplus_{k=n}^{\infty} a_k \otimes x^{\otimes(k-n)} \oslash \langle k-n \rangle_Q!$$

$$= \bigoplus_{k=0}^{\infty} a_{k+n} \otimes x^{\otimes k} \oslash \langle k \rangle_Q!$$

due to the formula, as well as the definition,

$$u\delta_Q^n(x^{\otimes k}) = \begin{cases} 0_\circ & (n > k) \\ \langle n \rangle_Q! & (n = k) \\ x^{\otimes(k-n)} \otimes \langle k \rangle_Q! \oslash \langle k-n \rangle_Q! & (n < k) \end{cases},$$

see an analogous formula for difference and q-difference operators in Appendix C below.

We see that the ultra-discrete equation $u\delta_Q(f(x)) = f(x)$ is now solved by a function $f(x)$ given by the max-plus series expansion (7.42) with $a_n = c$ for each $n \in \mathbb{Z}_{\geq 0}$, that is, the function

$$f(x) = \bigoplus_{k=0}^{\infty} c \otimes x^{\otimes k} \oslash \langle k \rangle_Q! = \max_{k \in \mathbb{Z}_{\geq 0}} \left\{ c + kx - \frac{k(k-1)}{2} Q \right\}$$

$$= c + \max_{k \in \mathbb{Z}_{\geq 0}} \left\{ kx - \frac{k(k-1)}{2} Q \right\} = c \otimes \bigoplus_{k=0}^{\infty} x^{\otimes k} \oslash \langle k \rangle_Q!,$$

which could be expressed by $f(x) = c \otimes u\exp_Q(x)$ as the definition of an ultra-discrete exponential function.

Note that if we concentrate only on tropical entire solutions to the equation $u\delta_Q f(x) = f(x)$, that is,

$$\max\{f(x), f(x - Q)\} - x = f(x - Q),$$

then the left-hand side is written as $f(x) - x$ so that the equation under consideration is of the form

$$f(x) = f(x - Q) + x \quad \text{or} \quad f(x) = (1_\circ \oplus x) \otimes f(x - Q),$$

which is the q-difference equation in the list to be given below in Appendix C. Recall that the two solutions in the list have infinite product forms $(z; q)_\infty$ and $\{x; Q\}_\infty$, respectively.

Now we have found a correspondence between 'Taylor-type' series expansions for the q-difference and ultra-discrete case:

q-difference case	ultra-discrete case
$\delta_q f(z) = f(qz) \quad (0 < |q| < 1)$	$u\delta_Q f(x) = f(x) \quad (Q = -\log|q| > 0)$
$f(z) = \displaystyle\sum_{k=0}^{+\infty} z^k / [k]_{1/q}!$	$f(x) = \displaystyle\bigoplus_{k=0}^{+\infty} x^{\otimes k} \oslash \langle k \rangle_Q!$

There are two known q-extensions of the exponential function that are expressed by a q-hypergeometric function as follows (see Ramis' paper [94], for example):

$$e_q(z) = {}_1\phi_0 \left(\begin{matrix} 0 \\ - \end{matrix} ; q, z \right) = \sum_{n=0}^{\infty} \frac{(0;q)_n}{(q;q)_n} z^n$$

and

$$E_q(z) = {}_0\phi_0 \left(\begin{matrix} - \\ - \end{matrix} ; q, -z \right) = \sum_{n=0}^{\infty} \frac{(-1)^k q^{n(n-1)/2}}{(q;q)_n} (-z)^n,$$

respectively. The latter is an entire function for $0 < |q| < 1$ and is rewritten as

$$E_q(z) = {}_1\phi_0 \left(\begin{matrix} 0 \\ - \end{matrix} ; 1/q, z \right) = \sum_{n=0}^{\infty} \frac{q^{n(n-1)/2} z^n}{(q;q)_n} = \sum_{n=0}^{\infty} \frac{(qz)^n}{(1/q; 1/q)_n},$$

so that as an ultra-discrete analogue of $E_q(z)$ we would have

$${}_1\Phi_0 \left(\begin{matrix} 0_\circ \\ - \end{matrix} ; Q, Q \otimes x \right) = \bigoplus_{n=0}^{\infty} \frac{(x \otimes Q)^{\otimes n}}{[Q;Q]_n} \oslash = \bigoplus_{n=0}^{\infty} \frac{x^{\otimes n}}{\langle n \rangle_Q!} \oslash,$$

with $[Q;Q]_n = \otimes_{j=1}^{n} (1_\circ \oplus Q^{\otimes j}) = \sum_{j=1}^{n} (jQ) = \frac{n(n+1)}{2} Q$ so that we see

$$[Q;Q]_n \oslash Q^{\otimes n} = \left(\frac{n(n+1)}{2} - n \right) Q = \frac{n(n-1)}{2} Q = \langle n \rangle_Q!.$$

7.8 Ultra-discrete hypergeometric function $_2\Phi_1$

As asked in Section 7.6, we consider again questions about the number $[-Q; -Q]_k$ and a suitable ultra-discretization of the entire function

$$f(z) = \sum_{k=0}^{\infty} \frac{q^{rk+k^2} z^k}{(q;q)_k^2}, \tag{7.43}$$

which is given by the q-hypergeometric series

$$2\phi_1 \left(\begin{matrix} 0,\ 0 \\ 1/q \end{matrix} ; 1/q,\ q^{r-1}z \right). \tag{7.44}$$

It was found as a solution to the three-term q-difference relation which appears to be related to q-Bessel equations. In fact, Ormerod's $J(z)$ function given in (7.38) seems to be a type of q-Bessel function $J_\nu^{(1)}(z)$ with parameter $\nu = 0$. This q-Bessel equation is given by

$$J_\nu^{(1)}(q^2 z) - (q^\nu + q^{-\nu})J_\nu^{(1)}(qz) + \left(1 + \frac{z^2}{4}\right)J_\nu^{(1)}(z) = 0$$

with the solution

$$J_\nu^{(1)}(z) = z^\nu {}_2\phi_1 \left(\begin{matrix} 0,\ 0 \\ q^{2\nu+2} \end{matrix} ; q^2,\ -\frac{z^2}{4} \right).$$

In the previous section we have seen that this $f(z)$ has the ultra-discrete counterpart

$$f(x) = \bigoplus_{k=0}^{\infty} (-Q)^{\otimes(rk+k^2)} \otimes x^{\otimes k}.$$

Thus we can say that we answer the questions above when its ultra-discrete hypergeometric series expression is given. In fact, it should be

$$2\Phi_1 \left(\begin{matrix} 0_\circ,\ 0_\circ \\ Q \end{matrix} ; Q,\ (-Q)^{\otimes(r-1)} \otimes x \right) \tag{7.45}$$

as the direct translation of (7.44) according to the rules found in this note. Recalling

$$2\phi_1 \left(\begin{matrix} 0,\ 0 \\ 1/q \end{matrix} ; 1/q,\ q^{r-1}z \right) = \sum_{k=0}^{\infty} \frac{(0; 1/q)_k \times (0; 1/q)_k \times (q^{r-1} \times z)^k}{(1/q; 1/q)_k \times (1/q; 1/q)_k}$$

$$= \sum_{k=0}^{\infty} \frac{q^{k(k+1)+k(r-1)} \times z^k}{(q; q)_k \times (q; q)_k}$$

$$= \sum_{k=0}^{\infty} \frac{q^{rk+k^2} \times z^k}{(q; q)_k^2},$$

the max-plus series $_2\Phi_1$ in (7.45) is

$$= \bigoplus_{k=0}^{\infty} \frac{[0_\circ;Q]_k \otimes [0_\circ;Q]_k \otimes \left((-Q)^{\otimes(r-1)} \otimes x\right)^{\otimes k}}{[Q;Q]_k \otimes [Q;Q]_k} \oslash$$

$$= \bigoplus_{k=0}^{\infty} \frac{(-Q)^{\otimes k(r-1)} \otimes x^{\otimes k}}{\frac{k(k+1)}{2}Q \otimes \frac{k(k+1)}{2}Q} \oslash$$

$$= \bigoplus_{k=0}^{\infty} \frac{(-Q)^{\otimes\left(k(r-1)+k(k+1)\right)} \otimes x^{\otimes k}}{[-Q;-Q]_k^{\otimes 2}} \oslash$$

$$= \bigoplus_{k=0}^{\infty} (-Q)^{\otimes(kr+k^2)} \otimes x^{\otimes k},$$

where we use the definition

$$[A;Q]_k = \bigotimes_{j=0}^{k-1} \left(1_\circ \oplus A \otimes Q^{\otimes j}\right), \quad A \in \mathbb{R}_\infty.$$

Thus we have

$$[0_\circ;Q]_k = \sum_{j=0}^{k-1} \max\left\{0, -\infty + Q^{\otimes j}\right\} = 0 = 1_\circ$$

and, since $Q > 0$,

$$[\pm Q; \pm Q]_k = \sum_{j=0}^{k-1} \max\left\{0, (\pm Q)^{\otimes(j+1)}\right\} = \begin{cases} \sum_{j=1}^{k} Q^{\otimes j} = \frac{k(k+1)}{2}Q, \\ \sum_{j=1}^{k} 0 = 1_\circ. \end{cases}$$

On the other hand, the direct translation of (7.43) is

$$f(x) = \bigoplus_{k=0}^{\infty} \frac{(-Q)^{\otimes(rk+k^2)} \otimes x^{\otimes k}}{[-Q;-Q]_k^{\otimes 2}} \oslash$$

$$= \bigoplus_{k=0}^{\infty} (-Q)^{\otimes(rk+k^2)} \otimes x^{\otimes k} \oslash 1_\circ^{\otimes 2}$$

$$= \bigoplus_{k=0}^{\infty} (-Q)^{\otimes(rk+k^2)} \otimes x^{\otimes k},$$

as we have already seen. Hence the validity of the introduction of $_2\Phi_1$ could be confirmed at least for a special choice of parameters. For general parameters we may propose to define

$$_2\Phi_1 \begin{pmatrix} A, B \\ C \end{pmatrix}; -Q, x \end{pmatrix}$$

by

$$= \bigoplus_{k=0}^{\infty} \frac{[A; -Q]_k \otimes [B; -Q]_k \otimes x^{\otimes k}}{[C; -Q]_k \otimes [-Q; -Q]_k} \oslash$$

$$= \max_{k \in \mathbb{Z}_{\geq 0}} \left\{ \sum_{j=0}^{k-1} \Big(\max\{0, A - jQ\} + \max\{0, B - jQ\} - \max\{0, C - jQ\} \Big) \right\}$$

$$= \max_{k \in \mathbb{Z}_{\geq 0}} \left\{ \sum_{j=0}^{k-1} \Big(\max\{jQ, A\} + \max\{jQ, B\} - \max\{jQ, C\} \Big) + \frac{k(k-1)}{2} Q \right\}.$$

Then

$$_2\Phi_1 \left(\begin{matrix} A, B \\ 0_\circ \end{matrix} ; -Q, x \right) = \bigoplus_{k=0}^{\infty} \frac{[A; -Q]_k \otimes [B; -Q]_k \otimes x^{\otimes k}}{[0_\circ; -Q]_k \otimes [-Q; -Q]_k} \oslash$$

$$= \bigoplus_{k=0}^{\infty} [A; -Q]_k \otimes [B; -Q]_k \otimes [Q; Q]_k \otimes x^{\otimes k}.$$

Finally we mention another approach via three-term relations with an analogue to the hypergeometric function. One knows that the $f(z)$ in (7.43) with (7.44),

$$f(z) = \sum_{k=0}^{\infty} \frac{q^{rk+k^2} z^k}{(q; q)_k^2} = {}_2\phi_1 \left(\begin{matrix} 0, \ 0 \\ 1/q \end{matrix} ; 1/q, \ q^{r-1} z \right)$$

satisfies the three-term relation

$$f(qz) - 2f(z) + f(z/q) = zq^r f(qz). \qquad (7.46)$$

On the other hand, its ultra-discrete analogue, say $f(x)$,

$$f(x) = \bigoplus_{k=0}^{\infty} (-Q)^{\otimes(rk+k^2)} \otimes x^{\otimes k} = {}_2\Phi_1 \left(\begin{matrix} 0, \ 0 \\ Q \end{matrix} ; Q, \ (-Q)^{\otimes(r-1)} \otimes x \right) \qquad (7.47)$$

satisfies

$$f(x - Q) \oplus f(x + Q) = 2 \otimes f(x) \oplus x \otimes (-Q)^{\otimes r} \otimes f(x - Q). \qquad (7.48)$$

In fact, putting

$$f(x) = \bigoplus_{k=0}^{\infty} a_k \otimes x^{\otimes k} = \max_{k \in \mathbb{Z}_{\geq 0}} \{kx + a_k\},$$

we can deduce that the left-hand side is

$$f(x + Q) = \max_{k \in \mathbb{Z}_{\geq 0}} \{kx + a_k + kQ\}$$

and the right-hand side becomes

$$\max\left\{\max_{k\in\mathbb{Z}_{\geq 0}}\{kx+a_k+2\}, \max_{k\in\mathbb{Z}_{\geq 0}}\{kx+a_{k-1}-(k+r-1)Q\}\right\}$$

with $a_{-1}:=0_\circ$. Noting that this identity holds for any $x\in\mathbb{R}$, we can compare the coefficients of the term kx in both sides to obtain

$$a_k+kQ=\max\{a_k+2,a_{k-1}-(k+r-1)Q\}$$

for each $k\in\mathbb{Z}_{\geq 0}$. It is shown that $a_k+2<a_{k-1}-(k+r-1)Q$ unless $k=2/Q$ and then

$$a_k=a_{k-1}-(2k+r-1)Q \quad (k\in\mathbb{Z}_{\geq 0}, k\neq 2/Q).$$

For the sake of convenience, we assume that for our base q, $|q|\neq e^{-m/2}$ for any non-negative integer m so that $Q\neq 2/m$. Then we have for each $k\in\mathbb{Z}_{\geq 0}$,

$$a_k=a_0+\big(k(k+1)+k(r-1)\big)(-Q)=a_0+(rk+k^2)(-Q)=a_0\otimes(-Q)^{\otimes(rk+k^2)}.$$

Then the solution to (7.48) has the max-plus series expansion

$$a_0\otimes\left\{\bigoplus_{k=0}^{\infty}(-Q)^{\otimes(rk+k^2)}\otimes x^{\otimes k}\right\}.$$

Here we just take $a_0=1_\circ$ to obtain our desired $f(x)$, since it satisfies $f(0_\circ)=(-Q)^{\otimes 0}=1_\circ$. It may be remarked that the opposite direction could be available, that is, when we start from the representations of $f(x)$ and its three-term relation (7.48), the translated three-term relation (7.46) can be solved by putting the power series expansion $f(z)=\sum_{k=0}^{\infty}a_k z^k$ in the parallel way to the above. It is a really classical way to deduce the recurrence relation

$$(q^k-2+q^{-k})a_k=q^{r+k-1}a_{k-1}, \quad k\in\mathbb{Z}_{\geq 0},$$

with $a_{-1}=0$ as well as $a_0=1$ and its solution

$$a_k=\prod_{j=1}^{k}\frac{q^{r+2k-1}}{(1-q^k)^2}=\frac{q^{rk+k(k+1)-k}}{(q;q)_k^2}=\frac{q^{rk+k^2}}{(q;q)_k^2}.$$

Appendix A

Classical Nevanlinna and Cartan theories

This section is devoted to presenting short presentations of the classical Nevanlinna and Cartan theories for the convenience of those readers who are not familiar with the theory in advance. We are restricting ourselves to recalling the basic notations, definitions and results only, omitting the proofs. However, we are giving references for the proofs, having tried to find most rigorous ones in each case. As for general references into Nevanlinna theory, we would like to point out the books by Hayman [48], and by Jank and Volkmann [57]. As for a general reference to the Cartan theory, we propose [38], [72] and [96] as the original Cartan paper [15] is not easy to find. Although the tropical versions of Nevanlinna and Cartan theories are surprisingly similar to the classical ones, certain differences appear anyway. We have made some effort to point out such differences in the course of the tropical theories above. In addition, the book [68] by Laine, and the paper [40] by Halburd and Korhonen are useful in comparing applications of the tropical theory to ultra-discrete equations to the corresponding reasoning in the classical complex differential and difference equations.

A.1 Classical Nevanlinna theory

The basic idea of Nevanlinna theory is to define two quantities, the proximity function and the counting function measuring, respectively, the affinity of a given meromorphic function f to an extended complex value α, and the number of the roots of $f = \alpha$ in a disk of radius r about the origin. The first key result, the first main theorem, tells us that the sum of the proximity function and the counting function is essentially independent of the extended complex value α to which the complete affinity is being measured in the sense just described.

To determine the two affinity functions, the proximity function and the counting function, let f be a meromorphic function in the complex plane. For each $r > 0$, let $n(r, f)$ denote the number of the poles of f in the disk $|z| \leq r$, each pole counted according to its multiplicity. Then counting function $N(r, f)$, that measures the average frequency of poles of f in the disk $|z| < r$, is now being defined as

$$N(r, f) := \int_0^r \frac{n(t, f) - n(0, f)}{t} dt + n(0, f) \log r.$$

The proximity function $m(r, f)$, that measures the average magnitude of f on the circle $|z| = r$, is now

$$m(r, f) := \frac{1}{2\pi} \int_0^{2\pi} \log^+ |f(re^{i\theta}| d\theta,$$

where $\log^+ x := \max\{\log x, 0\}$. In order to compute the affinities to a given complex value α, we just simply use $N(r, 1/(f - \alpha))$ and $m(r, 1/(f - \alpha))$ instead of $N(r, f)$ and $m(r, f)$.

The key definition in the Nevanlinna theory is the characteristic function $T(r, f)$, that is, the total affinity of f to poles:

$$T(r, f) := m(r, f) + N(r, f),$$

and the corresponding total affinity $T(r, 1/(f - \alpha))$ to a complex value α. Before starting our short introduction to Nevanlinna theory, we first list the following elementary inequalities of frequent use in what follows (here α, β are arbitrary complex numbers):

$$m(r, \alpha f + \beta g) \leq m(r, f) + m(r, g) + O(1),$$
$$m(r, fg) \leq m(r, f) + m(r, g),$$
$$N(r, \alpha f + \beta g) \leq N(r, f) + N(r, g),$$
$$N(r, fg) \leq N(r, f) + N(r, g),$$
$$T(r, \alpha f + \beta g) \leq T(r, f) + T(r, g) + O(1),$$
$$T(r, fg) \leq T(r, f) + T(r, g).$$

The starting point to Nevanlinna theory is the Poisson–Jensen formula: Given a meromorphic function g in the complex plane, let $(a_j), (b_k)$ be its zeros and poles, respectively, in the disk $|z| < r$, each of them being repeated according to multiplicity. Then

$$\log |g(z)| = \frac{1}{2\pi} \int_0^{2\pi} \log |g(re^{i\theta})| \frac{r^2 - |z|^2}{|re^{i\theta} - z|^2} d\theta$$

$$+ \sum_{|a_j|<r} \log \left| \frac{r(z-a_j)}{r^2 - \overline{a}_j z} \right| - \sum_{|b_k|<r} \log \left| \frac{r(z-b_k)}{r^2 - \overline{b}_k z} \right|.$$

As to the proof of the Poisson–Jensen formula, of the several options available, we propose to refer here to [57], p. 43–47.

Assuming now that $g(0) \neq 0, \infty$, we obtain the Jensen formula

$$\log |g(0)| = \frac{1}{2\pi} \int_0^{2\pi} \log |g(re^{i\theta})| d\theta + \sum_{|b_k|<r} \log \frac{r}{|b_k|} - \sum_{|a_j|<r} \log \frac{r}{|a_j|}. \quad (A.1)$$

Now, given a meromorphic function f, we may write $g(z) := z^{-p} f(z) = c_p + O(z)$, where $c_p \neq 0$, $p \in \mathbb{Z}$, to obtain the case where the value at the origin is finite and non-zero, in order to apply the Jensen formula. Making use of the simple identity $\log x = \log^+ x - \log \frac{1}{x}$, we easily obtain

$$m(r, f) + N(r, f) = m(r, 1/f) + N(r, 1/f) + \log |c_p|.$$

Replacing $f(z)$ here by $f(z) - \alpha$, where $\alpha \in \mathbb{C}$, we arrive at the first main theorem in Nevanlinna theory:

Theorem A.1. *For each meromorphic function f and each complex number α, it is true that*

$$T(r, 1/(f - \alpha)) = T(r, f) + O(1).$$

For a detailed proof of the first main theorem, one may look at [68], p. 21–24.

Remark A.2. Observe that the error term $O(1)$ in the first main theorem depends of α. To get rid of this dependence on α, it is possible to modify the definition of the proximity function. This results in the Ahlfors–Shimizu characteristic function, see e.g. [48], p. 10–13.

Remark A.3. The characteristic function obviously measures, in some sense, the complexity of the meromorphic function f under consideration. For example, $T(r, e^z) = \frac{r}{\pi}$, $T(r, e^{z^2}) = \frac{r^2}{\pi}$, $T(r, e^{z^n}) = \frac{r^n}{\pi}$, see e.g. [57]. In particular, it is an easy exercise to prove

Proposition A.4. *A meromorphic function f satisfies $T(r, f) = O(\log r)$ if and only if f is rational.*

A simple, rough characteristic for the complexity of meromorphic functions is the notion of their order, defined as

$$\rho(f) := \limsup_{r \to \infty} \frac{\log T(r, f)}{\log r}. \quad (A.2)$$

Hence, say, rationals are of order 0, the exponential function e^z of order 1, and e^{z^n} of order n. Going to more complex meromorphic functions, the notion of the iterated order may sometimes be useful. We restrict ourselves here to the notion of hyper-order

$$\rho_2(f) := \limsup_{r \to \infty} \frac{\log \log T(r, f)}{\log r}. \tag{A.3}$$

Indeed, when applying Nevanlinna theoretic considerations to differences of meromorphic functions, it appears that $\rho_2(f) = 1$ is the extremal growth complexity to obtain useful results, see below.

To obtain some description of the geometric nature of the characteristic function $T(r, f)$, we recall

Theorem A.5. *The characteristic function $T(r, f)$ is an increasing function and convex with respect to $\log r$.*

This is again an easy exercise to prove by making use of the Cartan identity

$$T(r, f) = \frac{1}{2\pi} \int_0^{2\pi} N(r, 1/(f - e^{i\theta})) d\theta + \log^+ |f(0)|. \tag{A.4}$$

A much more deep result than the first main theorem in Nevanlinna theory is the second main theorem, essentially based, in the classical theory, to the lemma for the logarithmic derivative:

Theorem A.6. *Suppose f is meromorphic in the plane such that $f(0) \neq 0, \infty$, and suppose that $1 < r < R$. Then*

$$m\left(r, \frac{f'}{f}\right) \leq \log^+ T(R, f) + \log^+\left(\frac{R}{r(R - r)}\right) + \log^+ |\log |f(0)|| + 7.$$

As for the proof, we refer to [16], p. 96–98. Observe that the key formula in this proof,

$$\frac{f'(z)}{f(z)} = \frac{1}{2\pi} \log |f(\rho e^{i\theta}| \frac{2\rho e^{i\theta}}{(\rho e^{i\theta} - z)^2} d\theta$$

$$+ \sum_{|a_j| < r} \left(\frac{\overline{a}_j}{\rho^2 - \overline{a}_j z} - \frac{1}{a_j - z}\right) + \sum_{|b_k| < r} \left(\frac{\overline{b}_k}{\rho^2 - \overline{b}_k z} - \frac{1}{b_k - z}\right)$$

follows by applying the differential operator $\partial/\partial x - i\partial/\partial y$ (with $z = x + iy$) to the Poisson–Jensen formula.

A relatively simple application of Borel lemma type reasoning results in

Theorem A.7. *Suppose f is meromorphic and non-constant in the complex plane. Then $m(r, f'/f) = S(r, f)$, where the notation $S(r, f)$ means that $S(r, f)/T(r, f)$ approaches to zero as $r \to \infty$ outside a possible exceptional set of finite linear measure. Moreover, $m(r, f'/f) = O(\log r)$ without an exceptional set, whenever f is of finite order.*

For a proof, look at [48], p. 40. More generally, we obtain

Theorem A.8. *Suppose f is meromorphic and non-constant in the complex plane and that $k \in \mathbb{N}$. Then $m(r, f^{(k)}/f) = S(r, f)$. Moreover, $m(r, f^{(k)}/f) = O(\log r)$ without an exceptional set, whenever f is of finite order.*

This last result is just a special case of a result originally due to H. Milloux, see here for [48], Theorem 3.1, and [48], p. 57–59, for a couple of related results. In particular, a special case of Milloux is that

$$T(r, f^{(k)}) \le (k + 1)T(r, f) + S(r, f)$$

for each $k \in \mathbb{N}$.

Theorem A.7, combined with the first main theorem, see e.g. [68], p. 44–46, may now easily be applied to obtain the second main theorem:

Theorem A.9. *Let f be a non-constant meromorphic functions, $q \ge 2$, and $\alpha_1, \cdots, \alpha_q$ be distinct complex numbers. Then*

$$m(r, f) + \sum_{j=1}^{q} m\left(r, \frac{1}{f - \alpha_j}\right) \le 2T(r, f) - N_{ram}(r, f) + S(r, f),$$

where $N_{ram}(r, f) := N(r, 1/f') + 2N(r, f) - N(r, f')$ is the integrated counting function for multiple points if f, meaning that each such multiple point (or a multiple pole) of multiplicity p is here to be counted $p - 1$ times.

Two easy variants of the second main theorem are:

Theorem A.10. *Let f be a non-constant meromorphic function, $q \ge 2$, and $\alpha_1, \ldots, \alpha_q$ be distinct complex numbers. Then*

$$(q - 2)T(r, f) - \sum_{j=1}^{q} N(r, 1/(f - \alpha_j)) + N_{ram}(r, f) = S(r, f).$$

Theorem A.11. *Let f be a non-constant meromorphic functions, $q \geq 2$, and $\alpha_1, \cdots, \alpha_q$ be distinct complex numbers. Then*

$$(q-1)T(r,f) \leq \overline{N}(r,f) + \sum_{j=1}^{q} \overline{N}(r, 1/(f-\alpha_j)) + S(r,f),$$

where $\overline{N}(r,f)$, resp. $\overline{N}(r, 1/(f-\alpha_j))$, stands for the counting function of distinct poles, resp. of distinct α_j-points of f.

As is well-known, second main theorem is a vast refinement of the classical Picard theorem stating that whenever a meromorphic function f omits three extended complex values, then f reduces to be a constant. This prompts to define what are called as deficiencies of meromorphic functions:

$$\delta(\alpha, f) := \liminf_{r\to\infty} \frac{m(r, 1/(f-\alpha))}{T(r,f)} = 1 - \limsup_{r\to\infty} \frac{N(r, 1/(f-\alpha))}{T(r,f)}, \qquad \alpha \in \mathbb{C},$$

$$\text{(A.5)}$$

and

$$\delta(\infty, f) := \liminf_{r\to\infty} \frac{m(r,f)}{T(r,f)} = 1 - \limsup_{r\to\infty} \frac{N(r,f)}{T(r,f)}. \qquad \text{(A.6)}$$

Indeed, if a meromorphic function f omits three values, we may assume, by a simple transformation that ∞ is one of these values. Then, by Theorem A.11, a contradiction $T(r,f) = S(r,f)$ readily follows, hence f has to be a constant.

Before proceeding to applications of Nevanlinna theory to differences of meromorphic functions, we recall a classical result whose counterpart appears in the tropical theory.

Theorem A.12. *Let f be a meromorphic function, and consider an irreducible rational function in f,*

$$R(z,f) = \frac{\sum_{j=0}^{p} a_j(z)f^j}{\sum_{j=0}^{q} b_j(z)f^j}$$

with meromorphic coefficients $a_j(z), b_j(z)$ small in the sense of $S(r,f)$. Then

$$T(r, R(z,f)) = dT(r,f) + S(r,f), \qquad d := \max\{p, q\}.$$

For a proof of this result, see [68], p.30–34. For the original proof, see [78]. For the convenience of the reader, we include here a table covering key notions and results in value distribution theory, to compare the classical and the tropical variant of the theory:

Classical Value Distribution	Tropical Value Distribution
A non-constant and meromorphic $f : \mathbb{C} \to \mathbb{C} \cup \{\infty\}$	A piecewise linear and continuous $f : \mathbb{R} \to \mathbb{R} \cup \{-\infty\}$
A finite measure set E of r	A finite logarithmic measure E of r
An angle $\theta \in [0, 2\pi)$	A signature $\sigma = \pm 1$
An average $\dfrac{1}{2\pi} \displaystyle\int_0^{2\pi} d\theta$	An average $\dfrac{1}{2} \displaystyle\sum_{\sigma = \pm 1}$
$\log^+ \lvert f(re^{i\theta}) \rvert = \max\{\log \lvert f(re^{i\theta}) \rvert, 0\}$	$f^+(\sigma r) = \max\{f(\sigma r), 0\}$
Nevanlinna characteristic function: $T(r, f) = \dfrac{1}{2\pi} \displaystyle\int_0^{2\pi} \log^+ \lvert f(re^{i\theta}) \rvert d\theta$ $\qquad\qquad + N(r, \infty, f)$	*Tropical characteristic function:* $T(r, f) = \dfrac{1}{2} \displaystyle\sum_{\sigma = \pm 1} f^+(\sigma r)$ $\qquad\qquad + N(r, \infty, f)$
Proximity function: $m(r, f) = \dfrac{1}{2\pi} \displaystyle\int_0^{2\pi} \log^+ \lvert f(re^{i\theta}) \rvert d\theta,$ $m(r, a, f) = m\big(r, 1/(f - a)\big)$ $= \dfrac{1}{2\pi} \displaystyle\int_0^{2\pi} \log^+ \left\lvert \dfrac{1}{f(re^{i\theta}) - a} \right\rvert d\theta$	*Tropical proximity function:* $m(r, f) = \dfrac{1}{2} \displaystyle\sum_{\sigma = \pm 1} f^+(\sigma r),$ $m(r, a, f) = m\big(r, 1_\circ \oslash (f \oplus a)\big)$ $= \dfrac{1}{2} \displaystyle\sum_{\sigma = \pm 1} \Big(1_\circ \oslash \big(f(\sigma r) \oplus a\big)\Big)^+$
Counting function $(f(0) \neq a, \infty)$: $N(r, a, f) = N\big(r, 1/(f - a)\big)$ $= \displaystyle\int_0^r n(t, a, f) \dfrac{dt}{t}$	*Tropical counting function:* $N(r, a, f) = N\big(r, 1_\circ \oslash (f \oplus a)\big)$ $= \dfrac{1}{2} \displaystyle\int_0^r n(t, a, f) dt$
Jensen Formula $(f(0) \neq 0, \infty)$: $\dfrac{1}{2\pi} \displaystyle\int_0^{2\pi} \log \lvert f(re^{i\theta}) \rvert d\theta - \log \lvert f(0) \rvert$ $= N(r, 1/f) - N(r, \infty, f)$	*Tropical Jensen Formula:* $\dfrac{1}{2} \displaystyle\sum_{\sigma = \pm 1} f(\sigma r) - f(0)$ $= N(r, 1_\circ \oslash f) - N(r, \infty, f)$

Classical Value Distribution	Tropical Value Distribution
First Main Theorem: $N(r,a,f)+m(r,a,f)$ $\quad = T(r,f)+O(1), \quad$ for $a \in \mathbb{C}$	*Tropical First Main Theorem:* $N(r,a,f)+m(r,a,f)$ $\quad - T(r,f)+O(1), \quad$ for $a < L_f$
Cartan theorem $(f(0) \neq \infty)$: $T(r,f) - \log^+ |f(0)|$ $\quad = \dfrac{1}{2\pi} \displaystyle\int_0^{2\pi} N(r,1,e^{i\varphi}f)d\varphi.$	*Tropical Cartan theorem:* $T(r,f) - f^+(0)$ $\quad = \displaystyle\bigoplus_{\tau=\pm 1} N(r,1_\circ,\tau f).$
$T(r,f)$ is a convex function of $\log r$	$T(r,f)$ is a convex function of r
Order and *hyper-order:* $\rho(f) = \limsup\limits_{r\to\infty} \dfrac{\log T(r,f)}{\log r},$ $\rho_2(f) = \limsup\limits_{r\to\infty} \dfrac{\log\log T(r,f)}{\log r}$	*Tropical order* and *hyper-order:* $\rho(f) = \limsup\limits_{r\to\infty} \dfrac{\log T(r,f)}{\log r},$ $\rho_2(f) = \limsup\limits_{r\to\infty} \dfrac{\log\log T(r,f)}{\log r}$
Lemma on logarithmic derivatives: $m\big(r,f'/f\big) = O(\log r)$ for all large $r > 0$, if $\rho(f) < \infty$. $m(r,f'/f) = O\big\{\log\big(rT(r,f)\big)\big\}$ for all $r \notin E$	*Lemma on tropical quotients:* $m\big(r,f(x+c)\oslash f(x)\big) = O\left(r^{\rho(f)-1}\right)$ for all $r \notin E$, if $\rho(f) < \infty$. $m\big(r,f(x+c)\oslash f(x)\big) = O\left\{\dfrac{T(r,f)}{r^\varsigma}\right\}$ for all $r \notin E$, if $0 < \varsigma < 1 - \rho_2(f)$.
Second Main Theorem: $(q-2)T(r,f) - \displaystyle\sum_{j=1}^q N(r,a_j,f)$ $+N_{\mathrm{ram}}(r,f) = O\big\{\log\big(rT(r,f)\big)\big\}$ for all $r \notin E$ and $\{a_j\}_{j=1}^q \subset \mathbb{P}^1(\mathbb{C})$	*Tropical Second Main Theorem:* $(q-1)T(r,f) - \displaystyle\sum_{j=1}^q N(r,a_j,f)$ $+N(r,1_\circ \oslash f) = O\big\{r^{-\varsigma}T(r,f)\big\}$ for all $r \notin E$ and $\max\{a_j\}_{j=1}^q < L_f,$ if $0 < \varsigma < 1 - \rho_2(f)$.

A.2   Difference variant of Nevanlinna theory

Difference variant of the Nevanlinna theory emerges from the idea of applying Nevanlinna type reasoning to complex difference equations. Foundations to the theory of complex difference equations had been made by Nörlund, Julia, Birkhoff et al. in early twentieth century, while Shimomura and Yanagihara were the first ones to apply Nevanlinna theory to investigate meromorphic solutions of complex difference equations, including some non-linear ones as well, in 1980's. During the last two decades, increasing interest to discrete Painlevé equations prompted the need to build up a more complete difference counterpart to Nevanlinna theory. Following the classical line of reasoning, it was natural to start with a difference counterpart to the lemma of logarithmic derivative. This had been made by Halburd and Korhonen in [40] and by Chiang and Feng [17], almost simultaneously and independently:

Theorem A.13 ([40]). *Let f be a non-constant meromorphic function of finite order, $c \in \mathbb{C}$ and $\delta < 1$. Then*

$$m\left(r, \frac{f(z+c)}{f(z)}\right) = o\left(\frac{T(r,f)}{r^{\delta}}\right)$$

for all r outside a possible exceptional set of finite logarithmic measure.

As for the proof in [40], observe that by Corollary 5.3, $T(r + |c|, f) = (1 + o(1))T(r, f)$ for all r outside a possible exceptional set of finite logarithmic measure, whenever f is of finite order. Therefore, the original version in [40] immediately implies this version. Theorem A.13 has been extended for meromorphic functions of hyper-order < 1 by Halburd, Korhonen and Tohge in [45].

Theorem A.14 ([17]). *Let η_1, η_2 be two complex numbers such that $\eta_1 \neq \eta_2$ and let f be a meromorphic function of finite order ρ. Then, for each $\varepsilon > 0$,*

$$m\left(r, \frac{f(z+\eta_1)}{f(z+\eta_2)}\right) = O(r^{\rho-1+\varepsilon}).$$

Similarly as in the classical theory, the previous lemmas are important in proving difference variants of the second main theorem. Such variants have been proved by Halburd and Korhonen in [40] for finite-order meromorphic functions. It was observed in [45] that key results in difference Nevanlinna theory can be extended to hold in the case where hyper-order is strictly less than one. In what follows we denote by $\widetilde{S}(r, f)$ a quantity such that $\widetilde{S}(r, f)/T(r, f)$ approaches to zero as $r \to \infty$ outside a possible exceptional set of finite logarithmic measure.

Theorem A.15. *Let $c \in \mathbb{C}$, and let f be a meromorphic function such that $\rho_2(f) < 1$ and $f(z)$ $f(z+c) \not\equiv 0$. Moreover, let $q \geq 2$, and $a_1(z), \ldots, a_q(z)$ be small periodic meromorphic functions with period c. Then*

$$m(r, f) + \sum_{j=1}^{q} m\left(r, \frac{1}{f - a_j}\right) \leq 2T(r, f) - N_{pair}(r, f) + \widetilde{S}(r, f),$$

where

$$N_{pair}(r, f) := 2N(r, f) - N(r, \Delta_c f) + N(r, 1/\Delta_c f).$$

A more general version may be proved by introducing what are called c-**separated a-pairs** of points, see again [40]:

Let f be a meromorphic function, and $c \in \mathbb{C}$. We denote by $n_c(r, a)$ the number of points z_0 in $|z| \leq r$ where $f(z_0) = a$ and $f(z_0 + c) = a$, counted according to the number of equal terms in the beginning of Taylor series expansions of $f(z) = a$ and $f(z + c) = a$ in a neighborhood of z_0. We call such points c-separated a-pairs of f in the disk $\{z : |z| \leq r\}$.

As an example, if in a neighborhood of z_0,

$$f(z) = a + c_1(z - z_0) + c_2(z - z_0)^2 + \alpha(z - z_0)^3 + O\left((z - z_0)^4\right)$$

and

$$f(z + c) = a + c_1(z - z_0) + c_2(z - z_0)^2 + \beta(z - z_0)^3 + O\left((z - z_0)^4\right)$$

where $\alpha \neq \beta$, then the point z_0 is counted three times in $n_c(r, a)$.

The integrated counting function is now defined in the usual way:

$$N_c\left(r, \frac{1}{f - a}\right) := \int_0^r \frac{n_c(t, a) - n_c(0, a)}{t} \, dt + n_c(0, a) \log r.$$

Similarly,

$$N_c(r, f) := \int_0^r \frac{n_c(t, \infty) - n_c(0, \infty)}{t} \, dt + n_c(0, \infty) \log r,$$

where $n_c(r, \infty)$ is the number of c-**separated pole pairs** of f, which are the same as the c-separated 0-pairs of $1/f$.

Difference analogue of $\overline{N}(r, f)$ now appears to be

$$\tilde{N}_c\left(r, \frac{1}{f-a}\right) := N\left(r, \frac{1}{f-a}\right) - N_c\left(r, \frac{1}{f-a}\right), \qquad (A.7)$$

which counts the number of those a-points (or poles) of f which are *not* in c-separated pairs. With this notation we obtain

Theorem A.16. *Let $c \in \mathbb{C}$, and let f be a meromorphic function such that $\rho_2(f) < 1$ and $f(z) - f(z+c) \not\equiv 0$. Let $q \geq 2$, and let $a_1(z)$, $a_2(z)$, $\cdots$, $a_q(z)$ be distinct small periodic functions with period c. Then*

$$(q-1)T(r, f) \leq \tilde{N}_c(r, f) + \sum_{k=1}^{q} \tilde{N}_c\left(r, \frac{1}{f-a_k}\right) + \tilde{S}(r, f). \qquad (A.8)$$

Observe that, contrary to the classical theory, $\tilde{N}_c(r, f)$ may be negative, even for all sufficiently large values of r, [40].

To close this section, we recall difference variants of two results, that have been frequently applied in the classical setting to complex differential equations. The first one is result due to Mohon'ko and Mohon'ko; this difference variant is due to Halburd and Korhonen in [40]:

Theorem A.17. *Let f be a non-constant meromorphic solution of $P(z, f) = 0$, where $\rho_2(f) < 1$ and $P(z, f)$ is a difference polynomial in f and its shifts. Assuming that $a(z)$ satisfies $T(r, a) = \tilde{S}(r, f)$, and that $P(z, a)$ does not vanish identically, then*

$$m\left(r, \frac{1}{f-a}\right) = \tilde{S}(r, f).$$

Finally, we recall two difference variants of the celebrated Clunie lemma. The first of these variants is due to Halburd and Korhonen in [40] as well:

Theorem A.18. *Let $f(z)$ be a non-constant meromorphic solution of*

$$f^n P(z, f) = Q(z, f),$$

where $P(z, f), Q(z, f)$ are difference polynomials in $f(z)$, $\delta < 1$ and $\varepsilon > 0$. If the degree of $Q(z, f)$ in f and its shifts is at most n, then

$$m(r, P(z, f)) = o\left(\frac{T(r, f)^{1+\varepsilon}}{r^\delta}\right)$$

outside of a possible exceptional set finite logarithmic measure.

An extension has been proved in [70]. In this extension, f^n on the left-hand side will be replaced by a difference polynomial $H(z, f)$:

Theorem A.19. *Let f be transcendental meromorphic solution of finite order of*

$$H(z, f)P(z, f) = Q(z, f),$$

where H, P, Q are difference polynomials (with small coefficients) such that the total degree n of $H(z, f)$ in f and its shifts is $\geq \deg Q(z, f)$. Assuming that $H(z, f)$ contains just one term of maximal total degree in f and its shifts, then

$$m(r, P(z, f)) = \widetilde{S}(r, f).$$

Observe that the proof of this theorem strongly relies on using Hölder type inequalities, while these are not needed to prove the tropical counterpart to this result.

A.3   Cartan's version of Nevanlinna theory

One can find several self-contained monographs about Cartan theory on holomorphic curves of the complex plane in projective spaces as well as its related fields, for example [24,66,72,83,96] and so on. Since Nevanlinna theory is concerned with the distribution of values of a holomorphic map in the complex projective line, Cartan extended the theory into the higher dimensional cases. As Nevanlinna did, Cartan introduced the proximity function and the counting function measuring the value distribution of those holomorphic curves. In order to see a relation between these two theories, let us show first so-called *Borel's theorem* as in Lang's monograph [72, Theorem 1.1, Chapter VI] (or *Borel's lemma* in Ru's monograph [96, Theorem $A3.3.2$]):

Theorem A.20 ([11]). *Let $h_0, \ldots, h_n$ be units, that is, entire functions without zeros. Suppose*

$$h_0 + \cdots + h_n = 0. \tag{A.9}$$

Define an equivalence $i \sim j$ if there exists a constant c (necessary $\neq 0$) such that $h_i = ch_j$. Let $\{S\}$ be the partition of $\{0, \ldots, n\}$ into equivalence classes. Then

$$\sum_{i \in S} h_i = 0.$$

If $n \leq 2$ then there is only one equivalence class.

This can be deduced as a corollary of Cartan's theorem below. There is one important idea for our discussion about the identity (A.9). Put $f_j = -h_j/h_0$ $(j = 1, \ldots, n)$ and rewrite it as

$$f_1 + \cdots + f_n = 1. \tag{A.10}$$

Then differentiating the both sides $(n-1)$ times as in [72], we have a system of linear equations

$$f_1 + \cdots + f_n = 1,$$
$$f_1^{(1)} + \cdots + f_n^{(1)} = 0,$$
$$\cdots$$
$$f_1^{(n-1)} + \cdots + f_n^{(n-1)} = 0,$$

so that it is natural to think of their Wronskian

$$W(f_1, \ldots, f_n) := \begin{vmatrix} f_1 & \cdots & f_n \\ f_1^{(1)} & \cdots & f_n^{(1)} \\ \vdots & \cdots & \vdots \\ f_1^{(n-1)} & \cdots & f_n^{(n-1)} \end{vmatrix}.$$

If we rewrite the system further to

$$f_1 + \cdots + f_n = 1,$$
$$\frac{f_1^{(1)}}{f_1} f_1 + \cdots + \frac{f_n^{(1)}}{f_n} f_n = 0,$$
$$\cdots$$
$$\frac{f_1^{(n-1)}}{f_1} f_1 + \cdots + \frac{f_n^{(j)}}{f_n} f_n = 0,$$

that is,

$$\begin{pmatrix} 1 & \cdots & 1 \\ \frac{f_1^{(1)}}{f_1} & \cdots & \frac{f_n^{(1)}}{f_n} \\ & \cdots & \\ \frac{f_1^{(n-1)}}{f_1} & \cdots & \frac{f_n^{(n-1)}}{f_n} \end{pmatrix} \begin{pmatrix} f_1 \\ f_2 \\ \vdots \\ f_n \end{pmatrix} = \begin{pmatrix} 1 \\ 0 \\ \vdots \\ 0 \end{pmatrix},$$

we naturally consider the determinant of this coefficient matrix,

$$L(f_1, \ldots, f_n) := \begin{vmatrix} 1 & \cdots & 1 \\ \frac{f_1^{(1)}}{f_1} & \cdots & \frac{f_n^{(1)}}{f_n} \\ & \cdots & \\ \frac{f_1^{(n-1)}}{f_1} & \cdots & \frac{f_n^{(n-1)}}{f_n} \end{vmatrix},$$

which is indeed the ratio $W(f_1, \ldots, f_n)/f_1 \cdots f_n$. Conversely, Cramer's rule says that each term f_j of the identity (A.10) can be extracted by using this $L(f_1, \ldots, f_n)$ and its variants

$$
L_i(f_1, \ldots, f_n) := \begin{vmatrix} 1 & \cdots & 1 & 1 & 1 & \cdots & 1 \\ \frac{f_1^{(1)}}{f_1} & \cdots & \frac{f_{i-1}^{(1)}}{f_{i-1}} & 0 & \frac{f_{i+1}^{(1)}}{f_{i+1}} & \cdots & \frac{f_n^{(1)}}{f_n} \\ \vdots & \cdots & \vdots & \vdots & \vdots & \cdots & \vdots \\ \frac{f_1^{(n-1)}}{f_1} & \cdots & \frac{f_{i-1}^{(n-1)}}{f_{i-1}} & 0 & \frac{f_{i+1}^{(n-1)}}{f_{i+1}} & \cdots & \frac{f_n^{(n-1)}}{f_n} \end{vmatrix},
$$

that is,

$$
f_i = \frac{L_i(f_1, \ldots, f_n)}{L(f_1, \ldots, f_n)}, \quad (j = 1, \ldots, n).
$$

Note that this discussion is only about linear algebra and requires us the linear independence of the f_i over the complex number field. It is natural that one needs to think of linear independence over the field of periodic meromorphic functions for a difference analogue of this discussion. Furthermore, for an ultra-discrete analogue of the result, one needs to introduce a suitable linear independence of tropical meromorphic functions on $\mathbb{R}$ over the max-plus algebra. Once this has been done, a similar reasoning to the above is permissible there and one can simply apply Nevanlinna's theory to this system of meromorphic functions following Cartan's idea. The symbols L and L_i are of course used to suggest logarithmic derivatives, since

$$
\frac{f^{(k)}}{f} = \prod_{m=0}^{k-1} \frac{(f^{(m)})'}{f^{(m)}}, \quad f^{(0)} := f.
$$

For further discussions, one will use many properties of Wronskian not only its multilinearity, which can be recalled in [68], for example. One of the important properties of $W(f_1, \ldots, f_n)$ and $L(f_1, \ldots, f_n)$ is their homogeneity of degree n and 0, respectively.

Proposition A.21 ([72, Proposition 1.5, p. 190]). *Let $f_1, \ldots, f_n$ be meromorphic. Then for any meromorphic function g,*

$$
W(gf_1, \ldots, gf_n) = g^n W(f_1, \ldots, f_n) \quad \text{and}
$$
$$
L(gf_1, \ldots, gf_n) = L(f_1, \ldots, f_n).
$$

Here let us refer to the paper [38] which G. G. Gundersen and W. K. Hayman dedicated to Henri Cartan on his 100th birthday and therefore taken over the formulation in his original papers [14,15], which is suitable also for our purpose to obtain an ultra-discrete analogue of his theory. Following their discussions and the notation there, we consider a system of p entire functions on $\mathbb{C}$ instead of a holomorphic curve of $\mathbb{C}$ to $\mathbb{P}^{p-1}(\mathbb{C})$. The full statement of their general form of Cartan's theorem in [14, 15] and [49] is as follows (see [38, Theorem 7.1]):

Theorem A.22 (Cartan-Gundersen-Hayman). *Let* $g_1, g_2, \ldots, g_p$ *be linearly independent entire functions, where* $p \geq 2$. *Suppose that for each complex number* z, *we have* $\max\{|g_1(z)|, |g_2(z)|, \ldots, |g_p(z)|\} > 0$. *For positive* r, *set*

$$T(r) = \frac{1}{2\pi} \int_0^{2\pi} u(re^{i\theta}) d\theta - u(0), \quad \text{where } u(z) = \sup_{1 \leq j \leq p} \log|g_j(z)|.$$

Let $f_1, f_2, \ldots, f_q$ *be* q *linear combinations of the* p *functions* $g_1, g_2, \ldots, g_p$, *where* $q > p$, *such that any* p *of the* q *functions* $f_1, f_2, \ldots, f_q$ *are linearly independent. Let* H *be the meromorphic functions defined by*

$$H = \frac{f_1 f_2 \cdots f_q}{W(g_1, g_2, \ldots, g_p)}, \tag{A.11}$$

where $W(g_1, g_2, \ldots, g_p)$ *is the Wronskian of* $g_1, g_2, \ldots, g_p$. *Then*

$$(q - p)T(r) \leq N(r, 0, H) - N(r, H) + S(r), \quad r > 0, \tag{A.12}$$

where $S(r)$ *is a quantity satisfying*

$$S(r) = O(\log T(r)) + O(\log r) \quad \text{as } r \to \infty \text{ n.e.}$$

We have

$$N(r, 0, H) \leq \sum_{j=1}^{q} N_{p-1}(r, 0, f_j), \tag{A.13}$$

and this gives

$$(q - p)T(r) \leq \sum_{j=1}^{q} N_{p-1}(r, 0, f_j) - N(r, H) + S(r), \quad r > 0,$$

If at least one of the quotients g_j/g_m *is a transcendental function, then*

$$S(r) = o(T(r)) \quad \text{as } r \to \infty \text{ n.e.},$$

whereas if all the quotients g_j/g_m *are rational functions, then*

$$S(r) \leq -\frac{1}{2}p(p-1)\log r + O(1) \quad \text{as } r \to \infty.$$

Furthermore, if all the quotients g_j/g_m *are rational functions, then there exist polynomials* $h_1, h_2, \ldots, h_p$, *and an entire function* ϕ, *such that*

$$g_j = h_j e^{\phi}, \quad j = 1, 2, \ldots, p.$$

Here $N(r, 0, H)$ and $N(r, H)$ denote the ordinary Nevanlinna counting functions of zeros and poles of the meromorphic function H, and $N_{p-1}(r, 0, f_j)$ in (A.13) is the **truncated** one in which a zero of f_j of multiplicity m is counted exactly $\min\{m, p - 1\}$ times.

The abbreviation **n.e.** for **nearly everywhere** means 'everywhere in $\mathbb{R}_{\geq 0}$, except possibly for a set of finite linear measure'.

Note that the Wronskian $W(g_1, g_2, \ldots, g_p)$ brings us some **ramification** data of the system $\{g_1, g_2, \ldots, g_p\}$ and the $(p-1)$-**truncation** in counting functions $N(r, 0, f_j)$.

As an application of this result, we can prove Theorem A.20, but here we prove its simpler statement.

Proposition A.23 ([72, Corollary 6.2, p. 223]). *Let $g_1, \ldots, g_n$ be entire functions without zeros (so units in the ring of entire functions). Suppose that*

$$g_1 + \cdots + g_n = 1.$$

Then $g_1, \ldots, g_n$ are linearly dependent if $n \geq 2$.

Proof. On contrary we assume that $g_1, \ldots, g_n$ are linearly independent over $\mathbb{C}$, so each ratio g_j/g_m $(j \neq m)$ is transcendental as a non-constant unit. We apply Theorem A.22 with $p = n$, $q = n + 1$ and

$$f_j = g_j \quad (1 \leq j \leq n) \quad \text{and} \quad j_{n+1} = g_1 + \cdots + g_n.$$

Since these f_j are all zero-free entire functions so that

$$N(r, 0, f_j) \equiv 0 \quad (1 \leq j \leq n + 1),$$

we have a contradiction:

$$T(r) \leq S(r) = o\big(T(r)\big) \quad \text{as} \quad r \to \infty, \quad \text{n.e.}$$

$\square$

This statement is, of course, a generalization of Picard's little theorem saying that all holomorphic mappings $f : \mathbb{C} \to \mathbb{P}^1(\mathbb{C}) \setminus \{a, b, c\}$ are constants and therefore Theorem A.22 is a generalization of Nevanlinna's second main theorem. In fact, Picard's theorem is stated by using this identity. By a Möbius transformation we may now assume that $\{a, b, c\} = \{0, 1, \infty\}$, then the assumption that $f(z)$ omits these three values is equivalent to the statement that the identity $g_1 + g_2 = 1$ holds on $\mathbb{C}$ and $g_1 = f$ and

$g_2 = 1 - f$ are units. The linear independence of g_1 and g_2 is equivalent to the statement that f is a constant. Results that correspond to the Fermat type identity or holomorphic maps to a Fermat variety are also found in the paper [36] by M. L. Green, for example, as well as in [72, Chapter VII, §4], [24, §3.4] or [96, Theorem $A.3.2.8$] as an application of Cartan's theorems such as Theorem A.22 or 'Generalized ABC Theorem' [96, Theorem $A.3.2.6$] and so on.

Here we simply observe the case when $n = 2$ in Theorem A.22 so that $f := g_2/g_1$ is a non-constant meromorphic function on $\mathbb{C}$. Assume the q functions f_j are given by $a_j g_1 + b_j g_2$ with $a_j \neq 0$ $(1 \leq j \leq q)$. Then the statement of Cartan's second main theorem coincides that of Nevanlinna's. Note that for the Wronskian of g_1, g_2 we have

$$W(g_1, g_2) = g_1 g_2' - g_1' g_2 = \left(\frac{g_2}{g_1}\right)' g_1^2$$

and thus for the function H in Theorem A.22, we have

$$\frac{1}{H} = \frac{W(g_1, g_2)}{f_1 f_2 \cdots f_q} = \frac{f'}{C g_1^{q+2} \prod_{j=1}^q (f - \alpha_j)},$$

with $\alpha_j := -b_j/a_j \in \mathbb{C}$ and $C := \prod_{j=1}^q a_j$. The quantity $N(r, 0, H) - N(r, H)$ in (A.12) corresponds to

$$N(r, f) + \sum_{j=1}^q N(r, (1/(f - \alpha_j))) - \left(N(r, 1/f') + 2N(r, f) - N(r, f')\right)$$

in the second main theorem of Nevanlinna as we have seen in the previous section.

This observation on a system of entire functions coincides an observation on a linearly non-degenerate holomorphic curve with reduced representation $(f_1, f_2, \ldots, f_p)$ and a family of q hyperplanes in $\mathbb{P}^{p-1}(\mathbb{C})$ located in **general position** as in [24, 72, 96].

Definition A.24. A collection of hyperplanes in $\mathbb{P}^{p-1}(\mathbb{C})$ is in general position if the intersection of any m of them has dimension $p - m - 1$ for each $m \leq p - 1$, and if the intersection of any p of them reduces to empty.

Cartan's generalization of the second main theorem [15] is a strong result in value distribution of holomorphic curves in the n-dimensional complex projective space $\mathbb{P}^n(\mathbb{C})$ [24, 66, 72, 96], as well as an efficient tool for certain problems in the complex plane $\mathbb{C}$ [38, 49] or for the study of Gauss maps of minimal surfaces in $\mathbb{R}^3$, hyperbolic complex spaces [66] or Diophantine approximation [83, 96].

Recall that the n-dimensional complex projective space $\mathbb{P}^n(\mathbb{C})$ is the quotient space $(\mathbb{C}^{n+1} - \{\mathbf{0}\})/ \sim$, where $(a_1, a_2, \ldots, a_{n+1}) \sim (b_1, b_2, \ldots, b_{n+1})$ in $\mathbb{C}^{n+1}$ if and only if $(a_1, a_2, \ldots, a_{n+1}) = \lambda(b_1, b_2, \ldots, b_{n+1})$ for some $\lambda \in \mathbb{C} - \{0\}$. As we have seen in Theorem A.22, the Cartan characteristic function $T_f(r)$ of a holomorphic curve

$$f = [f_0 : f_1 : \cdots : f_n] : \mathbb{C} \to \mathbb{P}^n(\mathbb{C}),$$

or its associated system of $n + 1$ entire functions f_j,

$$F := (f_0, f_1, \ldots, f_n) : \mathbb{C} \to \mathbb{C}^{n+1} \setminus \{\mathbf{0}\},$$

is defined by

$$T_f(r) = T(r, F) = \int_0^{2\pi} u(re^{i\theta}) \frac{d\theta}{2\pi} - u(0),$$

where $r > 0$ and

$$u(z) = \max_{0 \le j \le n} \log |f_j(z)|.$$

Here $f_0, f_1, \ldots, f_n$ are entire functions such that for all complex numbers z the quantity $\max_{0 \le j \le n} |f_j(z)|$ is non-zero, that is, all the f_j have no common zeros in the whole $\mathbb{C}$. We call the holomorphic map $F : \mathbb{C} \to \mathbb{C}^{n+1} \setminus \{\mathbf{0}\}$ a **reduced representation** of the curve f. Then we see that such a reduced representation can be obtained up to the multiplication of a common unit. Note that for a fixed reduced representation of the curve f we will also define its proximity functions and counting functions for hyperplanes, but all these functions are given naturally to be independent of the choice of the representation. Instead of considering each hyperplane H_j in $\mathbb{P}^n(\mathbb{C})$ itself which is defined by a linear equation $\hat{H}_j(w) := \sum_{k=0}^n h_{jk} w_k = 0$ $(1 \le j \le q)$, we will mainly observe its representing vector $\boldsymbol{h}_j = (h_{j0}, \cdots, h_{jn}) \in \mathbb{C}^{n+1}$. Here $w = [w_0 : \cdots : w_n]$ is a homogeneous coordinate system of $\mathbb{P}^n(\mathbb{C})$. Then it is convenient to use a symbol $(\bullet, \bullet)$ to denote the inner product on $\mathbb{C}^{n+1}$ given by $(\boldsymbol{h}_j, w) := \sum_{i=0}^n h_{ji} w_i = \hat{H}_j(w)$, which is the representing linear form of the hyperplane H_j. For a reduced representation of a curve $f = [f_0 : \cdots : f_n] : \mathbb{C} \to \mathbb{P}^n(\mathbb{C})$, the product $(\boldsymbol{h}_j, f(z)) := \sum_{i=0}^n h_{ji} f_i(z) = \hat{H}_j(f(z))$ is an entire function on $\mathbb{C}$ for every j. Since f is supposed to be linearly non-degenerate, the Wronskian of the $f_j(z)$ $(0 \le j \le n)$ does not vanish. Of course,

$(h_j, f(z)) = \hat{H}_j(f(z)) \not\equiv 0$ on $\mathbb{C}$ for every j $(1 \leq j \leq q)$ under this assumption.

Definition A.25. The proximity function for the hyperplane H_k of f is

$$m_f(r, H_k) = \int_0^{2\pi} \lambda_{H_j}\left(f(re^{i\theta})\right) \frac{d\theta}{2\pi}$$

with the Weil function

$$\lambda_{H_j}(w) = -\log \frac{\|\hat{H}_j(w)\|}{\|w\|\|h_j\|} = -\frac{1}{2}\log \frac{|\sum_{i=0}^n h_{ji}w_i|^2}{(\sum_{i=0}^n |w_i|^2)(\sum_{i=0}^n |h_{ji}|^2)}$$

of f with respect to the hyperplane H_j. The counting function for H_j of f is

$$N_f(r, H_j) = \int_0^r \left\{ n\big(t, 0, \hat{H}_j(f)\big) - n\big(0, 0, \hat{H}_j(f)\big) \right\} \frac{dt}{t} + n\big(0, 0, \hat{H}_j(f)\big) \log r,$$

where $n\big(t, 0, \hat{H}_j(f)\big)$ denotes the number (with multiplicity counted) of zeros of the entire function $\hat{H}_j(f)$ in $|z| < t$. The truncated counting function $N_f^{[n]}(r, H_j)$ is defined similarly.

Note that $\lambda_{H_j}\big(f(z)\big)$, $m_f(r, H_j)$, $N_f(r, H_j)$ and $N_f^{[n]}(r, H_j)$ are only defined under the assumption $f(\mathbb{C}) \not\subset H_j$, that is, $\hat{H}_j\big(f(z)\big) \not\equiv 0$. Furthermore, these definitions are independent of the choice of h_j and $\hat{H}_j(w)$ defining H_j as well as the choice of a reduced representation of f as we mentioned.

The Cartan characteristic function $T_f(r)$, or the **height** of f may be also defined as the sum $m_f(r, H) + N_f(r, H)$ for any hyperplane H not containing the image of f. The first main theorem holds also in this context, so that it is shown that Cartan's characteristic function also depends on H only up to a bounded function. (See, for example [96, Theorem A3.1.1 (The First Main Theorem)].)

Recalling $\mathbb{P}^1(\mathbb{C})$ is identical to $\mathbb{C} \cup \{\infty\}$ in a natural way, one sees that the definitions of $T_f(r)$, $m_f(r, H_j)$ and $N_f(r, H_j)$ coincide with Nevanlinna's functions for meromorphic functions on $\mathbb{C}$ which are defined by means of the chordal distance.

In the above settings, Cartan's second main theorem can be stated as in the form (see [72, Theorem 6.1, p.220] or [96, Theorem A3.1.7]):

Theorem A.26. *Let* $f = [f_0 : f_1 : \ldots : f_n] : \mathbb{C} \to \mathbb{P}^n(\mathbb{C})$ *be a holomorphic map, with* $f_0, f_1, \ldots, f_n$ *entire functions without common zeros. Assume* the image of f is not contained in any hyperplane. Let $H_1, \ldots, H_q$ be hyperplanes in general position. Let $W := W(f_0, \ldots, f_n)$ be the Wronskian. Then

$$\sum_{k=1}^{q} m_f(r, H_k) + N(r, 0, W) \leq (n+1)T_f(r) + O\big(\log r + \log^+ T_f(r)\big), \quad n.e.$$

By Cartan's defect relation [15] we mean the following statement:

Theorem A.27. *Let $f : \mathbb{C} \to \mathbb{P}^n(\mathbb{C})$ be a linearly non-degenerate holomorphic curve and $\{H_j\}_{j=1}^q$ be a family of hyperplanes $H_j \subset \mathbb{P}^n(\mathbb{C})$ in general position. Then we have*

$$\sum_{j=1}^{q} \delta_f^{[n]}(H_j) \leq n + 1. \tag{A.14}$$

Here $\delta_f^{[n]}(H_j)$ denotes the n-truncated deficiency for the hyperplane H_j of the curve f, which is defined by

$$\delta_f^{[n]}(H_j) := \delta^{[n]}\big(0, (\boldsymbol{h}_j, f)\big) = 1 - \limsup_{r \to \infty} \frac{N^{[n]}\big(r, 0, (\boldsymbol{h}_j, f)\big)}{T_f(r)}, \tag{A.15}$$

where $N^{[n]}(r, 0, \bullet)$ is the n-truncated counting function with respect to the zeros of the entire function $\bullet$. (See for example, [24], [38], [72] or [96].)

A key estimate is the following relation between ramification and truncations:

Lemma A.28 ([96, Lemma A3.2.1]). *Let $H_1, \ldots, H_q$ be the hyperplanes in $\mathbb{P}^n(\mathbb{C})$, located in general position. Then*

$$\sum_{j=1}^{q} N_f(r, H_j) - N(r, 0, W) \leq \sum_{j=1}^{q} N_f^{[n]}(r, H_j).$$

As another natural generalization of Picard's theorem, for example, Fujimoto [22] and Green [34] interpret Borel's theorem in the context of holomorphic curves of $\mathbb{C}$ into $\mathbb{P}^n(\mathbb{C})$ and show

Theorem A.29. *Let $f : \mathbb{C} \to \mathbb{P}^n(\mathbb{C})$ be a holomorphic curve. Assume that the image of f lies in the complement of $n + p$ hyperplanes in general position, that is, f omits those hyperplanes, where $p \in \{1, \ldots, n+1\}$. Then the image of f is contained in a linear subspace at most of dimension $[n/p]$.*

In particular, by taking $p = n + 1$ it follows that if the image of a holomorphic curve $f : \mathbb{C} \to \mathbb{P}^n(\mathbb{C})$ lies in the complement of $2n + 1$ hyperplanes in general position, then f must be a constant.

Further extensions of Picard's theorem for holomorphic curves missing hyperplanes can be found in papers [23,35,36] and in monographs [24,66,72,96] and so forth.

In Ru's monograph [96] for example, one also finds some more refinements such as 'A General SMT with a Good Error Term' (Theorem $A3.1.3$), 'Product to the sum estimate' (Theorem $A3.1.6$) or 'Cartan's Second Main Theorem with Truncated Counting Functions' (Theorem $A3.2.2$).

The Cartan characteristic function has many useful properties [38,72,96]. Let us cite those references to summarize some of the essential properties in a little more concrete form for the purpose of our applications.

First, we can confirm that it is independent of the choice of reduced representations in the following way. For a holomorphic curve $f : \mathbb{C} \to \mathbb{P}^n(\mathbb{C})$, consider two reduced representations of f such as $(f_0, \ldots, f_n)$ and $(F_0, \ldots, F_n)$. Then, since the f_j's and F_j's are entire and

$$\sup_{j=0,\ldots,n} |f_j(z)| \neq 0 \quad \text{and} \quad \sup_{j=0,\ldots,n} |F_j(z)| \neq 0,$$

it follows that there exists a nowhere vanishing entire function h such that

$$F_j(z) = h(z)f_j(z)$$

for all $z \in \mathbb{C}$ and $j \in \{0, \ldots, n\}$. By writing f as $F = [F_0 : \cdots : F_n]$ and defining

$$T_F(r) = \int_0^{2\pi} U(re^{i\theta}) \frac{d\theta}{2\pi} - U(0), \quad U(z) = \sup_{k \in \{0,\ldots,n\}} \log |F_k(z)|,$$

we have the identity

$$T_F(r) = T_f(r) + \frac{1}{2\pi} \int_0^{2\pi} \log |h(re^{i\theta})| d\theta - \log |h(0)|.$$

However, since $h(z)$ is entire and nowhere zero so that $\log |h(z)|$ is harmonic, we have of course

$$\log |h(0)| = \frac{1}{2\pi} \int_0^{2\pi} \log |h(re^{i\theta})| d\theta.$$

Hence $T_F(r) = T_f(r)$ holds for the two reduced representations of f.

Second, the difference is only a bounded factor $O(1)$ as $r \to \infty$ between $T_f(r)$ and so-called **Ahlfors' characteristic function** which is equally given by

$$\int_0^{2\pi} \log \|F(re^{i\theta})\| \frac{d\theta}{2\pi} - \log \|F(0)\|,$$

where $F = (f_0, \ldots, f_n)$ is a reduced representation of f and $\|F(z)\| = (|f_0(z)|^2 + \cdots + |f_n(z)|^2)^{1/2}$.

Finally, as one finds in [96, Theorem A3.1.2] for example, we have

$$T\left(r, \frac{f_i}{f_j}\right) - O(1) \le T_f(r) \le \sum_{\substack{1 \le \mu \le n \\ (\mu \ne \nu)}} T\left(r, \frac{f_\mu}{f_\nu}\right) + O(1), \qquad \text{(A.16)}$$

for any $j, \nu \in \{0, 1, \ldots, n\}$ with $f_j(z)f_\nu(z) \not\equiv 0$. This shows that $T_f(r)/\log r \to \infty$ as $r \to \infty$, when at least one quotient f_i/f_j is a transcendental function. Thus it is natural to call the curve f **transcendental** then.

Cartan's characteristic appears to indicate an alternative way of defining the order of growth of a holomorphic curve f as

$$\rho(f) = \limsup_{r\to\infty} \frac{\log^+ T_f(r)}{\log r}, \quad \rho_2(f) = \limsup_{r\to\infty} \frac{\log^+ \log^+ T_f(r)}{\log r}. \qquad \text{(A.17)}$$

As we mentioned, these quantities $\rho(f)$ and $\rho_2(f)$ in definition (A.17) are both independent of a reduced representation of $f = [f_0 : f_1 : \cdots : f_n]$. In fact, inequality (A.16) implies that $\rho(f) = \rho^*(f)$ and $\rho_2(f) = \rho_2^*(f)$ with

$$\rho^*(f) = \max_{k,j \in \{0,1,\ldots,n\}} \limsup_{r\to\infty} \frac{\log^+ T(r, f_k/f_j)}{\log r},$$

and

$$\rho_2^*(g) = \max_{k,j \in \{0,1,\ldots,n\}} \limsup_{r\to\infty} \frac{\log^+ \log^+ T(r, f_k/f_j)}{\log r}.$$

A.4   Difference variant of Cartan theory

In a similar way as Cartan's value distribution theory extends Nevanlinna theory, there is a natural generalization of difference variant of Nevanlinna theory for holomorphic curves in the complex projective space. We will describe this difference variant of Cartan's theory in this section.

Let $g(z)$ be a meromorphic function and let $c \in \mathbb{C}$. For a fixed c, we will use the short notation

$$g(z) \equiv g = \bar{g}^{[0]}, \quad g(z+c) \equiv \bar{g}, \quad g(z+2c) \equiv \bar{\bar{g}}, \quad \ldots, \quad g(z+nc) \equiv \bar{g}^{[n]}$$

to suppress the explicit z-dependence of $g(z)$. The **Casorati determinant** of $g_1, \ldots, g_n$ $(n \in \mathbb{N})$ is then defined by

$$C(g_1, \ldots, g_n) = \begin{vmatrix} g_1 & g_2 & \cdots & g_n \\ \bar{g}_1 & \bar{g}_2 & \cdots & \bar{g}_n \\ \vdots & \vdots & \ddots & \vdots \\ \bar{g}_1^{[n-1]} & \bar{g}_2^{[n-1]} & \cdots & \bar{g}_n^{[n-1]} \end{vmatrix}$$

$$= \begin{vmatrix} g_1 & g_2 & \cdots & g_n \\ \Delta_c g_1 & \Delta_c g_2 & \cdots & \Delta_c g_n \\ \vdots & \vdots & \ddots & \vdots \\ \Delta_c^{n-1} g_1 & \Delta_c^{n-1} g_2 & \cdots & \Delta_c^{n-1} g_n \end{vmatrix}$$

with the difference operator $\Delta_c g = \bar{g} - g$ and it does not vanish identically on $\mathbb{C}$ if and only if the functions $g_1, \ldots, g_n$ are linearly independent over the field $\mathcal{P} := \mathcal{P}(c)$ of periodic functions with the period c. Note that by the transformation $z \mapsto cz$ of the independent variable we can take $c = 1$ in the definitions of the operator Δ_c and the Casorati determinant without loss of generality. Note also that Wronskians and Casoratians naturally share the properties followed with basic calculation rules of determinants. The only difference is in relation to the concept of 'multiplicity', in each case. For example with $c = 1$, the multiplicity of the zero of an entire function $h(z)$ at the origin is defined as the exponent of the leading term of its power series expansion $h(z) = \sum_{j=m_p}^{\infty} a_j z^j$ or its binomial series expansion $h(z) = \sum_{j=m_b}^{\infty} b_j z(j)$ with

$$z(j) = z(z-1) \cdots (z-j+1), \quad z(0) = 1,$$

respectively. One sees that $m_p \neq m_b$ in general with such simple examples as $z^2 = z(1) + z(2)$ and $z(2) = -z + z^2$ both having a zero at $z = 0$ which is double and simple in the ordinary sense but simple and double in the sense of the difference Δ_1, respectively.

Let us observe the correspondence of some properties for Wronskians and Casoratians now. Let $g_1, \ldots, g_n, h$ be meromorphic functions and $c_1, \ldots, c_n$ be constants for $n \in \mathbb{N}$ (see [68, Proposition 1.4.3], for example).

Wronski determinants	Casorati determinants
$W(c_1 g_1, \ldots, c_n g_n)$ $= c_1 \cdots c_n W(g_1, \ldots, g_n)$	$C(c_1 g_1, \ldots, c_n g_n)$ $= c_1 \cdots c_n C(g_1, \ldots, g_n)$
$W\left(z^0, z^1, \ldots, \frac{z^{n-1}}{n!}, h\right) = h^{(n)}$	$C\left(z(0), z(1), \ldots, \frac{z(n-1)}{n!}, g\right) = \Delta_1^n g$
$W(g_1, \ldots, g_n, 1)$ $= (-1)^n W\left((g_1)', \ldots, (g_n)'\right)$	$C(g_1, \ldots, g_n, 1)$ $= (-1)^n C\left(\Delta_c g_1, \ldots, \Delta_c g_n\right)$
$W(h g_1, \ldots, h g_n)$ $= h^n W(g_1, \ldots, g_n)$	$C(h g_1, \ldots, h g_n)$ $= h \overline{h} \cdots \overline{h}^{[n-1]} C(g_1, \ldots, g_n)$
$W(g_1, \ldots, g_n)/g_1^n$ $= W\left(\left(\frac{g_2}{g_1}\right)', \ldots, \left(\frac{g_n}{g_1}\right)'\right)$	$C(g_1, \ldots, g_n)/\left(g_1 \overline{g_1} \cdots \overline{g_1}^{[n-1]}\right)$ $= C\left(\Delta_c\left(\frac{g_2}{g_1}\right), \ldots, \Delta_c\left(\frac{g_n}{g_1}\right)\right)$

These observations make it possible to treat the identity (A.9) or (A.10) with difference operators. In fact, we may consider a system of linear equations

$$
\begin{pmatrix}
1 & \cdots & 1 \\
\frac{\Delta_c f_1}{f_1} & \cdots & \frac{\Delta_c f_n}{f_n} \\
& \cdots & \\
\frac{\Delta_c^{n-1} f_1}{f_1} & \cdots & \frac{\Delta_c^{n-1} f_n}{f_n}
\end{pmatrix}
\begin{pmatrix}
f_1 \\ f_2 \\ \vdots \\ f_n
\end{pmatrix}
=
\begin{pmatrix}
1 \\ 0 \\ \vdots \\ 0
\end{pmatrix}
$$

as well as the determinant of the coefficient matrix

$$
H(f_1, \ldots, f_n) :=
\begin{vmatrix}
1 & \cdots & 1 \\
\frac{\Delta_c f_1}{f_1} & \cdots & \frac{\Delta_c f_n}{f_n} \\
& \cdots & \\
\frac{\Delta_c^{n-1} f_1}{f_1} & \cdots & \frac{\Delta_c^{n-1} f_n}{f_n}
\end{vmatrix}
= \frac{C(f_1, \ldots, f_n)}{f_1 \cdots f_n}.
$$

Application of Cramer's rule again shows that each term f_j of the identity (A.10) is given as

$$
f_i = \frac{L_i(f_1, \ldots, f_n)}{L(f_1, \ldots, f_n)}, \quad (j = 1, \ldots, n)
$$

by $H(f_1, \ldots, f_n)$ and its variants

$$H_i(f_1, \ldots, f_n) := \begin{vmatrix} 1 & \cdots & 1 & 1 & 1 & \cdots & 1 \\ \frac{\Delta_c f_1}{f_1} & \cdots & \frac{\Delta_c f_{i-1}}{f_{i-1}} & 0 & \frac{\Delta_c f_{i+1}}{f_{i+1}} & \cdots & \frac{\Delta_c f_n}{f_n} \\ \vdots & \cdots & \vdots & \vdots & \vdots & \cdots & \vdots \\ \frac{\Delta_c^{n-1} f_1}{f_1} & \cdots & \frac{\Delta_c^{n-1} f_{i-1}}{f_{i-1}} & 0 & \frac{\Delta_c^{n-1} f_{i+1}}{f_{i+1}} & \cdots & \frac{\Delta_c^{n-1} f_n}{f_n} \end{vmatrix},$$

which is completely parallel to the case of Borel's theorem. In fact, it was noted that this discussion only requires us the linear independence of the f_i over the field $\mathcal{P}_c$ of c-periodic functions in $\mathbb{C}$ as a difference analogue of the previous discussion. As we learned in Chapter 3, there is a difference analogue of the lemma on logarithmic derivatives. Thus, we again need to consider a product of **logarithmic differences** of the form

$$\frac{\Delta_c^k f}{f} = \prod_{m=0}^{k-1} \frac{\Delta_c(\Delta_c^m f)}{\Delta_c^m f}, \quad \Delta_c^{(0)} f := f,$$

but thanks to the relation $\Delta_c(f)/f = \overline{f}/f - 1$, we may concentrate on a product of **logarithmic shifts** of the form

$$\frac{\overline{f}^{[k]}}{f} = \prod_{m=0}^{k-1} \frac{\overline{f}^{[m]}}{\overline{f}^{[m]}}, \quad \overline{f}^{[0]} := f,$$

instead. Of course, we have seen the homogeneity of Wronskian and Casoratian in the above list, while the second identity in Proposition A.21, that is,

$$L(gf_1, \ldots, gf_n) = L(f_1, \ldots, f_n)$$

has its counterpart such as

$$H(gf_1, \ldots, gf_n) = H(f_1, \ldots, f_n) \prod_{j=1}^{n-1} \frac{\overline{g}^{[j]}}{g}$$

since

$$\frac{C(gf_1, \ldots, gf_n)}{(gf_1) \cdots (gf_n)} = \frac{C(f_1, \ldots, f_n)}{f_1 \cdots f_n} \prod_{j=1}^{n-1} \frac{\overline{g}^{[j]}}{g}.$$

Without our expected identity of the form $H(gf_1, \ldots, gf_n) = H(f_1, \ldots, f_n)$, we can proceed our discussion in a parallel way to Cartan's discussion, since the lemma on logarithmic shifts shows the growth of the quantity $m\left(r, \prod_{j=1}^{n-1} \frac{\overline{g}^{[j]}}{g}\right)$ is indeed *negligible* with respect to the growth of our holomorphic curve if it is of hyper-order less than one. Furthermore, by

Theorem 3.15 in Chapter 3 it follows that any non-decreasing continuous function $T : [0, +\infty) \to [0, +\infty)$ such that

$$\limsup_{r \to \infty} \frac{\log \log T(r)}{\log r} = \rho_2 < 1,$$

satisfies

$$T(r + s) = T(r) + o\left(\frac{T(r)}{r^\delta}\right), \tag{A.18}$$

where r runs to infinity outside of a set of finite logarithmic measure, $s > 0$ and $\delta \in (0, 1 - \rho_2)$. The asymptotic identity (A.18) can be used to estimate counting functions such as $N\left(r, 0, \overline{f_j}^{[j-1]}\right)$ by means of $N(r, 0, f_j)$ up to the error term $o\left(\frac{T_g(r)}{r^\delta}\right)$ for example, so that we can deduce the estimate

$$N(r, 0, \tilde{L}) - N(r, \tilde{L}) \leq N(r, 0, L) - N(r, L) + o\left(\frac{T_g(r)}{r^\delta}\right)$$

with $\delta \in (0, 1 - \rho_2)$ for such an auxiliary function $\tilde{L}$ as

$$\tilde{L} = \frac{f_0 f_1 \overline{f_q} \cdots \overline{f_p}^{[p-1]} f_{p+1} \cdots f_q}{C(g_1, g_2, \ldots, g_p)}$$

together with L in Theorem A.30 below.

In Cartan's generalization of the second main theorem the ramification term is expressed in terms of the Wronski determinant of a set of linearly independent entire functions. Therefore, it seems quite natural that Halburd, Korhonen and Tohge in [45] proved the following theorem as a difference analogue of Cartan's result where the ramification term has been replaced by a quantity expressed in terms of the Casorati determinant of entire functions which are linearly independent over the field $\mathcal{P}_c^1$ of c-periodic meromorphic functions of hyper-order less than 1.

Theorem A.30 ([45, Theorem 2.1]). *Let $n \geq 1$, and let $g_0, \ldots, g_n$ be entire functions, linearly independent over $\mathcal{P}_c^1$, such that $\max\{|g_0(z)|, \ldots, |g_n(z)|\} > 0$ for each $z \in \mathbb{C}$, and*

$$\rho_2 := \rho_2(g) < 1, \quad g := [g_0 : \cdots : g_n].$$

Let $\varepsilon > 0$. If $f_0, \ldots, f_q$ are $q + 1$ linear combinations of the $n + 1$ functions $g_0, \ldots, g_n$ over $\mathbb{C}$, where $q > n$, such that any $n + 1$ of the $q + 1$ functions $f_0, \ldots, f_q$ are linearly independent over $\mathcal{P}_c^1$, and

$$L := \frac{f_0 f_1 \cdots f_q}{C(g_0, g_1, \ldots, g_n)}, \tag{A.19}$$

then

$$(q - n)T_g(r) \leq N\left(r, \frac{1}{L}\right) - N(r, L) + o\left(\frac{T_g(r)}{r^{1-\rho_2-\varepsilon}}\right) + O(1),$$

where r approaches infinity outside of an exceptional set E of finite logarithmic measure (i.e. $\int_E dt/t < \infty$).

On the first glance it may appear that natural requirement for the linear independence of $g_0, \ldots, g_n$ in Theorem A.30 would be over the field $\mathcal{P}_c$ of all c-periodic meromorphic functions. However, when working under the condition that the holomorphic curve defined by the coordinate functions $g_0, \ldots, g_n$ is of hyper-order less than one, it is sufficient to impose the weaker condition given in Theorem A.30, where the corresponding field is $\mathcal{P}_c^1 \subset \mathcal{P}_c$. This fact is due to the following lemma.

Lemma A.31 ([45, Lemma 3.2]). *If the holomorphic curve $g = [g_0 : \cdots : g_n]$ satisfies $\rho_2(g) < 1$ and if $c \in \mathbb{C}$, then $C(g_0, \ldots, g_n) \equiv 0$ if and only if the entire functions $g_0, \ldots, g_n$ are linearly dependent over the field $\mathcal{P}_c^1$.*

A special case of Theorem A.30 has been also given, independently, by P.-M. Wong, H.-F. Law and P. P. W. Wong [111] for the class of finite-order holomorphic curves with finite-order coordinate functions. Their result is also an extension of the results by Halburd and Korhonen in [41] to holomorphic curves into $\mathbb{P}^n(\mathbb{C})$.

A difference analogue of Green and Fujimoto's Theorem A.29 is proved as follows:

Theorem A.32 ([45, Theorem 1.1]). *Let $f : \mathbb{C} \to \mathbb{P}^n(\mathbb{C})$ be a holomorphic curve such that $\rho_2(f) < 1$, let $c \in \mathbb{C}$ and let $p \in \{1, \ldots, n+1\}$. If $n+p$ hyperplanes in general position have forward invariant preimages under f with respect to the translation $\tau(z) = z+c$, then the image of f is contained in a projective linear subspace over $\mathcal{P}_c^1$ of dimension $\leq [n/p]$.*

An example such as [45, Example 1.2] shows that the growth condition $\rho_2(f) < 1$ in Theorem A.32 cannot be replaced by $\rho_2(f) \leq 1$.

Examples obtained by a similar methods to Green's in [34] confirm the sharpness of the upper bound $[n/p]$ in Theorem A.32 as we see in [45, Section 7]. As an immediate consequence of Theorem A.32, we also have

Corollary A.33 ([45, Corollary 1.3]). *Let $f : \mathbb{C} \to \mathbb{P}^n(\mathbb{C})$ be a holomorphic curve such that $\rho_2(f) < 1$, and let $c \in \mathbb{C}$. If $2n + 1$ hyperplanes*

H_j $(j = 0, 1, \ldots, 2n)$ in $\mathbb{P}^n(\mathbb{C})$ *are located in general position and satisfy the condition*

$$\tau\left(f^{-1}H_j\right) \subset f^{-1}H_j$$

about their preimages under f with respect to the translation $\tau_c(z) = z + c$, then f is periodic with period c.

This result tells us that f reduces to a constant if the preimages $f^{-1}H_j$ under f are all empty so that the above condition for **forward invariance** is satisfied with respect to $\tau_c(z)$ for any $c \in \mathbb{C}$. Hence one may also regard Corollary A.33 a generalization of Green's Picard-type theorem in [34] for holomorphic curves of hyper-order less than one. It is not difficult here again to see that the growth condition $\rho_2(f) < 1$ in Corollary A.33 cannot be weakened by examples as in [45].

We have observed that in their paper [41] Halburd and Korhonen obtained an analogue of Nevanlinna's second main theorem for the difference operator Δ_c. Let us now show that their result for constant targets follows from Theorem A.30 in a similar way which we used to prove that Nevanlinna's second main theorem follows from Cartan's result.

Let w be a meromorphic function in $\mathbb{C}$ of hyper-order $\rho_2(w) < 1$. Then there exist linearly independent entire functions g_0 and g_1 with no common zeros such that $w = g_0/g_1$, and, according to (A.16), it follows that $\rho_2(g) < 1$ for $g = [g_0 : g_1]$. Note that both of the entire functions g_0 and g_1 themselves may be of hyper-order greater or equal to one in general, while their ratio w as well as the map g must be of hyper-order strictly less than one. Let $a_j \in \mathbb{C}$ be distinct numbers for $j = 0, \ldots, q-1$, and denote $f_j = g_0 - a_j g_1$ and $f_q = g_1$. Then, by Theorem A.30, it follows that

$$(q-1)T_g(r) \le N\left(r, \frac{1}{L}\right) - N(r, L) + o(T_g(r)) \qquad \text{(A.20)}$$

where

$$L = \frac{f_0 f_1 \cdots f_{q-1} f_q}{g_0 \overline{g}_1 - \overline{g}_0 g_1}.$$

Since $\rho_2(w) < 1$, it follows by Theorem 3.15 that $N(r, \overline{w}) \le N(r + |c|, w) = N(r, w) + o(N(r, w))$ outside of a possible exceptional set of finite logarithmic measure. Therefore, by interpreting (A.20) in terms of the counting functions of the form (A.7), and using (A.16) we have

$$(q-1)T(r, w) \le \widetilde{N}(r, w) + \sum_{j=0}^{q-1} \widetilde{N}\left(r, \frac{1}{w - a_j}\right) - N_0\left(r, \frac{1}{\Delta_c w}\right) + o(T(r, w))$$

where $N_0(r, 1/\Delta_c w)$ counts the number of those zeros of $\Delta_c w$ which do not coincide with any of the a_j-points or poles of w, and r runs to infinity outside of a set of finite logarithmic measure. This is an extension of [41, Theorem 2.5] as we desired.

Appendix B

Introduction to ultra-discrete Painlevé equations

Integrability is a central common feature of all Painlevé equations, whether they are continuous, discrete or ultra-discrete. In what follows we approach the topic of ultra-discrete Painlevé equations by using continuous and discrete Painlevé equations as reference points. We will briefly cover the concept integrability, and some ideas used for integrability testing of these equations. The idea is that we introduce these notions by using the well-known case of (differential) Painlevé equations as an example, and then continue to the details of analogous concepts for discrete and ultra-discrete equations.

B.1 Painlevé equations

When talking about *Painlevé equations*, one can think of this term as a general name for all equations - continuous, discrete or ultra-discrete - that are in some sense of Painlevé type. However, a 'Painlevé equation' more commonly refers to an equation in the list of six original non-linear differential equations

$$w'' = 6w^2 + z \qquad (P_I)$$

$$w'' = 2w^3 + zw + \alpha \qquad (P_{II})$$

$$w'' = \frac{(w')^2}{w} - \frac{1}{z}w' + \frac{1}{z}(\alpha w^2 + \beta) + \gamma w^3 + \frac{\delta}{w} \qquad (P_{III})$$

$$w'' = \frac{(w')^2}{2w} + \frac{3}{2}w^3 + 4zw^2 + 2(z^2 - \alpha)w + \frac{\beta}{w} \qquad (P_{IV})$$

$$w'' = \left\{ \frac{1}{2w} + \frac{1}{w-1} \right\} (w')^2 - \frac{w'}{z}$$
$$+ \frac{(w-1)^2}{z^2 w}(\alpha w^2 + \beta) + \frac{\gamma w}{z} + \frac{\delta w(w+1)}{w-1} \tag{P_V}$$
$$w'' = \frac{1}{2}\left(\frac{1}{w} + \frac{1}{w-1} + \frac{1}{w-z} \right)(w')^2 - \left(\frac{1}{z} + \frac{1}{z-1} + \frac{1}{w-z} \right)w'$$
$$+ \frac{w(w-1)(w-z)}{z^2(z-1)^2}\left(\alpha + \beta \frac{z}{w^2} + \gamma \frac{z-1}{(w-1)^2} + \delta \frac{z(z-1)}{(w-z)^2} \right), \tag{P_{VI}}$$

where $\alpha, \beta, \gamma, \delta \in \mathbb{C}$, discovered by Painlevé [88, 89], Fuchs [21] and Gambier [25] over a century ago. Painlevé and his colleagues went through an extensive classification problem by sieving through the class

$$w'' = F(z, w, w') \tag{B.1}$$

of second-order ordinary differential equations, where F is rational in w and w', and rejecting those equations which did not have the **Painlevé property**. An equation has the Painlevé property if the only movable singularities of its solutions are poles. By rejecting all those equations which do not possess the Painlevé property, Painlevé, Gambier and Fuchs obtained a list of 50 equations, 44 of which could be solved in terms of known functions, or transformed into another equation in the list. The remaining six equations (P_I)–(P_{VI}) are now known as Painlevé equations. A question which immediately arises is that does the implication go in the opposite direction? Namely, do the Painlevé equations indeed possess the Painlevé property? Painlevé himself drafted a method of proof to show that this is indeed the case. The proof contained gaps, however, and the matter was only relatively recently settled by Hinkkanen, Laine and others in a series of articles, see [37] and the references therein. They proved using a complex analytic method that all solutions of equations (P_I), (P_{II}) and (P_{IV}) are meromorphic, and that (P_{III}) and (P_V) can be transformed using exponential transformation of the independent variable $z = \exp(\zeta)$ into modified equations which have only meromorphic solutions. For (P_{VI}) no such transformation is possible, but the equation still satisfies the Painlevé property [37]. Another method for proving that the Painlevé equations possess the Painlevé property is based on the underlying isomonodromy problems and Riemann-Hilbert techniques, see the the work of Miwa [77] and Malgrange [75].

Since all solutions of (P_I), (P_{II}) and (P_{IV}), and of modified equations (P_{III}) and (P_V) are meromorphic, Nevanlinna theory is readily applicable to study the solutions. It has been shown by Steinmetz [102] and Shimo-mura [97] that all solutions of (P_I), (P_{II}) and (P_{IV}) are of finite order (and the possible orders are now known as well, see [51,52,99]). Also, Shimomura has proved that the modified versions of (P_{III}) and (P_V) have solutions of finite hyper-order and no other solutions [98]. This jump in the growth of solutions of modified equations compared to the non-modified equations is explained by the exponential transformation of the independent variable used to obtain them. The growth of solutions of Painlevé equations appears to be - in a broad sense - of 'finite order'. There is a direct analogue of this growth behavior of solutions in the discrete and ultra-discrete cases, to which we will go into soon.

Painlevé equations have been extensively studied since their discovery. They possess vast amount of special properties and applications that are specific to integrable equations. Before continuing any further along this line of thought, we should pause to explain what exactly do we mean by integrability, since this concept has several definitions depending on the context. By an integrable equation we mean an equation which is solvable either explicitly in terms of special functions, or indirectly by using an associated linear problem together with the inverse scattering transform techniques, or more generally inverse spectral methods [2]. Linear and linearizable equations are considered to be integrable as well. All six Painlevé equations can be expressed as a compatibility condition

$$\Psi_{z\zeta} = \Psi_{\zeta z} \tag{B.2}$$

of a linear iso-monodromy problem

$$\begin{cases} \Psi_z = A(z,\zeta)\Psi \\ \Psi_\zeta = B(z,\zeta)\Psi \end{cases} \tag{B.3}$$

for suitable choices of matrix-valued functions A and B. Here z is the independent variable of the Painlevé equation and ζ is a spectral (or iso-monodromy) parameter, which is external to the Painlevé equation. For example the Painlevé II equation (P_{II}) is the compatibility condition for the system

$$\begin{cases} \Psi_z = \begin{pmatrix} -i\zeta & w \\ w & i\zeta \end{pmatrix} \Psi \\ \Psi_\zeta = \begin{pmatrix} -i(4\zeta^2 + z + 2w^2) & 4\zeta w + 2iw_z + \alpha\zeta^{-1} \\ 4\zeta w - 2iw_z + \alpha\zeta^{-1} & i(4\zeta^2 + z + 2w^2) \end{pmatrix} \Psi \end{cases}.$$

The fact that Painlevé equations can be represented in this form implies that they are integrable, since they can now be solved using inverse scattering transform techniques. Painlevé equations are characteristic examples of equations satisfying this definition of integrability without being solvable explicitly in terms of known functions. This makes them a useful starting point when trying to develop methods to find new integrable equations.

B.2   Integrability of Painlevé equations and integrability testing

One of the central themes we have adopted in this appendix is integrability testing. By this we mean that can we find an easily tested simple criterion, which implies reliably that the equation under testing is integrable without actually proving the fact? This type of tests are useful in the attempts to find new integrable equations, independently of whether we are exploring the realms of differential, discrete or ultra-discrete equations. In the case of ordinary differential equations, the Painlevé property has turned out to be an extremely reliable test for integrability. For instance, in the case of the class of second-order equations (B.1), this test singles out exactly integrable equations within (B.1). There are so far no known counterexamples to show that the test would fail in the case of ordinary differential equations. In the case of discrete and ultra-discrete equations the situation is not as clear and there are many proposed tests for integrability.

In the first-order case Nevanlinna theory can be used as a particularly effective tool in testing integrability. The following example is actually a century old theorem by Malmquist [76], reproven by Yosida [115] in 1930's using Nevanlinna theory. We choose to consider this example here because the idea of applying Nevanlinna theory to rule out classes of equations extends in a natural way to discrete and ultra-discrete situations, also in the higher order cases, as we will see later on.

Example B.1. If

$$f' = R(z, f), \tag{B.4}$$

where $R(z, f)$ is rational in both arguments, admits a non-rational meromorphic solution f, then

$$f' = a_2 f^2 + a_1 f + a_0,$$

i.e., equation (B.4) reduces into the Riccati equation.

Proof. Since

$$2T(r, f) + S(r, f) \geq T(r, f') = T(r, R(z, f))$$
$$= \deg_f(R(z, f))T(r, f) + S(r, f),$$

it follows that $\deg_f(R(z, f)) \leq 2$. Substituting $f = 1/g$ into

$$f' = \frac{a_2 f^2 + a_1 f + a_0}{b_2 f^2 + b_1 f + b_0} \tag{B.5}$$

we have

$$g' = -\frac{a_0 g^2 + a_1 g + a_2}{\frac{b_2}{g^2} + \frac{b_1}{g} + b_0} =: F(z, g).$$

Suppose now that $a_0 b_0 \not\equiv 0$. Then $\deg_g(F(z, g)) > 2$ unless $b_2 \equiv b_1 \equiv 0$. But then

$$2T(r, g) \geq T(r, g') = T(r, F(z, g)) = \deg_g(F(z, g))T(r, g) + S(r, g)$$
$$> 2T(r, g) + S(r, g).$$

Thus $b_2 \equiv b_1 \equiv 0$, and

$$f' = \frac{a_2}{b_0} f^2 + \frac{a_1}{b_0} f + \frac{a_0}{b_0}.$$

The case $a_0 b_0 \not\equiv 0$ can be treated in a similar way using a slightly modified transformation. □

B.3 Discrete Painlevé equations

There are many ordinary non-linear discrete equations which possess a striking number of special properties that are analogous to the properties of Painlevé equations. Based on this similarity, such equations are referred to as discrete Painlevé equations. Some of the discrete equations now considered to be of Painlevé type, have been known in the mathematical literature long before people started to call them discrete Painlevé equations. For example, equation

$$x_{n+1} + x_n + x_{n-1} = \frac{an + b}{x_n} + c, \tag{B.6}$$

where a, b and c are constant parameters, have appeared in studies on the coefficients in the three-point recurrence relations of certain classes of orthogonal polynomials dating at least back to Shohat in 1930's [101]. The discrete equation (B.6) also appears in studies of a partition function in quantum gravity, see, e.g., [10, 12, 90]. From these studies and others like them it became clear that there is something very special about equation (B.6). Further studies on the equation (B.6) showed, for instance, that it is the compatibility condition for a linear problem [55], and it has a simple continuum limit to the first Painlevé equation (P_I) [12, 19]. It became a standard convention to call the equation (B.6) as a discrete Painlevé I, based on the continuum limit.

Other discrete equations with similar special properties were soon found as reductions of integrable lattice equations [80] and by looking for discrete isomonodromy problems [20, 58]. Some examples of discrete Painlevé equations are

$$x_{n+1} + x_n + x_{n-1} = \frac{\alpha n + \beta + \gamma(-1)^n}{x_n} + \delta \qquad (\text{d-}P_I)$$

$$x_{n+1} + x_{n-1} = \frac{\alpha n + \beta}{x_n} + \frac{\gamma + \delta(-1)^n}{x_n^2} \qquad (\text{d-}P_I)$$

$$x_{n+1} + x_{n-1} = \frac{(\alpha n + \beta)x_n + \gamma + \delta(-1)^n}{1 - x_n^2} \qquad (\text{d-}P_{II})$$

$$x_{n+1}x_{n-1} = \frac{\alpha x_n^2 + \beta \lambda^n x_n + \gamma \lambda^{2n}}{(x_n - 1)(x_n - \alpha)} \qquad (\text{d-}P_{III})$$

$$\begin{cases} x_{n+1} + x_n = \frac{\alpha + (\beta + \delta n)y_n}{y_n^2 - 1} \\ y_n + y_{n-1} = \frac{a + (\beta - \delta/2 + \delta n)x_n}{x_n^2 - 1} \end{cases} \qquad (A_{III}^c)$$

where α, β, γ and δ are constant parameters. This is by no means nowhere near of being a complete list of discrete Painlevé equations. In fact, there are many more discrete Painlevé equations compared to their differential counterparts. A more comprehensive but still partial list of discrete Painlevé equations can be found in [32], for example.

Ablowitz, Halburd and Herbst [1] suggested to embed discrete Painlevé equations in the complex plane and to look at their meromorphic solutions. This way obtained complex analytic difference analogues of Painlevé equations can be studied using complex analytic methods, such as Nevanlinna theory. For example, the discrete Painlevé equation (d-P_I) becomes then

$$y(z + 1) + y(z - 1) = \frac{\pi_1 z + \kappa_1}{y(z)} + \frac{\pi_2}{y(z)^2}, \qquad (\text{B.7})$$

where $\pi_1(z)$ and $\kappa_1(z)$ are arbitrary meromorphic periodic functions of period 1, and $\pi_2(z)$ is a meromorphic periodic function of period 2. We will elaborate on this approach at the end of Section B.4 below.

Similarly as in the case of differential Painlevé equations, difference Painlevé equations can also be written as compatibility conditions of linear systems. (The same is true for purely discrete versions of Painlevé equations, and this is the more commonly used context in the literature.) For instance, a difference analogue of (B.2) and (B.3) applicable for discrete Painlevé equations arising as Bäcklund transformations of the (continuous) Painlevé equations is given by

$$E_z \partial_\zeta \Psi(z, \zeta) = \partial_\zeta E_z \Psi(z, \zeta) \tag{B.8}$$

and

$$\begin{cases} \Psi(z+1, \zeta) = U(z, \zeta)\Psi(z, \zeta) \\ \partial_\zeta \Psi(z, \zeta) = V(z, \zeta)\Psi(z, \zeta), \end{cases} \tag{B.9}$$

respectively, where $E_z \Psi(z, \zeta) := \Psi(z+1, \zeta)$ is the z-shift operator. Condition (B.9) is equivalent to

$$\partial_\zeta U(z, \zeta) = V(z+1, \zeta)U(z, \zeta) - U(z, \zeta)V(z, \zeta). \tag{B.10}$$

We will now go through the details of a particular example studied in the discrete case by Fokas, Grammaticos and Ramani [20] (see [44] for the details in the complex analytic case). In (B.10), by choosing

$$U(z, \zeta) := \begin{pmatrix} 2\zeta/\Omega(z+1) & -\Omega(z)/\Omega(z+1) \\ 1 & 0 \end{pmatrix}, \tag{B.11}$$

for some function Ω, and by writing

$$\Psi(z, \zeta) = \begin{pmatrix} \psi_1(z, \zeta) \\ \psi_2(z, \zeta) \end{pmatrix}$$

it follows by (B.9) that

$$\Omega(z+1)\psi_1(z+1, \zeta) + \Omega(z)\psi_1(z-1, \zeta) = 2\zeta\psi_1(z, \zeta).$$

If the matrix V is chosen such that its trace vanishes, then (B.10) implies that

$$V(z, \zeta) = \begin{pmatrix} a(z, \zeta) & -\frac{\Omega(z)}{2\zeta}[a(z, \zeta) + a(z+1, \zeta)] \\ \frac{\Omega(z)}{2\zeta}[a(z, \zeta) + a(z-1, \zeta)] & -a(z, \zeta) \end{pmatrix},$$

where

$$\zeta[a(z+1, \zeta) - a(z, \zeta)] - \Omega(z+1)^2 \frac{a(z+1, \zeta) + a(z+2, \zeta)}{4\zeta} \\ + \Omega(z)^2 \frac{a(z-1, \zeta) + a(z, \zeta)}{4\zeta} = 1. \tag{B.12}$$

Suppose now that (B.12) has a solution of a special type given by

$$a(z, \zeta) = \phi(z)/\zeta + 2\beta(z)\zeta.$$

Then,

$$\phi(z) = \beta\Omega^2(z) + z + \alpha, \qquad \beta(z) = \beta,$$

where α and β are constants, and

$$[\phi(z-1) + \phi(z)]\,\Omega(z)^2 + [\phi(z+1) + \phi(z+2)]\,\Omega(z+1)^2 = 0. \qquad \text{(B.13)}$$

By multiplying (B.13) side by side with $\phi(z) + \phi(z+1)$, we have

$$\begin{aligned}
(\phi(z) + \phi(z+1))\,[\phi(z+1) + \phi(z+2)]\,\Omega(z+1)^2 \\
+ (\phi(z) + \phi(z+1))\,[\phi(z-1) + \phi(z)]\,\Omega(z)^2 = 0,
\end{aligned} \qquad \text{(B.14)}$$

which implies that

$$(\phi(z) + \phi(z+1))(\phi(z) + \phi(z-1)) = \frac{\gamma}{\phi(z) - z - \alpha}, \qquad \text{(B.15)}$$

where γ is an arbitrary periodic function with period one. If we choose γ as a constant and substitute $y(z) = 1/(\phi(z) + \phi(z-1))$ into equation (B.15), then we have

$$y(z+1) + y(z-1) = \frac{1 - (2z + 2\alpha - 1)y(z)}{\gamma y^2(z)}. \qquad \text{(B.16)}$$

Equation (B.16) is a special case of a difference Painlevé equation (B.7).

B.4 Discrete Painlevé equations and integrability testing

In the early 90's the list of newly discovered discrete Painlevé equations started to expand quickly: more and more of them were found using different methods. There were clearly many discrete Painlevé equations to be found, and so there was a need to develop systematic methods to find them. Since the Painlevé property is a reliable enough test in finding integrable ordinary differential equations, it was natural to look into finding a discrete analogue of the Painlevé property which would be similarly applicable in finding new discrete integrable equations.

The first suggested discrete analogue for the Painlevé property, which was used in a systematic way to find new discrete Painlevé equations, is the **singularity confinement** method by Grammaticos, Ramani and Papageorgiou [33]. The basic idea in the singularity confinement test is that one considers a sequence of iterates $(x_n)_{n \in \mathbb{Z}}$ of a discrete equation, say (B.6), by choosing initial conditions in such a way that they lead the solution into a singularity. For instance, one can take

$$x_0 = k$$
$$x_1 = 0$$

as the initial condition for (B.6), which imply that $x_2 = \infty$. In order to continue the iteration through the singularity, it is convenient to look at a perturbation

$$
\begin{aligned}
x_0 &= k \\
x_1 &= \varepsilon,
\end{aligned}
\tag{B.17}
$$

where $\varepsilon > 0$. Then, by iterating (B.6) further, we obtain

$$
\begin{aligned}
x_0 &= k \\
x_1 &= \varepsilon \\
x_2 &= \frac{a+b}{\varepsilon} + c - k - \varepsilon \\
x_3 &= -\frac{a+b}{\varepsilon} + k + \varepsilon \cdot \frac{2a+b}{a+b} + O(\varepsilon^2) \\
x_4 &= -\varepsilon \cdot \frac{4a+b}{a+b} + O(\varepsilon^2) \\
x_5 &= \frac{2ac+ak+bk}{b+4a} + O(\varepsilon)
\end{aligned}
$$

which by letting $\varepsilon \to 0$ becomes

$$
\begin{aligned}
x_0 &= k, \quad x_1 = 0, \quad x_2 = \infty, \\
x_3 &= \infty, \quad x_4 = 0, \quad x_5 = \frac{2ac+ak+bk}{b+4a}.
\end{aligned}
\tag{B.18}
$$

In general such iteration would lead into a sequence of iterates with seemingly infinitely many singularities. According to singularity confinement test, if the iterates become finite after a finite number of steps and retain sufficient information on the initial conditions (B.17), then the singularities are considered to be confined and the equation should be integrable. This is exactly what happens in the case (B.18). However, if we change the equation (B.6) slightly, for example into

$$
x_{n+1} + x_n + x_{n-1} = \frac{an^2 + b}{x_n} + c,
\tag{B.19}
$$

(which is no longer a discrete Painlevé equation) then the corresponding iteration sequence becomes

$$
\begin{aligned}
x_0 &= k, \quad x_1 = 0, \quad x_2 = \infty, \quad x_3 = \infty, \quad x_4 = 0, \\
x_5 &= \infty, \quad x_6 = \infty, \quad x_7 = \infty, \ldots
\end{aligned}
$$

which is not confined.

The singularity confinement test has been very successful in finding new discrete equations of Painlevé type [93], and it also extends to the case of lattice equations [33]. It is not perfect, however. Hietarinta and Viallet [50] found an example of a non-integrable equation, which passes the singularity confinement test. They suggested to improve the singularity confinement method by demanding that in addition to passing the singularity confinement test, the solutions should also have zero algebraic entropy. This means roughly speaking that the degrees of the iterates should grow slowly as a function of the initial conditions.

Ablowitz, Halburd and Herbst [1] suggested a purely complex analytic difference analogue for the Painlevé property. They embedded the considered discrete equation into the complex plane, and looked at it as a complex delay equation. By this assumption, the discrete Painlevé equation (B.6), for instance, becomes the difference equation

$$y(z+1) + y(z) + y(z-1) = \frac{az+b}{y(z)} + c, \qquad (B.20)$$

where $z \in \mathbb{C}$. According to the approach by Ablowitz, Halburd and Herbst, if a difference equation such as (B.20) has sufficiently many finite-order meromorphic solutions, then it should be integrable, or of Painlevé type. The following example, which is a result due to Yanagihara [112] with an alternate proof relying to the new methods obtained in [40, 41], is an exact difference analogue of Example B.1.

Example B.2. If

$$y(z+1) = R(z, y(z)), \qquad (B.21)$$

where $R(z, y(z))$ is rational in both arguments, admits a non-rational meromorphic solution y of finite order, then

$$y(z+1) = \frac{a_1(z)y(z) + a_0(z)}{b_1(z)y(z) + b_0(z)},$$

where $a_1 b_0 \not\equiv a_0 b_1$.

Proof. Since y is of finite order, it follows by Theorem A.13 that

$$m(r, y(z+1)) \le m(r, y(z)) + m\left(r, \frac{y(z+1)}{y(z)}\right) = m(r, y(z)) + \widetilde{S}(r, y),$$

and similarly that $m(r, y(z)) \le m(r, y(z+1)) + \widetilde{S}(r, y)$. Thus we have the asymptotic identity $m(r, y(z)) = m(r, y(z+1)) + \widetilde{S}(r, y))$. An application of Theorem 3.15 to the function $N(r, y)$ together with the inequalities

$$N(r-1, y(z)) \le N(r, y(z+1)) \le N(r+1, y(z))$$

yields $N(r, y(z)) = N(r, y(z+1)) + \widetilde{S}(r, y)$. Hence by applying Theorem A.12, we have

$$T(r, y(z)) + \widetilde{S}(r, y) = T(r, y(z+1)) = T(r, R(z, y(z)))$$
$$= \deg_y(R(z, y(z)))T(r, y(z)) + \widetilde{S}(r, y),$$

and so $\deg_y(R(z, y(z))) = 1$. □

In the same way as Example B.1 singled out the Riccati equation out of a large class of first-order differential equations, Example B.2 shows that the only difference equation out of the analogous class (B.21) of first-order difference equations that can have finite-order meromorphic solutions, is the difference Riccati equation. Therefore, at least in the first-order case, the difference analogue of the Painlevé property proposed by Ablowitz, Halburd and Herbst works exactly as it should. The next theorem from [43] gives strong evidence that the same conclusion is valid also in the second-order case.

Before stating the theorem we need to recall some notation from classical Nevanlinna theory. If y is a meromorphic function, then $\mathcal{S}(y)$ denotes the field of small functions with respect to y. A non-rational meromorphic solution y of a difference equation is said to be **admissible** if all coefficients of the considered equation belong to the class $\mathcal{S}(y)$, or in other words are small with respect to y. This means, for instance, that all transcendental solutions of a difference equation with rational coefficients are admissible by this definition. Admissible meromorphic solutions can be defined in an analogous way for differential, q-difference or ultra-discrete equations. The concept was originally introduced for differential equations, see, e.g., [68].

Theorem B.3. *If the equation*

$$y(z+1) + y(z-1) = R(z, y(z)), \tag{B.22}$$

where $R(z, y(z))$ is rational in $y(z)$ and meromorphic in z, has an admissible meromorphic solution of finite order, then either $y(z)$ satisfies a difference Riccati equation

$$y(z+1) = \frac{p(z+1)y(z) + q(z)}{y(z) + p(z)}, \tag{B.23}$$

where $p, q \in \mathcal{S}(y)$, or equation (B.22) can be transformed by a linear change in $y(z)$ to one of the following equations:

$$y(z+1) + y(z) + y(z-1) = \frac{\pi_1 z + \pi_2}{y(z)} + \kappa_1 \tag{B.24}$$

$$y(z+1) + y(z-1) = \frac{\pi_1 z + \pi_3}{y(z)} + \pi_2 \tag{B.25}$$

$$y(z+1) + y(z-1) = \frac{\pi_1 z + \kappa_1}{y(z)} + \frac{\pi_2}{y(z)^2} \tag{B.26}$$

$$y(z+1) + y(z-1) = \frac{(\pi_1 z + \kappa_1)y(z) + \pi_2}{\exp(-i\pi z) - y(z)^2} \tag{B.27}$$

$$y(z+1) + y(z-1) = \frac{(\pi_1 z + \kappa_1)y(z) + \pi_2}{1 - y(z)^2} \tag{B.28}$$

$$y(z+1)y(z) + y(z)y(z-1) = p \tag{B.29}$$

$$y(z+1) + y(z-1) = py(z) + q \tag{B.30}$$

where $\pi_k, \kappa_k \in \mathcal{S}(y)$ *are arbitrary finite-order periodic functions with period* k.

Equation (B.23) is a special case of the difference Riccati equation, and equations (B.24), (B.25), (B.26) and (B.28) are known difference Painlevé equations. Equation (B.30) is a second order linear difference equation and (B.29) is linearizable by a simple transformation. The purely discrete form of (B.27) is transformable into the discrete version of (B.28), but in the complex plane no such transformation is known. Therefore the above list consists exactly of Painlevé difference equations within the class (B.22), and it contains no other equations. Hence, at least when operating within the class (B.22), the approach by Ablowitz, Halburd and Herbst gives exactly the right answers.

In the differential case, the Painlevé equations actually possess the Painlevé property. What about the difference case? Do the difference Painlevé equations above possess finite-order meromorphic solutions? In the special case, when the coefficients of these equations are constants, the answer in general to this question is yes: the autonomous version of difference Painlevé equations can be generically integrated in terms of elliptic functions, see, e.g., [44]. The following example, see, e.g., [44], makes use of addition laws of elliptic functions to construct a two parameter family of finite-order meromorphic solutions of an autonomous difference Painlevé equation.

Example B.4. Suppose that ω_1 and ω_2 are complex constants such that the imaginary part of ω_2/ω_1 is non-zero. The Weierstrass function is defined by

$$\wp(z) = \frac{1}{z^2} + \sum_{m,n}' \left\{ \frac{1}{(z - \Omega_{mn})^2} - \frac{1}{\Omega_{mn}^2} \right\}, \tag{B.31}$$

where the sum $\sum'_{m,n}$ runs over all integers m and n such that $(m, n) \neq (0, 0)$, and

$$\Omega_{mn} = 2(m\omega_1 + n\omega_2).$$

The function $\wp$ is meromorphic with periods $2\omega_1$ and $2\omega_2$, it is an even function satisfying the differential equation

$$\wp'^2 = 4\wp^3 - g_2\wp - g_3, \tag{B.32}$$

where

$$g_2 = 60\sum_{m,n}{}'\Omega_{mn}^{-4} \quad \text{and} \quad g_3 = 140\sum_{m,n}{}'\Omega_{mn}^{-6}.$$

The function $\wp$ is of order 2, and it satisfies the addition law

$$\wp(z_1 + z_2) = \frac{1}{4}\left\{\frac{\wp'(z_1) - \wp'(z_2)}{\wp(z_1) - \wp(z_2)}\right\}^2 - \wp(z_1) - \wp(z_2). \tag{B.33}$$

By combining the addition law (B.33), equation (B.32) and the fact that $\wp$ is even, it can be seen that $y(z) := \wp(h(z - z_0)) - \wp(h)$ satisfies

$$y(z+1) + y(z-1) = \frac{\alpha y(z) + \beta}{y^2(z)}, \tag{B.34}$$

where h is a fixed constant such that $\wp(h)$ is finite, $\alpha = (12\wp^2(h) - g_2)/2$ and $\beta = \wp'^2(h)$. Therefore, equation (B.34) admits a two-parameter family of finite-order meromorphic solutions, where the parameters are z_0 and g_3.

In the general non-autonomous case the question of existence of finite-order meromorphic solutions remains open in its full generality. Shimomura [100] has shown that under some conditions on the parameter values, equations (B.25), (B.26) and (B.28) have meromorphic solutions, but Shimomura's method does not give information on the growth of these solutions.

It was recently shown that weakening the finite-order condition in Theorem B.3 to 'hyper-order < 1' gives the same list of equations [45]. This condition cannot be weakened any further without altering the assertions of the theorems. For instance

$$y(z) = \exp(2^z)$$

is an entire solution of hyper-order exactly one for the second-order non-linear difference equation

$$y(z+1)y(z-1)^2 = y(z)^4,$$

which is not a difference Painlevé equation and is not considered to be integrable.

B.5 Ultra-discrete Painlevé equations

An important feature, which is common to many integrable equations, is the existence of a continuous limit to another integrable equation or equations. The idea is that by making a suitable change of variables to the dependent variable of the equation, depending on an arbitrary small parameter ϵ, and by allowing $\epsilon \to 0$, the original equation reduces to another integrable equation. Continuous limits can be used, for example, to form coalescence cascades of Painlevé equations, both differential and discrete. They were used also as a guideline in determining which subclasses of discrete equations to concentrate the search, when trying to find new discrete Painlevé equations by the singularity confinement method [93].

As an example of a continuous limit we consider the Lotka-Volterra equation

$$\frac{d}{dt}x_n(t) = x_n(t)\left(x_{n+1}(t) - x_{n-1}(t)\right), \tag{B.35}$$

which is an integrable discretization of the famous KdV equation

$$\frac{\partial}{\partial t}u(x,t) = \frac{\partial^3}{\partial x^3}u(x,t) + u(x,t)\frac{\partial}{\partial x}u(x,t). \tag{B.36}$$

By making a change of variables

$$x_n(t) = 1 + \frac{1}{6}\epsilon^2 u\left((n+2t)\epsilon, \frac{\epsilon^3 t}{3}\right),$$

and letting $\epsilon \to 0$, the Lotka-Volterra equation (B.35) is mapped into the KdV equation (B.36) [105]. Similar limits can be used to transform discrete Painlevé equations into differential Painlevé equations. For instance, an application of the transformation

$$
\begin{aligned}
w &= -1/2 + \varepsilon^2 u \\
\delta &= -3 \\
\alpha n + \beta + \gamma(-1)^n &= -(3 + 2\varepsilon^4 t)/4 \\
\epsilon &\to 0
\end{aligned}
\tag{B.37}
$$

to the discrete Painlevé I equation

$$w_{n+1} + w_n + w_{n-1} = \frac{\alpha n + \beta + \gamma(-1)^n}{w_n} + \delta$$

yields the Painlevé I equation (P_I) [20].

Continuous limits such as (B.37) have been used in both finding and naming new discrete Painlevé equations [93]. Another type of limit, the so called ultra-discrete limit, can be used in a similar way in identifying new ultra-discrete Painlevé equations. This method was introduced by Tokihiro et al. [105], and the basic idea is as follows. For the dependent variable v of the equation a new variable V is introduced by defining

$$v = \exp\left(\frac{V}{\epsilon}\right),$$

where ϵ is an arbitrary positive parameter. The ultra-discrete limit is taken by allowing ϵ tend to zero and by applying the identity

$$\lim_{\epsilon \to +0} \epsilon \log\left(\exp\left(\frac{A}{\epsilon}\right) + \exp\left(\frac{B}{\epsilon}\right)\right) = \max\{A, B\} \tag{B.38}$$

to the equation. Consider as an example the following ultra-discrete limit by Grammaticos et al. [31]. By writing the discrete Painlevé I equation

$$x_{n+1}x_{n-1} = \frac{\alpha\lambda^n}{x_n} + \frac{1}{x_n^2}, \tag{B.39}$$

where α and $\lambda > 0$ are constant parameters, in the form

$$x_{n+1}x_n^2 x_{n-1} = \alpha\lambda^n x_n + 1,$$

and defining [31]

$$x_n = \exp\left(\frac{X_n}{\epsilon}\right), \qquad n \in \mathbb{Z},$$
$$\lambda = \exp\left(\frac{1}{\epsilon}\right), \qquad \alpha = \exp\left(\frac{a}{\epsilon}\right), \tag{B.40}$$

it follows that

$$X_{n+1} + 2X_n + X_{n-1} = \epsilon \log\left(\exp\left(\frac{n + a + X_n}{\epsilon}\right) + \exp(0)\right). \tag{B.41}$$

By applying formula (B.38) to the right-hand side of (B.41), we obtain

$$X_{n+1} + 2X_n + X_{n-1} = \max\{X_n + n + a, 0\}, \tag{B.42}$$

which is a known ultra-discrete form of Painlevé I equation.

Note that making transformations such as (B.40) is only possible for real variables and parameters if they are all positive. Therefore the process of ultra-discretization is only applicable for multiplicative discrete Painlevé equations, such as (B.39). For this type of equations the solution sequences remain positive by choosing positive initial conditions and restricting all parameters values to be positive. Several examples of discrete Painlevé equations for which the ultra-discretization is possible are listed in [31], and these equations are given by

$$x_{n+1}x_{n-1} = \alpha\lambda^n + \frac{1}{x_n},$$

$$x_{n+1}x_{n-1} = \alpha\lambda^n x_n + 1,$$

$$x_{n+1}x_{n-1} = \frac{\lambda^n(x_n + \alpha\lambda^n)}{x_n(1 + x_n)}$$

$$x_{n+1}x_{n-1} = \frac{x_n + \alpha\lambda^n}{1 + \beta x_n\lambda^n}$$

$$x_{n+1}x_{n-1} = \frac{(x_n + \alpha\lambda^n)(x_n + \beta\lambda^n)}{(1 + \gamma x_n\lambda^n)(1 + \delta x_n\lambda^n)}$$

where the parameter values of α, β, γ, δ and λ are positive to make sure that the ultra-discretization process can be performed. By taking the ultra-discrete limit these equations become

$$X_{n+1} + X_n + X_{n-1} = \max\{X_n + n + a, 0\},$$

$$X_{n+1} + X_{n-1} = \max\{X_n + n + a, 0\},$$

$$X_{n+1} + X_{n-1} = n + \max\{n + a - X_n, 0\} - \max\{X_n, 0\},$$

$$X_{n+1} - X_n + X_{n-1} = \max\{n + a - X_n, 0\} - \max\{X_n + n + b, 0\},$$

$$X_{n+1} - 2X_n + X_{n-1} = \max\{n + a - X_n, 0\} + \max\{n + b - X_n, 0\}$$
$$- \max\{X_n + n + c, 0\} - \max\{X_n + n + d, 0\},$$

respectively. The first work with ultra-discrete forms of all multiplicative type of Painlevé equations (corresponding to affine Weyl groups A_1 to E_8) has been recently published by Ormerod and Yamada [87].

It is justified to ask why (or when) ultra-discrete equations obtained by the method described above can be considered to be of Painlevé type? As mentioned in [31], one of the main integrability criteria for discrete and continuous Painlevé equations is the existence of a Lax pair. This is the case also for ultra-discrete equations. We will next follow the reasoning due to Joshi, Nijhoff and Ormerod [60] to give an example of a Lax pair of an ultra-discrete Painlevé equation. The discrete equation

$$x_{n+1}x_{n-1} = \frac{\alpha\lambda^n + x_n^2}{1 + \alpha\lambda^n x_n^2}, \tag{B.43}$$

with α and λ being real constants such that $\alpha = (\lambda - 1)^2/\lambda$ and $x_n = x(\lambda^n)$, is one of the known discrete forms of Painlevé III equation. It can be expressed as the compatibility condition of

$$\begin{cases} \phi(\lambda x, k) = \begin{pmatrix} \frac{x_{n+1}}{x_n} & \frac{(\lambda-1)k^2 x}{x_n} \\ (\lambda-1)xx_{n+1} & 1 \end{pmatrix} \phi(x,k) \\ \phi(x, \lambda k) = \begin{pmatrix} \frac{x_{n+1}}{x_n} + \alpha x x_n x_{n+1} & \frac{(\lambda-1)k^2 x}{x_n} + \left(1 - \frac{1}{\lambda}\right)x_n \\ \frac{(\lambda-1)}{\lambda^2} \cdot \frac{1}{k^2 x_n} + \left(1-\frac{1}{\lambda}\right)xx_{n+1} & \frac{1}{\lambda}\left(\frac{x_n}{x_{n+1}} + \frac{\alpha x}{x_n x_{n+1}}\right) \end{pmatrix} \phi(x,k) \end{cases}$$

$$\text{(B.44)}$$

where $k = k_0 \lambda^m$ and $x = x_0 \lambda^m$. The ultra-discretization of (B.43) results in

$$X_{n+1} + X_{n-1} = \max\{2X_n, A + nQ\} - \max\{2X_n + A + nQ, 0\}, \quad \text{(B.45)}$$

where the ultra-discrete variables A, Q and X_n are defined by $\alpha = \exp(A/\epsilon)$, $\lambda = \exp(Q/\epsilon)$ and $x_n = \exp(X_n/\epsilon)$. Equation (B.45) follows from (B.43) by taking $\epsilon \to +0$.

Ultra-discretizing the Lax pair of (B.43) yields an ultra-discrete Lax pair of (B.45). This can be seen by denoting

$$\phi(x,k) = \phi(\lambda^n, \lambda^m) = \begin{pmatrix} u_n^m \\ v_n^m \end{pmatrix},$$

and, after some algebraic manipulation, writing the compatibility condition (B.44) in a simplified form as

$$\begin{cases} \phi(\lambda x, k) = \begin{pmatrix} \frac{x_{n+1}}{x_n} & \frac{(\alpha\lambda)^{1/2}k^2 x}{x_n} \\ (\alpha\lambda)^{1/2}xx_{n+1} & 1 \end{pmatrix} \phi(x,k) \\ \phi(x, \lambda k) = \begin{pmatrix} \frac{1}{(\alpha\lambda)^{1/2}\lambda^2 k^2 x_{n+1}} & \frac{\left(\frac{\alpha}{\lambda}\right)^{1/2} x_n}{\lambda x_{n+1}} \end{pmatrix} \phi(x,k) \end{cases} . \qquad \text{(B.46)}$$

By applying the ultra-discretization procedure with

$$\alpha = \exp(A/\epsilon), \quad \lambda = \exp(Q/\epsilon), \quad x_n = \exp(X_n/\epsilon),$$
$$u_n^m = \exp(U_n^m/\epsilon), \quad v_n^m = \exp(V_n^m/\epsilon)$$

to (B.46) it follows that

$$\begin{cases} U_{n+1}^m = \max\left\{U_n^m + X_{n+1} - X_n, V_n^m + \frac{A}{2} + \left(2m + n + \frac{1}{2}\right)Q - X_n\right\} \\ V_{n+1}^m = \max\left\{U_n^m + \frac{A}{2} + \left(n + \frac{1}{2}\right)Q + X_{n+1}, V_n^m\right\} \\ U_n^{m+1} = \max\left\{U_{n+1}^m, V_n^m + \frac{A}{2} - \frac{Q}{2} + X_n\right\} \\ V_n^{m+1} = \max\left\{U_{n+1}^m + \frac{A}{2} - \left(2m + \frac{3}{2}\right)Q - X_{n+1}, V_{n+1}^m + X_n - X_{n+1} - Q\right\} \end{cases} .$$

$$\text{(B.47)}$$

This may be written, using the definition of the standard ultra-discrete multiplication, as

$$\begin{cases} \phi((n+1)Q, mQ) = L(nQ, mQ) \otimes \phi(nQ, mQ) \\ \phi(nQ, (m+1)Q) = M(nQ, mQ) \otimes \phi((n+1)Q, mQ) \end{cases}, \qquad (B.48)$$

where

$$\begin{cases} L(nQ, mQ) = \begin{pmatrix} X_{n+1} - X_n & \frac{A}{2} + (2m + n + \frac{1}{2})Q - X_n \\ \frac{A}{2} + (n + \frac{1}{2})Q + X_{n+1} & 0 \end{pmatrix} \\ M(nQ, mQ) = \begin{pmatrix} 0 & \frac{A}{2} - \frac{Q}{2} + X_n \\ \frac{A}{2} - (2m + \frac{3}{2})Q - X_{n+1} & X_n - X_{n+1} - Q \end{pmatrix} \end{cases}.$$

$$(B.49)$$

It can now be verified that equation (B.45) can indeed be expressed as a compatibility condition of (B.48), and so (B.48) forms its Lax pair. This, as in [60], can be done by writing the system (B.48) as

$$L(nQ, (m+1)Q) \otimes M(nQ, mQ) = M((n+1)Q, mQ) \otimes L((n+1)Q, mQ), \qquad (B.50)$$

which is a two-by-two tropical matrix equation. By comparing the entries on the left and the right had sides of these tropical matrixes on row 1 and column 1, it follows that

$$X_{n+2} + X_n = \max\{2X_{n+1}, A + (n+1)Q\} - \max\{A + (n+1)Q + 2X_{n+1}, 0\}$$

after some algebraic manipulation. This equation is exactly (B.45) where the index n has been replaced by $n + 1$.

All the other entries of the tropical matrix equation (B.50) yield the same equation, or no further information. Namely, by looking at the entries on the second column of the second row, it follows that

$$\max\{A + nQ + X_{n+1} + X_n, X_n - X_{n+1} - Q\}$$
$$= \max\{A + nQ - X_{n+2} - X_{n+1}, X_{n+1} - X_{n+2} - Q\},$$

which is again equivalent to (B.45). The second row first column and first row second column entries lead to vacuous equations where the left and the right-hand sides are trivially the same.

B.6   Ultra-discrete Painlevé equations and integrability testing

In the previous sections we have seen that finite-order meromorphic solutions play an important role in testing for differential and difference

Painlevé equations. Is there an ultra-discrete analogue of this phenomenon that can be applied to search for new ultra-discrete equations of Painlevé type? Grammaticos et al. [31] suggest to apply an analogous low growth/complexity criterion which was considered by Arnol'd [5] and Veselov [107] in the case of discrete mappings. Grammaticos et al. considered the class

$$X_{n+1} + \sigma X_n + X_{n-1} = \max\{X_n + \phi_n, 0\}, \qquad (\text{B.51})$$

where σ is a constant and ϕ_n is an arbitrary function of n, and showed that the only integer values of σ which may prevent solutions of (B.51) growing exponentially towards $\pm\infty$ are $\sigma = -1, 0, 1, 2$. The restriction to integer values is motivated by the fact that ultra-discrete equations with integer parameter values and integer initial conditions describe generalizations of ultra-discrete cellular automata [105]. The values $-1, 0, 1, 2$ of σ correspond to known ultra-discrete Painlevé equations within the class (B.51). However, the method in [31] does not give information on the exact form of the function ϕ_n: Equations within the class (B.51) are of Painlevé type only for some special choices of ϕ_n even in the case where $\sigma = -1, 0, 1, 2$.

Joshi and Lafortune [59] have proposed an ultra-discrete analogue of the singularity confinement method, which can pin down the coefficient ϕ_n in (B.51). A difficulty in finding such an analogue is that solutions of ultra-discrete equations do not have singularities in the classical sense for finite index values. Joshi and Lafortune overcame this problem by considering differentiability of solutions at certain critical situations. Consider, as an example, the discrete equation

$$x_{n+1}x_{n-1} = \frac{1}{x_n} + k, \qquad (\text{B.52})$$

where k is a constant parameter. Equation (B.52) is actually an autonomous equation and it is a special case of the QRT-class [91] known to be integrable. By ultra-discretizing (B.52), we obtain

$$X_{n+1} + X_n + X_{n-1} = \max\{X_n + K, 0\}, \qquad (\text{B.53})$$

where K is a positive constant. The fundamental idea in the singularity confinement method for discrete equations is that one considers a nonautonomous class of equations, and obtains, by demanding that singularity confinement property holds for the equation in question, a restriction for the coefficients. For instance, looking at

$$x_{n+1}x_{n-1} = \frac{1}{x_n} + \psi_n, \qquad (\text{B.54})$$

where ψ_n is a function of n, and running the singularity confinement argument, yields

$$x_{n+1}x_{n-1} = \frac{1}{x_n} + k\lambda^n,$$

where k and $\lambda > 0$ are constants [30]. In the ultra-discrete singularity confinement by Joshi and Lafortune the basic idea is to look at the behavior of iterates under the initial conditions such as

$$X_n = -K + \varepsilon, \tag{B.55}$$

where ε is assumed to be a non-zero constant such that $|\varepsilon|$ is small, and to compare the cases where ε is positive and negative. If we substitute (B.55) into (B.53) and, in addition, we assume that

$$X_{n-1} > 2|K|, \tag{B.56}$$

it follows that

$$X_{n+1} = K - X_{n-1} - \varepsilon + \max\{\varepsilon, 0\}. \tag{B.57}$$

Therefore the exact form of X_{n+1} now depends on the sign of ε, since we are working under the assumption (B.56). If $\varepsilon > 0$, we have

$$X_{n+1} = K - X_{n-1},$$

while if $\varepsilon < 0$, it follows that

$$X_{n+1} = K - X_{n-1} - \varepsilon.$$

Behavior of the iterate in these two cases is different, and by using the terminology of [59], it follows that X_{n+1} is not differentiable at $X_n = -K$. This corresponds to a singularity in the usual discrete case. By continuing the iteration further, we have under the assumption that $\varepsilon > 0$,

$$\begin{aligned}
X_n &= -K + \varepsilon \\
X_{n+1} &= K - X_{n-1} \\
X_{n+2} &= X_{n-1} - \varepsilon \\
X_{n+3} &= X_{n-1} \\
X_{n+4} &= K - X_{n-1} + \varepsilon \\
X_{n+5} &= -K - \varepsilon \\
X_{n+6} &= X_{n-1}.
\end{aligned} \tag{B.58}$$

Similarly, by assuming that $\varepsilon < 0$, we have

$$X_n = -K + \varepsilon$$
$$X_{n+1} = K - X_{n-1} - \varepsilon$$
$$X_{n+2} = X_{n-1}$$
$$X_{n+3} = X_{n-1} + \varepsilon \qquad \text{(B.59)}$$
$$X_{n+4} = K - X_{n-1}$$
$$X_{n+5} = -K - \varepsilon$$
$$X_{n+6} = X_{n-1}.$$

By comparing the iteration sequences (B.58) and (B.59) we can see that the iterates X_{n+1}, X_{n+2}, X_{n+3} and X_{n+4} are 'incompatible', or non-differentiable, under the initial conditions (B.55) and (B.56), but the iterates X_{n+5} and X_{n+6} are again 'compatible', or differentiable. This corresponds to a singularity being confined in the usual singularity confinement method. Now, by looking at a non-autonomous generalization

$$X_{n+1} + X_n + X_{n-1} = \max\{X_n + \phi_n, 0\}$$

of (B.53), where ϕ_n is an arbitrary function of n, and demanding that

$$X_{n-1} > \max\{\phi_n + \phi_{n+1}, -\phi_{n+2}, -\phi_{n+3} - \phi_{n+2} + \phi_n, \phi_{n+3} + \phi_{n+4}\} \quad \text{(B.60)}$$

instead of (B.56), we obtain

$$X_n = -\phi_n + \varepsilon$$
$$X_{n+1} = \phi_n - X_{n-1}$$
$$X_{n+2} = X_{n-1} - \varepsilon$$
$$X_{n+3} = X_{n-1} - \phi_n + \phi_{n+2}$$
$$X_{n+4} = \phi_{n+3} - X_{n-1} + \varepsilon$$
$$X_{n+5} = -\phi_{n+3} - \phi_{n+2} + \phi_n - \varepsilon$$

when $\varepsilon > 0$, and

$$X_n = -\phi_n + \varepsilon$$
$$X_{n+1} = \phi_n - X_{n-1} - \varepsilon$$
$$X_{n+2} = X_{n-1}$$
$$X_{n+3} = X_{n-1} - \phi_n + \phi_{n+2} + \varepsilon$$
$$X_{n+4} = \phi_{n+3} - X_{n-1}$$
$$X_{n+5} = -\phi_{n+3} - \phi_{n+2} + \phi_n - \varepsilon$$

if $\varepsilon < 0$. Now again the iterate X_{n+5} is compatible with the change of sign of ε, but X_{n+6} is differentiable only if ϕ_n satisfies

$$\phi_{n+5} - \phi_{n+3} - \phi_{n+2} + \phi_n = 0$$

for all n. This equation can be solved to obtain

$$\phi_n = \alpha + \beta n + \gamma(-1)^n + \delta \cos\left(\frac{2\pi n}{3}\right) + \omega \sin\left(\frac{2\pi n}{3}\right),$$

where α, β, γ, δ and ω are arbitrary constants. This corresponds to a known form of an ultra-discrete Painlevé equation [59].

We have seen that finite-order meromorphic solutions have a special role both in the case of continuous Painlevé and discrete Painlevé equations. In the continuous case the solutions appear to be of finite order, at least in a broad sense, while in the discrete case the existence of finite-order meromorphic solutions is a good complex analytic candidate for the discrete Painlevé property. With this background in mind, Halburd and Southall [46] suggested that the existence of sufficiently many piecewise linear solutions with integer slopes, or tropical meromorphic solutions, of finite order of ultra-discrete equations should be considered as an ultra-discrete analogue of the Painlevé property. We consider the first-order case in the following example, which is analogous to Examples B.1 and B.2.

Example B.5. If

$$y(x+1) = P(x, y(x)), \tag{B.61}$$

where P is tropical difference polynomial, has an admissible tropical entire solution of finite order, then

$$\deg_y(P(x, y(x))) = 1.$$

In the proofs of Examples B.1 and B.2, the key idea is that Nevanlinna characteristic is applied to both sides of the first-order equation, and then appropriate estimates on the characteristic yield the result. An identity due to Valiron and Mohon'ko play a key role in these proofs, and a similar estimate would certainly be very useful here as well. Unfortunately there is not so far an ultra-discrete version of the Valiron-Mohon'ko identity, which would extend to rational functions as well. However, there is an analogue due to Laine and Yang [71], which is applicable for tropical difference polynomials of $y(x)$, as opposed to just polynomials of $y(x)$. Hence, in the polynomial case (B.61) with entire solutions the ultra-discrete analogue is actually more general than its differential and difference counterparts. However, a rational extension of (B.61) remain as an open problem, as well as the case where solutions are tropical meromorphic.

The following proof of Example B.5 is actually a special case of Theorem 7.15, but we include it here to illustrate the connection to Examples B.1 and B.2.

Proof. By applying the tropical proximity function to both sides of (B.61), it follows that

$$m(r, y(x+1)) = m(r, P(x, y(x))). \tag{B.62}$$

But now, by the tropical analogue of the lemma on the logarithmic derivatives, Theorem 3.27, it follows that

$$m(r, y(x+1)) \leq m(r, y(x)) + m\left(r, \frac{y(x+1)}{y(x)} \oslash\right) = m(r, y(x)) + \widetilde{S}(r, y),$$

and similarly

$$m(r, y(x)) \leq m(r, y(x+1)) + \widetilde{S}(r, y),$$

and thus

$$m(r, y(x+1)) = m(r, y(x)) + \widetilde{S}(r, y). \tag{B.63}$$

On the other hand, by Theorem 4.3, we have

$$m(r, P(x, y(x))) = \deg_y(P(x, y(x)))m(r, y(x)) + \widetilde{S}(r, y). \tag{B.64}$$

Since y is tropical entire, we have $m(r, y(x)) \equiv T(r, y(x))$. Hence the assertion follows by combining (B.62), (B.63) and (B.64). $\square$

In the case of difference Painlevé equations, a type of complex analytic interpretation of singularity confinement appears in the proofs of the main results [43]. This 'complex confinement' tells the number of poles of solutions compared to certain special values of solutions, which depend on the form of the equation. If the pole density is not in balance with the densities of these special values, then all admissible meromorphic solutions are of infinite order. This raises the question of whether there is a similar connection in the ultra-discrete case? The method of Halburd and Southall has not so far been applied to obtain precise information on the coefficients of the ultra-discrete equations. On the other hand, Joshi's and Lafortune's ultra-discrete singularity confinement works under certain conditions on the coefficients, such as (B.60). Perhaps by a suitable interpretation of the ultra-discrete singularity confinement for the case of tropical meromorphic solutions, one can obtain a broad classification result, such as an ultra-discrete analogue of Theorem B.3, thus capturing new ultra-discrete equations of Painlevé type.

B.7   Hypergeometric solutions to Painlevé equations

One of the key properties common to most (but not all) continuous, discrete and ultra-discrete Painlevé equations is the existence of special solutions. In their paper [64] as well as [63], K. Kajiwara, T. Masuda, M. Noumi, Y. Ohta and Y. Yamada presented a procedure to construct hypergeometric solutions through the linearization of discrete Riccati equations and then gave the list of q-Painlevé equations and their hypergeometric solutions, together with the data necessary for their construction along the procedure. The detailed procedure is as follows:

Step 0: Assume that we have a q-difference Riccati equation

$$w(qz) = \frac{A(z)w(z) + B(z)}{C(z)w(z) + D(z)}, \qquad (B.65)$$

where the coefficients $A(z)$, $B(z)$, $C(z)$ and $D(z)$ are functions of z.

Step 1: Put an ansatz

$$w(z) = \frac{F(z)}{G(z)} \qquad (B.66)$$

and linearize the equation (B.65) into

$$\begin{pmatrix} F(qz)/H(z) \\ G(qz)/H(z) \end{pmatrix} = \begin{pmatrix} A(z) & B(z) \\ C(z) & D(z) \end{pmatrix} \begin{pmatrix} F(z) \\ G(z) \end{pmatrix} \qquad (B.67)$$

where $H(z)$ is an arbitrary decoupling factor.

Step 2: Under the assumption that $B(z) \not\equiv 0$, eliminate $G(z)$ from (B.67) and obtain the three-term relation for F:

$$F(qz) + c_1(z)F(z) + c_2(z)F(z/q) = 0, \qquad (B.68)$$

with

$$c_1(z) = -\frac{H(z)}{B(z/q)}\big(A(z)B(z/q) + B(z)D(z/q)\big),$$

$$c_2(z) = \frac{B(z)}{B(z/q)}H(z)H(z/q)\big(A(z/q)D(z/q) - B(z/q)D(z/q)\big).$$

Step 3: Recall that the three-term relation for a basic hypergeometric series often takes the form

$$V_1(z)\big(\Phi(qz) - \Phi(z)\big) + V_2(z)\Phi(z) + V_3(z)\big(\Phi(z/q) - \Phi(z)\big) = 0, \qquad (B.69)$$

where the coefficients $V_1(z)$, $V_2(z)$ and $V_3(z)$ are factorized into binomials involving the independent variable z and some parameters.

Step 4: Compare (B.68) and (B.69) and have
$$\frac{V_2(z)}{V_1(z)} = 1 + c_1(z) + c_2(z), \quad \frac{V_3(z)}{V_1(z)} = c_2(z). \tag{B.70}$$

Step 5: Look for the decoupling factor $H(z)$ so that these quantities factorize *by trial and error with computer algebra*.

Step 6: Identify the three-terms relation with that for appropriate hypergeometric function which appears for q-Painlevé equations, for example.

Step 7: Repeat the steps 2 to 5 for $G(z)$ instead of $F(z)$, under the assumption that $C(z) \not\equiv 0$.

For example for the case of $(\mathcal{A}_1 + \mathcal{A}_1')^{(1)}$ (see [64, Section 3, 5.1, p.1461]), a q-Painlevé equation
$$\big(w(qz)w(z) - 1\big)\big(w(z)w(z/q) - 1\big) = \frac{\alpha z^2 w(z)}{w(z) + z}$$

with the parameter $\alpha = q$ is considered, and the hypergeometric solution
$$w(z) = \frac{\Phi(qz)}{\Phi(z)}, \quad \Phi(z) = {}_1\phi_1\left(\begin{array}{c} 0 \\ -q \end{array}; q, -qz\right)$$

is deduced. As the data for the procedure, Kajiwara et. al recorded (1) q-difference Riccati equation, (2) three-term relation and (3) identification as follows:

(1) $\quad w(qz) = \dfrac{1}{w(z)} - qz,$

(2) $\quad \Phi(qz) + z\Phi(z) = \Phi(z/q),$

(3) $\quad w(z) = \dfrac{F(z)}{G(z)}, \quad F(z) = \Phi(qz), \quad G(z) = \Phi(z).$

In this case there is no need to introduce decoupling and other factors at all. This type of the q-Painlevé equation, indeed a q-difference analogue of the Painlevé II equation (see [92]), is also studied by T. Hamamoto, K. Kajiwara and N. S. Witte [47, Lemma 2.1], for example. As the general solution of the above three-term relation for $\Phi(z)$ with the same parameter, they give
$$\Phi(z) = A(z){}_1\phi_1\left(\begin{array}{c} 0 \\ -q \end{array}; q, -qz\right) + B(z)e^{\pi i(\log z)/(\log q)}{}_1\phi_1\left(\begin{array}{c} 0 \\ -q \end{array}; q, qz\right),$$

where $A(z)$ and $B(z)$ are arbitrary q-periodic functions. In fact the entire function
$$\mathrm{Ai}_q(z) := {}_1\phi_1\left(\begin{array}{c} 0 \\ -q \end{array}; q, -z\right) = \sum_{k=0}^{\infty} \frac{q^{k(k-1)/2}}{(-q;q)_k (q;q)_k} z^k$$

is known as a q-Airy function, since this solves the q-Airy equation

$$u(q^2 z) + zu(qz) - u(z) = 0$$

with $u(0) = 1$. Also the entire function $\,_1\phi_1\left(\begin{matrix} 0 \\ -q \end{matrix}; q,\, z\right)$ satisfies the q-difference equation

$$v(q^2 z) - zv(z) - v(z) = 0.$$

This equation follows when we put $u(z) = e^{\pi i (\log z)/(\log q)} v(z)$ in the q-Airy equation for the purpose to deduce a solution $u(z)$ satisfying the initial conditions $u(0) = 0$ and $u'(0) = 1$. But if we restrict our interest into (single-valued) entire solutions on $\mathbb{C}$, both A and B reduce to constants and $B = 0$ in addition, since $e^{\pi i (\log z)/(\log q)} = (-1)^{\log_q z}$ is not accepted as our desired function then.

In his survey article [85, Table 2, p. 54], Noumi gives a list of the Weyl group symmetries which characterize the q-Painlevé equations including $(A_1 + A_1')^{(1)}$, relevant q-hypergeometric functions with the terminating cases. According to the list, $\,_1\phi_1\left(\begin{matrix} 0 \\ -q \end{matrix}; q,\, z\right)$ is the hypergeometric function relevant to $(A_1 + A_1')^{(1)}$. Hypergeometric functions relevant to q-P_{II} and q-P_V are

$$\,_1\phi_1\left(\begin{matrix} 0 \\ -q \end{matrix}; q,\, -qz\right) \quad \text{and} \quad \,_2\phi_1\left(\begin{matrix} a,\ b \\ 0 \end{matrix}; q,\, z\right),$$

respectively.

Then one might ask whether the ultra-discrete limit procedure or a direct translation from 'q-difference' to 'ultra-discrete' is available to deduce a special max-plus transcendental solution to ultra-discrete analogue u-P_{II} and u-P_V to the above q-P_{II} and q-P_V. In other words, is it possible to give a correct series expansion to each of

$$\,_1\Phi_1\left(\begin{matrix} 0_\circ \\ Q \end{matrix}; -Q,\, Q \otimes x\right) \quad \text{and} \quad \,_2\Phi_1\left(\begin{matrix} A,\ B \\ 0_\circ \end{matrix}; -Q,\, x\right),$$

for example? Let us postpone considering the latter to the later section and concentrate on the former, that is, to find an ultra-discrete analogue of $\mathrm{Ai}_q(qz)$. As mentioned, we know the series expansion

$$\mathrm{Ai}_q(qz) = \,_1\phi_1\left(\begin{matrix} 0 \\ -q \end{matrix}; q,\, -qz\right) = \sum_{k=0}^{\infty} \frac{q^{k(k+1)/2}}{(-q; q)_k (q; q)_k} z^k.$$

Let us follow the procedure to deduce this expression. Now we consider an ultra-discrete analogue of the q-Airy equation, or more concretely that of the above three term relation (2),

$$\Psi\big((x \otimes (-Q))\big) \oplus x \otimes \Phi(x) = \Psi\big(x \oslash (-Q)\big),$$

i.e. with x replaced by $x + Q$,

$$\max\Big\{\Psi(x), \Psi(x + Q) + x + Q\Big\} = \Psi(x + 2Q) \qquad (B.71)$$

for $Q = -\log|q|$, and further, as its solution, let us observe a max-plus series

$$\Psi(x) = \bigoplus_{k=0}^{\infty} a_k \otimes x^{\otimes k} = \max_{k \in \mathbb{Z}_{\geq 0}} \{kx + a_k\}$$

for $a_k \in \mathbb{R}_\infty$ ($k \in \mathbb{Z}_{\geq 0}$). Then we obtain

$$\Psi(x + Q) + x + Q = \max_{k \in \mathbb{Z}_{\geq 0}} \big\{(k + 1)x + (k + 1)Q + a_k\big\}$$

$$= \max_{k \in \mathbb{N}} \big\{kx + kQ + a_{k-1}\big\},$$

$$\Psi(x + 2Q) = \max_{k \in \mathbb{Z}_{\geq 0}} \{kx + 2kQ + a_k\},$$

and therefore putting $a_{-1} = 0_\circ$, we rewrite (B.71) into the identity

$$\max_{k \in \mathbb{Z}_{\geq 0}} \big\{kx + \max(a_k, kQ + a_{k-1})\big\} = \max_{k \in \mathbb{Z}_{\geq 0}} \{kx + 2kQ + a_k\},$$

which holds on the $\mathbb{R}$ entirely. Hence we can compare the coefficients and deduce that

$$\max\{a_k, kQ + a_{k-1}\} = 2kQ + a_k \qquad (B.72)$$

is true for each $k \in \mathbb{Z}_{\geq 0}$. If $a_k \geq kQ + a_{k-1}$ for some $k \in \mathbb{Z}_{\geq 0}$, this identity (B.72) implies $a_k = 2kQ + a_k$ so that $k = 0$. Hence we have always

$$kQ + a_{k-1} = 2kQ + a_k, \quad k \in \mathbb{N},$$

and therefore

$$a_k = \sum_{j=1}^{k} (-jQ) + a_0 = -\frac{k(k + 1)}{2}Q + a_0 = a_0 \oslash [Q; Q]_k.$$

The solution $\Psi(x)$ to the 'u-Airy equation' (B.71) with $\Psi(0) = a_0 = 1_\circ$ has the max-plus series expansion

$$\Psi(x) = \bigoplus_{k=0}^{\infty} x^{\otimes k} \oslash [Q; Q]_k.$$

Now we wonder if

$$_1\Phi_1\left(\begin{matrix}0_\circ\\Q\end{matrix}; -Q, Q\otimes x\right)$$

was correct or not. Let us return to the series expansion of

$$\mathrm{Ai}_q(qz) = \sum_{k=0}^{\infty}\frac{q^{k(k+1)/2}}{(-q;q)_k(q;q)_k}z^k,$$

which can be written also by

$$\mathrm{Ai}_q(qz) = \sum_{k=0}^{\infty}\frac{(1/q)^{k(k+1)/2}}{(-1/q;1/q)_k(1/q;1/q)_k}(-z)^k = {}_1\phi_1\left(\begin{matrix}0\\-1/q\end{matrix}; 1/q, -z\right).$$

The ultra-discrete translation of this is now

$$_1\Phi_1\left(\begin{matrix}0_\circ\\-1_\circ\oslash(-Q)\end{matrix}; 1_\circ\oslash(-Q), x\right) = {}_1\Phi_1\left(\begin{matrix}0_\circ\\-Q\end{matrix}; Q, x\right)$$

$$:= \bigoplus_{k=0}^{\infty}\frac{[0_\circ;Q]_k\otimes x^{\otimes k}}{[-Q;Q]_k\otimes[Q;Q]_k}\oslash$$

$$= \bigoplus_{k=0}^{\infty}\frac{x^{\otimes k}}{1_\circ\otimes[Q;Q]_k}\oslash$$

which coincides with the representation for the $\Psi(x)$. Then this could give a special solution by means of the ultra-discrete hypergeometric function to the ultra-discrete analogue of $(\mathcal{A}_1+\mathcal{A}_1')^{(1)}$ or its modified equation.

Another question might arise about an elliptic solution to discrete Painlevé equations. It is known that d-P_3 has an elliptic solution and q-P_{III} has a basic hypergeometric solution, and also that u-P_{III} has an ultra-discrete hypergeometric solution. Is there any hierarchy relation among these solutions? We have learned that Nobe gave an ultra-discretization of Jacobi's elliptic functions by using an ultra-discretized elliptic theta function, and Ormerod gave an ultra-discretized hypergeometric solution to u-P_{III} from the q-analogue of a hypergeometric solution to q-P_{III}. Do they give any information on the above question?

These questions seem to be waiting for a systematic study.

Appendix C

Some operators in complex analysis

In this appendix we observe several similarities among the properties of the q-difference operator for entire functions on $\mathbb{C}$ and those of the ultra-discrete operator for tropical entire functions on $\mathbb{R}$.

C.1   Logarithmic order and type

In Chapter 2, we studied how to measure the growth of a tropical entire function by means of the coefficients of its max-plus series expansion. This was given as a complete translation of the results for entire functions on the complex plane. In fact, it is well-known in classical function theory that the order $\rho(f)$ of a transcendental entire function

$$f(z) = \sum_{n=0}^{\infty} a_{m_n} z^{m_n}, \quad a_{m_n} \in \mathbb{C} \setminus \{0\}, \tag{C.1}$$

is related with its Taylor expansion coefficients via the equality

$$\limsup_{r \to \infty} \frac{\log \log M(r, f)}{\log r} = \limsup_{n \to \infty} \frac{\log m_n}{\log \sqrt[m_n]{1/|a_{m_n}|}}, \tag{C.2}$$

where $M(r, f)$ is the maximum modulus function of f. Moreover, if the order $\rho = \rho(f)$ is positive and finite, then also the type $\tau(f) := \limsup_{r \to \infty} \frac{\log M(r, f)}{r^\rho}$ is related with the Taylor coefficients a_{m_n} as follows:

$$\limsup_{r \to \infty} \frac{\log M(r, f)}{r^\rho} = \frac{1}{e\rho} \limsup_{n \to \infty} \frac{m_n}{\left(\sqrt[m_n]{1/|a_{m_n}|} \right)^\rho}. \tag{C.3}$$

As is well-known, $f(x)$ is of minimal, mean, or maximal type, if the value is zero, positive and finite, or infinite, respectively.

At this point, we call attention to a relation between the two equalities (C.2) and (C.3) such that

the ratio $\dfrac{\log Y}{\log X}$ in (C.2) is replaced by the ratio $\dfrac{Y}{X^\rho}$ in (C.3)

on each side, that is to put

$$(X, Y) = \big(r,\ \log M(r, f)\big)$$

on the left-hand side and

$$(X, Y) = \big(\sqrt[m_n]{1/|a_{m_n}|},\ \log m_n\big)$$

on the right-hand side respectively. Here it is required, however, to associate a weight multiplier $1/(e\rho)$ with the ratio on the right-hand side of (C.3).

A similar situation also holds for the logarithmic order or q-order and the logarithmic type or q-type of entire functions. These are defined by replacing r by $\log r$ in the denominator of the left-hand sides of (C.2) and (C.3), respectively, that is to define

$$\limsup_{r \to \infty} \frac{\log\log M(r, f)}{\log\log r} =: \rho_{\log} \quad \text{and} \quad \limsup_{n \to \infty} \frac{\log M(r, f)}{(\log r)^{\rho_{\log}}} =: \lambda_{\log},$$

provided $1 < \rho_{\log} < +\infty$. This means that $\log T(r, f)$ behaves asymptotically like

$$\big\{\lambda_{\log} + o(1)\big\}(\log r)^{\rho_{\log}}$$

on suitable intervals over $\mathbb{R}_{>0}$. In fact, the following two relations on the logarithmic order and type, in terms of Taylor coefficients, are obtained by Juneja, Kapoor and Bajpai [61, 62]. See also the paper [9] by Berg and Pedersen with an appendix due to Hayman:

$$\limsup_{r \to \infty} \frac{\log\log M(r, f)}{\log\log r} = 1 + \limsup_{n \to \infty} \frac{\log m_n}{\log\log \sqrt[m_n]{1/|a_{m_n}|}} \tag{C.4}$$

and

$$\limsup_{r \to \infty} \frac{\log M(r, f)}{(\log r)^{\rho_{\log}}} = \frac{(\rho_{\log} - 1)^{\rho_{\log} - 1}}{(\rho_{\log})^{\rho_{\log}}} \limsup_{n \to \infty} \frac{m_n}{\left(\log \sqrt[m_n]{1/|a_{m_n}|}\right)^{\rho_{\log} - 1}}, \tag{C.5}$$

provided $1 < \rho_{\log} < +\infty$.

Studying (C.4) and (C.5) in detail admits a modification as follows:

Theorem C.1. *Let $f(z)$ be an entire function with (C.1). Then we have*

$$\limsup_{r \to \infty} \frac{\log\log M(r, f)}{\log\log r} = \limsup_{n \to \infty} \frac{\log\log(1/|a_{m_n}|)}{\log\log \sqrt[m_n]{1/|a_{m_n}|}} \tag{C.6}$$

and

$$\limsup_{r \to \infty} \frac{\log M(r,f)}{(\log r)^{\rho_{\log}}} = \frac{1}{\left(1 + \frac{1}{\rho_{\log}-1}\right)^{\rho_{\log}-1} \cdot \rho_{\log}} \limsup_{n \to \infty} \frac{\log\left(1/|a_{m_n}|\right)}{\left(\log \sqrt[m_n]{1/|a_{m_n}|}\right)^{\rho_{\log}}}$$

(C.7)

provided $1 < \rho_{\log} < +\infty$.

As for the proof of this theorem, see [104, p. 119–125]. The proof of the identities (C.6) and (C.7) for $f(z)$ given by (C.1) on $\mathbb{C}$ is completely parallel to the proof of the similar relations on a tropical entire function on $\mathbb{R}$ with max-plus series expansion.

Observe that we here have a scaling relation again with

$$(X, Y) := \left(\log r, \ \log M(r, f)\right)$$

on the left, while

$$(X, Y) := \left(\log \sqrt[m_n]{1/|a_{m_n}|}, \ \log\left(1/|a_{m_n}|\right)\right)$$

with the logarithmic order $\rho_{\log}$ instead of the order ρ and with the multiplier $e(\rho_{\log}) \cdot \rho_{\log}$ $(1 < \rho_{\log} < +\infty)$. Here we have

$$e(t) := \left(1 + \frac{1}{t-1}\right)^{t-1}, \quad 1 < t < +\infty,$$

which tends to Euler's constant e as $t \to 1+0$ and as $t \to +\infty$, respectively. Therefore we can consider the relation (C.7) to be also valid, even if $\rho_{\log} = 1$ and $f(z)$ is not a polynomial. In fact, $f(z)$ is a transcendental entire function when and only when

$$\lim_{r \to \infty} \frac{\log M(r,f)}{\log r} = +\infty.$$

On the other hand,

$$\frac{\log(1/|a_{m_n}|)}{\log \sqrt[m_n]{1/|a_{m_n}|}} = m_n \to +\infty, \quad (n \to \infty),$$

when and only when the Taylor series $\sum_{n=0}^{\infty} a_{m_n} z^{m_n}$ of $f(z)$ is infinite.

Actually, the relations above motivate the reader to explore for some other possibilities. Indeed, a similar relation may be found for entire functions of several variables in $\mathbb{C}^n$ based on some results proved in the monograph [73] by P. Lelong and L. Gruman. In fact, they proved the following result in terms of the order ρ and type τ:

If $f(z) = \sum_{q=0}^{\infty} P_q(z)$ is the expression of an entire function in homogeneous polynomials and

$$C_q := \sup_{p(z) \leq 1} |P_q(z)|$$

with a norm $p(z)$, then the order ρ and for $\rho > 0$ the type σ of f with respect to $p(z)$ are given by

$$\rho := \limsup_{r \to \infty} \frac{\log M_{f,p}(r)}{\log r} = \limsup_{q \to \infty} \frac{q \log q}{-C_q} = \limsup_{q \to \infty} \frac{\log q}{\log\left(\sqrt[q]{1/C_q}\right)}$$

and

$$\tau := \limsup_{r \to \infty} \frac{M_{f,p}(r)}{r^{\rho}} = \frac{1}{e\rho} \limsup_{q \to \infty} \frac{q}{\left(\sqrt[q]{1/C_q}\right)^{\rho}},$$

respectively.

Of course, Lelong and Gruman introduce them as the order ρ and type τ for those of an entire function $f : \mathbb{C}^n \to \mathbb{C}$ with respect to a norm $p(z)$. Their proof of this result can be done similarly as in the case of $n = 1$. Now we can also define the logarithmic order and logarithmic type of the entire function f of several complex variables with respect to a norm $p(z)$ in the same way as we chose to do above and deduce the same relations among its order, type, logarithmic order and logarithmic type as we found for an entire function on $\mathbb{C}$ or a tropical entire function on $\mathbb{R}$.

C.2   Some operators and related series expansions in complex analysis

Let us start this section by making a list of correspondence between an entire function $f(z)$ on $\mathbb{C}$ and its ultra-discrete counterpart $f(x)$ on $\mathbb{R}$:

q-difference case	ultra-discrete case
base: $q \in \mathbb{C}$ $(0 < \|q\| < 1)$	*shift:* $Q := \log(1/\|q\|)$ (> 0)
$(1 - z)f(qz) = f(z),$ with $f(0) = 1$ $(z \in \mathbb{C})$	$(1_\circ \oplus x) \otimes f(x - Q) = f(x),$ with $f(0) = 1_\circ$ $(x \in \mathbb{R})$
$f(z) = \displaystyle\prod_{j=0}^{\infty}(1 - q^j z)$ $=: (z; q)_\infty$	$f(x) = \displaystyle\bigotimes_{j=0}^{\infty}\left(1_\circ \oplus (-Q)^{\otimes j} \otimes x\right)$ $=: \{x; Q\}_\infty$
$f(z) = \displaystyle\sum_{n=0}^{\infty} \frac{(z/q)^n}{\prod_{k=1}^{n}\{1 - (1/q)^k\}}$ $= \displaystyle\sum_{n=0}^{\infty} \frac{q^{n(n-1)/2} \times z^n}{\prod_{k=1}^{n}(q^k - 1)}$	$f(x) = \displaystyle\bigoplus_{n=0}^{\infty} \frac{(x \otimes Q)^{\otimes n}}{\bigotimes_{k=1}^{n}(1_\circ \oplus Q^{\otimes k})} \oslash$ $= \displaystyle\bigoplus_{n=0}^{\infty} \frac{(-Q)^{\otimes n(n-1)/2} \otimes x^{\otimes n}}{\bigotimes_{k=1}^{n}((-Q)^{\otimes k} \oplus 1_\circ)} \oslash$
$\log M(r, f)$ $= \dfrac{(\log r)^2 \times \{1 + o(1)\}}{\log(1/\|q\|)^2}$	$\log M(r, f)$ $= \dfrac{(\log r)^{\otimes 2} \otimes \{1_\circ \oplus o(1)\}}{\log Q^{\otimes 2}} \oslash$
roots: $q^{-j} = (1/q)^j \to \infty$ $(j \to \infty)$ *multiplicity:* simple for each j	*roots:* $jQ = Q^{\otimes j} \to \infty$ $(j \to \infty)$ *multiplicity:* $\omega_f(jQ) = 1$ for each j
$\displaystyle\limsup_{r \to \infty} \frac{\log \log M(r, f)}{\log \log r} = 2$	$\displaystyle\limsup_{r \to \infty} \frac{\log M(r, f)}{\log r} = 2$
$\displaystyle\limsup_{r \to \infty} \frac{\log M(r, f)}{(\log r)^2} = \frac{1}{2 \log(1/\|q\|)}$	$\displaystyle\limsup_{r \to \infty} \frac{M(r, f)}{r^2} = \frac{1}{2Q}$

Motivated by this list, we may think of a possible relation between a differential, difference, q-difference, ultra-discrete operator and corresponding series expansions, respectively.

Let us begin with the differential operator $d/dz = {}'$ and the Taylor series expansion of an analytic function

$$f(z) = \sum_{n=0}^{\infty} \frac{a_n}{n!} z^n, \quad a_n \in \mathbb{C}, \tag{C.8}$$

about the origin. Then we obtain $a_n = f^{(n)}(0) = \frac{d^n f}{dz^n}(0)$ for each n by termwise differentiation

$$\frac{d}{dz} f(z) = \sum_{n=0}^{\infty} \frac{a_{n+1}}{n!} z^n$$

and

$$\frac{d^n}{dz^n} f(z) = \sum_{k=n}^{\infty} \frac{a_k}{(k-n)!} z^{k-n} = \sum_{k=0}^{\infty} \frac{a_{k+n}}{k!} z^k.$$

Note that the differential equation $\frac{d}{dz} f(z) = f(z)$ is solved by $f(z) = ce^z$ ($c \in \mathbb{C}$) since $a_n = c$ for any $n \in \mathbb{Z}_{\geq 0}$.

Now, it is natural to ask whether a similar reasoning is true for other operators, that is, for difference operator, q-difference operator and ultra-discrete operators together with some series expansion corresponding to the Taylor series expansion above. Analogous ultra-discrete operator has already been considered in Chapter 7 above.

C.2.1   *The case of difference operator*

First, we consider the difference operator Δ defined by

$$\Delta f(z) = f(z+1) - f(z),$$

and the function $f(z)$ given formally by a binomial series (or factorial series) expansion about the origin, which is to be written in the form (C.9) below. In fact, following [54], and defining

$$z(0) = 1, \ z(1) = z, \ z(n) = z(z-1)\cdots(z-n+1) = n!\binom{z}{n} \quad (n \geq 2),$$

consider a function $f(z)$ given formally by

$$f(z) = \sum_{n=0}^{\infty} \frac{a_n}{n!} z(n), \quad a_n \in \mathbb{C}, \tag{C.9}$$

possibly near $z = 0$. Now we recall that for $n \geq 2$,

$$\Delta z(n) = nz(n-1), \quad \Delta z(0) \equiv 0 \quad \text{and} \quad \Delta z(1) \equiv 1$$

and we obtain $a_n = \Delta^n f(0)$ for each $n \in \mathbb{Z}_{\geq 0}$ under the assumption that we can operate Δ to the series (C.9) termwise, that is,

$$\Delta f(z) = \sum_{n=1}^{\infty} \frac{a_n}{n!} \Delta z(n) = \sum_{n=1}^{\infty} \frac{a_n}{(n-1)!} z(n-1) = \sum_{n=0}^{\infty} \frac{a_{n+1}}{n!} z(n).$$

In fact, we have

$$f(0) = a_0, \quad \text{and} \quad \Delta f(0) = a_1$$

and the general term $a_n = \Delta^n f(0)$ is obtained by

$$\Delta^n f(z) = \sum_{k=n}^{\infty} \frac{a_k}{k!} \Delta^n z(k) = \sum_{k=n}^{\infty} \frac{a_k}{(k-n)!} z(k-n) = \sum_{k=0}^{\infty} \frac{a_{k+n}}{k!} z(k)$$

due to the formula

$$\Delta^n z(k) = \begin{cases} 0 & (n > k) \\ n! & (n = k) \\ \frac{k!}{(k-n)!} z(k-n) & (n < k) \end{cases},$$

which is very similar to that for $\frac{d^n}{dz^n} z^k$.

Note that the difference equation $\Delta f(z) = f(z)$ is solved by $f(z) = c2^z = ce^{z \log 2}$ $(c \in \mathbb{C})$ since $a_n = c$ for any $n \in \mathbb{Z}_{\geq 0}$ and therefore

$$f(z) = c \sum_{n=0}^{\infty} \frac{z(n)}{n!} = c \sum_{n=0}^{\infty} \binom{z}{n} = c(1+1)^z = c2^z.$$

C.2.2 The case of q-difference operator

Second, we observe the q-difference or q-derivative operator δ_q for $q \in \mathbb{C}$ with $0 < |q| < 1$, which is defined by

$$\delta_q f(z) = \begin{cases} \dfrac{f(qz) - f(z)}{(q-1)z}, & (z \neq 0) \\ f'(0), & (z = 0), \end{cases}$$

and the function $f(z)$ given formally by a q-series expansion about the origin of the form (C.10) below. Following [94], we use the notations $[n]_q = \frac{q^n - 1}{q - 1}$ for $n \in \mathbb{N}$ and $[n]_q!$ defined by

$$[n]_q! = \prod_{k=1}^{n} [k]_q = \prod_{k=1}^{n} \frac{q^k - 1}{q - 1} = \frac{(q; q)_n}{(1 - q)^n} \quad (n \in \mathbb{N})$$

with $(q; q)_n = \prod_{m=1}^{n} (1 - q^m)$. We put $(q; q)_0 = 1$ and $[0]_q! = 1$ in addition. In general, we recall the notations for a nonzero complex number q with $|q| \neq 1$ and $a \in \mathbb{C}$,

$$(a; q)_0 = 1, \quad (a; q)_n = \prod_{m=0}^{n-1} (1 - aq^m) \quad (n \in \mathbb{N})$$

and

$$(a;q)_\infty = \prod_{n=0}^{\infty}(1 - aq^n)$$

if $|q| < 1$. Consider a function $f(z)$ given formally by

$$f(z) = \sum_{n=0}^{\infty}\frac{a_n}{[n]_q!}z^n, \quad a_n \in \mathbb{C} \quad (n \in \mathbb{Z}_{\geq 0}) \tag{C.10}$$

near $z = 0$. Now we have, for $n \geq 2$

$$\delta_q z^n = \frac{q^n z^z - z^n}{(q-1)z} = [n]_q z^{n-1}, \quad \delta_q z^0 \equiv 0 \quad \text{and} \quad \delta_q z \equiv 1$$

and obtain $a_n = \delta_q^n f(0)$ for each $n \in \mathbb{Z}_{\geq 0}$ under the assumption that we can operate δ_q to (C.10) termwise, that is,

$$\delta_q f(z) = \sum_{n=1}^{\infty}\frac{a_n}{[n]_q!}\delta_q z^n = \sum_{n=1}^{\infty}\frac{a_n}{[n-1]_q!}z^{n-1} = \sum_{n=0}^{\infty}\frac{a_{n+1}}{[n]_q!}z^n.$$

In fact,

$$f(0) = a_0, \quad \text{and} \quad \delta_q f(0) = a_1$$

and the general term $a_n = \delta_q^n f(0)$ is obtained by

$$\delta_q^n f(z) = \sum_{k=n}^{\infty}\frac{a_k}{[k]_q!}\delta_q^n z^k = \sum_{k=n}^{\infty}\frac{a_k}{[k-n]_q!}z^{k-n} = \sum_{k=0}^{\infty}\frac{a_{k+n}}{[k]_q!}z^k$$

due to the formula

$$\delta_q^n z^k = \begin{cases} 0 & (n > k) \\ n! & (n = k) \\ \frac{[k]_q!}{[k-n]_q!}z^{k-n} & (n < k) \end{cases},$$

which is very similar to those for both $\frac{d^n}{dz^n}z^k$ and $\Delta^n z(k)$. This coincides with the q-counterpart of Taylor series introduced by Jackson [56] (see also [4], for example)

$$f(z) = \sum_{n=0}^{\infty}\frac{(1-q)^n}{(q;q)_n}\delta_q^n f(a)[z-a]_n = \sum_{n=0}^{\infty}\frac{\delta_q^n f(a)}{[n]_q!}[z-a]_n$$

with

$$[z-a]_n = (z-a)(z-qa)\cdots(z-q^{n-1}a), \; n \in \mathbb{N}, \quad [z-a]_0 = 1,$$

when $a = 0$. As observed in [94], we see that the difference equation $\delta_q f(z) = f(z)$ is solved by $f(z) = c \exp_q(z) = c e_q((1-q)z)$ $(c \in \mathbb{C})$ since $a_n = c$ for any $n \in \mathbb{Z}_{\geq 0}$ and therefore

$$f(z) = c \sum_{n=0}^{\infty} \frac{z^n}{[n]_q!} = c e_q((1-q)z) = c \exp_q(z).$$

But since $|q| < 1$, this is holomorphic only in the unit disk $|z| < 1$. We are now interested in entire functions so that the radius of convergence of the series above need to be infinite. For this purpose, recall

Proposition C.2 ([94, Proposition 5.2]). *Let q be a non-zero complex number, with $|q| \neq 1$. Consider the q-difference equation*

$$f(qz) - (1-z)f(z) = 0. \tag{1_q}$$

i) *This equation admits a unique formal power series solution $\widehat{e}_q$ such that $\widehat{e}_q(0) = 1$;*

$$\widehat{e}_q(z) = \sum_{n=0}^{+\infty} \frac{z^n}{(q;q)_n}.$$

If $|q| > 1$ this series converges in all the complex plane and defines an entire function e_q. If $|q| < 1$ the radius of converges of this series is one and its sum defines a holomorphic function e_q in the open unit disk.

ii) *If $|q| > 1$*

$$e_q(z) = \sum_{n=0}^{+\infty} \frac{z^n}{(q;q)_n} = \prod_{n=1}^{+\infty} (1 - p^n z) = (pz;p)_\infty.$$

The function e_q has zeros at $q, q^2, \ldots, q^n, \ldots$ (these zeros are simple).

iii) *If $|q| < 1$*

$$e_q(z) = \sum_{n=0}^{+\infty} \frac{z^n}{(q;q)_n} = \frac{1}{\prod_{n=0}^{+\infty}(1 - q^n z)} = \frac{1}{(z;q)_\infty}$$

for $|z| > 1$. This infinite product $(z;q)_\infty$ converges in all the complex plane and e_q admits a meromorphic extension to all $\mathbb{C}$ (that we denote also by e_q). The function e_q has no zeros and poles only at $1, q, \ldots, q^n, \ldots$ (these poles are simple).

Bibliography

[1] Ablowitz, M. J., Halburd, R. G. and Herbst, B. (2000). On the extension of the Painlevé property to difference equations, *Nonlinearity* **13**, pp. 889–905.

[2] Ablowitz, M. J. and Segur, H. (1977). Exact linearization of a Painlevé transcendent, *Phys. Rev. Lett.* **38**, 20, pp. 1103–1106.

[3] Akian, M., Gaubert, S. and Guterman, A. E. (2009). *Linear independence over tropical semirings and beyond, Contemp. Math.*, Vol. 495 (Amer. Math. Soc.), pp. 1–38.

[4] Annaby, M. H. and Mansour, Z. S. (2008). q-Taylor and interpolation series for Jackson q-difference operators, *J. Math. Anal. Appl.* **344**, 1, pp. 472–483.

[5] Arnol'd, V. I. (1990). Dynamics of complexity of intersections, *Bol. Soc. Brasil. Mat. (N.S.)* **21**, 1, pp. 1–10.

[6] Bergweiler, W. and Hayman, W. K. (2003). Zeros of solutions of a functional equation, *Comput. Methods Funct. Theory* **3**, pp. 55–78.

[7] Bergweiler, W., Ishizaki, K. and Yanagihara, N. (1998). Meromorphic solutions of some functional equations, *Methods Appl. Anal.* **5**, pp. 248–258.

[8] Bergweiler, W., Ishizaki, K. and Yanagihara, N. (2002). Growth of meromorphic solutions of some functional equations. I., *Aequationes Math.* **63**, pp. 140–151.

[9] Berg, C. and Pedersen, H. L. (2007). Logarithmic order and type of indeterminate moment problems (With an appendix by Walter Hayman), in *Difference equations, special functions and orthogonal polynomials. Proceedings of the International Conference on Difference Equations and Applications held in Munich, July 25-30, 2005.* (World Sci. Publ.), pp. 51–79.

[10] Bessis, D., Itzykson, C. and Zuber, J.-B. (1980). Quantum field theory techniques in graphical enumeration, *Adv. Appl. Math.* **1**, pp. 109–157.

[11] Bloch, A. (1926). Sur les systèmes de fonctions holomorphes àvariétés linéaires lacunaires, *Ann. Sci. École Norm. Sup.* **43**, pp. 309–362.

[12] Brézin, E. and Kazakov, V. A. (1990). Exactly solvable field theories of closed strings, *Phys. Lett. B.* **236**, pp. 144–150.

[13] Butkovič, P. (2010). *Max-linear Systems: Theory and Applications*

(Springer-Verlag, London).

[14] Cartan, H. (1928). Sur les systèmes de fonctions holomorphes àvariétés linéaires lacunaires et leurs applications, *Ann. Sci. École Norm. Sup.* **45**, pp. 255–346.

[15] Cartan, H. (1933). Sur les zéros des combinaisons linéaires de p fonctions holomorphes données, *Mathematica Cluj* **7**, pp. 5–31.

[16] Cherry, W. and Ye, Z. (2001). *Nevanlinna's Theory of Value Distribution. Second main theorem and its error terms* (Springer-Verlag, Berlin).

[17] Chiang, Y.-M. and Feng, S.-J. (2008). On the Nevanlinna characteristic of $f(z+\eta)$ and difference equations in the complex plane, *Ramanujan J.* **16**, pp. 105–129.

[18] Clunie, J. (1962). On integral and meromorphic functions, *J. London Math. Soc.* **37**, pp. 17–27.

[19] Douglas, M. R. and Shenker, S. H. (1990). Strings in less than one dimension, *Nucl. Phys. B* **335**, pp. 635–654.

[20] Fokas, A. S., Grammaticos, B. and Ramani, A. (1993). From continuous to discrete Painlevé equations, *J. Math. Anal. Appl.* **180**, pp. 342–360.

[21] Fuchs, L. (1905). Sur quelques équations différentielles linéares du second ordre, *C. R. Acad. Sci. Paris* **141**, pp. 555–558.

[22] Fujimoto, H. (1972). On holomorphic maps into a taut complex space, *Nagoya Math. J.* **46**, pp. 49–61.

[23] Fujimoto, H. (1974). On meromorphic maps into the complex projective space, *J. Math. Soc. Japan* **26**, pp. 272–288.

[24] Fujimoto, H. (1993). *Value distribution theory of the Gauss map of minimal surfaces in $\mathbb{R}^m$, Aspects of Mathematics*, Vol. E21 (Friedr. Vieweg & Sohn, Braunschweig).

[25] Gambier, B. (1993). Sur les équations différentielles du second ordre et du premier degré dont l'intégrale générale est à points critiques fixes, *Acta Math.* **33**, pp. 1–55.

[26] Gol'dberg, A. and Ostrovskii, I. (2008). *Value Distribution of Meromorphic Functions, Translations of Mathematical Monographs*, Vol. 236 (Amer. Math. Soc.)

[27] Gasper, G. and Rahman, M. (2004). *Basic Hypergeometric Series*, 2nd edn., *Encyclopedia of Mathematics and its Applications*, Vol. 96 (Cambridge University Press, Cambridge).

[28] Gondran, M. and Minoux, M. (1979). *Graphes et algorithmes, Collection de la Direction des Études et Recherches d'Électricité de France*, Vol. 37 (Éditions Eyrolles, Paris).

[29] Gondran, M. and Minoux, M. (1984). Linear algebra in dioids: a survey of recent results, in *Algebraic and combinatorial methods in operations research North-Holland Math. Stud.*, Vol. 95 (North-Holland, Amsterdam), pp. 147–163.

[30] Grammaticos, B., Nijhoff, F. W. and Ramani, A. (1999). Discrete Painleve equations, in *The Painlevé property, One century later, CRM Ser. Math. Phys.* (Springer, New York), pp. 413–516.

[31] Grammaticos, B., Ohta, Y., Ramani, A., Takahashi, D. and Tamizhmani,

K. M. (1997). Cellular automata and ultra-discrete Painlevé equations, *Phys. Lett. A* **226**, 1-2, pp. 53–58.

[32] Grammaticos, B. and Ramani, A. (2000). The hunting for the discrete Painlevé equations, *Regul. Chaotic Dyn.* **5**, pp. 53–66.

[33] Grammaticos, B., Ramani, A. and Papageorgiou, V. (1991). Do integrable mappings have the Painlevé property?, *Phys. Rev. Lett.* **67**, pp. 1825–1828.

[34] Green, M. L. (1972). Holomorphic maps into complex projective space omitting hyperplanes, *Trans. Amer. Math. Soc.* **169**, pp. 89–103.

[35] Green, M. L. (1974). On the functional equation $f^2 = e^{2\phi_1} + e^{2\phi_2} + e^{2\phi_3}$ and a new Picard theorem, *Trans. Amer. Math. Soc.* **195**, pp. 223–230.

[36] Green, M. L. (1975). Some Picard theorems for holomorphic maps to algebraic varieties, *Amer. J. Math.* **97**, 43–75.

[37] Gromak, V. I., Laine, I. and Shimomura, S. (2002). *Painlevé differential equations in the complex plane* (Walter de Gruyter, Berlin).

[38] Gundersen, G. and Hayman, W. (2004). The strength of Cartan's version of Nevanlinna theory, *Bull. London Math. Soc.* **36**, pp. 433–454.

[39] Halburd, R. G. (2015). *Private communication*.

[40] Halburd, R. G. and Korhonen, R. (2006). Difference analogue of the lemma of the logarithmic derivative with applications to difference equations, *J. Math. Anal. Appl.* **314**, pp. 477–487.

[41] Halburd, R. G. and Korhonen, R. (2006). Nevanlinna theory for the difference operator, *Ann. Acad. Sci. Fenn. Math.* **31**, pp. 463–478.

[42] Halburd, R. G. and Korhonen, R. (2012). Nondecreasing functions, exponential sets and generalized Borel lemmas, *J. Aust. Math. Soc.* **88**, pp. 353–361.

[43] Halburd, R. G. and Korhonen, R. (2007). Finite-order meromorphic solutions and the discrete Painlevé equations, *Proc. London Math. Soc.* **94**, pp. 443–474.

[44] Halburd, R. G. and Korhonen, R. (2007). Meromorphic solutions of difference equations, integrability and the discrete Painlevé equations, *J. Phys. A: Math. Theor.* **40**, pp. R1–R38.

[45] Halburd, R. G., Korhonen, R. and Tohge, K. (2014). Holomorphic curves with shift-invariant hyperplane preimages, *Trans. Amer. Math. Soc.* **366**, pp. 4267–4298.

[46] Halburd, R. G. and Southall, N. (2009). Tropical Nevanlinna theory and ultra-discrete equations, *Int. Math. Res. Not. IMRN* **2009**, pp. 887–911.

[47] Hamamoto, T., Kajiwara, K., Witte, N. S. (2006). Hypergeometric solutions to the q-Painlevé equation of type $(A_1 + A_1')^{(1)}$, *Int. Math. Res. Not. IMRN* **2006**, pp. 1–26.

[48] Hayman, W. (1964). *Meromorphic Functions* (Clarendon Press, Oxford).

[49] Hayman, W. (1985). Warings Problem für analytische Funktionen (German), in *Bayer. Akad. Wiss. Math.-Natur. Kl. Sitzungsber. 1984*, pp. 1–13.

[50] Hietarinta, J. and Viallet, C.-M. (1998). Singularity confinement and chaos in discrete systems, *Phys. Rev. Lett.* **81**, pp. 325–328.

[51] Hinkkanen, A. and Laine, I. *preprint*.

[52] Hinkkanen, A. and Laine, I. (2004). Growth results for Painlevé transcen-

dents, *Math. Proc. Cambridge Philos. Soc.* **137**, 3, pp. 645–655.

[53] Ismail, M. E. H. (2005). *Classical and Quantum Orthogonal Polynomials in One Variable, Encyclopedia of Mathematics and its Applications*, Vol. 98 (Cambridge University Press, Cambridge).

[54] Ishizaki, K. and Yanagihara, N. (2004). Wiman-Valiron method for difference equations, *Nagoya Math. J.* **175**, pp. 75–102.

[55] Its, A. R., Kitaev, A. V. and Fokas, A. S. (1990). The isomonodromy approach in the theory of two-dimensional quantum gravitation, *Russ. Math. Surv.* **45**, 6, pp. 155–157.

[56] Jackson, F. H. (1909). q-form of Taylor's theorem, *Messenger Math.* **39**, pp. 62–64.

[57] Jank, G. and Volkmann, L. (1985). *Einführung in die Theorie der ganzen und meromorphen Funktionen mit Anwendungen auf Differentialgleichungen* (Birkhäuser, Basel–Boston).

[58] Joshi, N., Burtonclay, D. and Halburd, R. (1992). Nonlinear nonautonomous discrete dynamical systems from a general discrete isomonodromy problem, *Lett. Math. Phys.* **26**, pp. 123–131.

[59] Joshi, N. and Lafortune, S. (2005). How to detect integrability in cellular automata, *J. Phys. A* **38**, 28, pp. L499–L504.

[60] Joshi, N., Nijhoff, F. W. and Ormerod, C. (2004). Lax pairs for ultra-discrete Painlevé cellular automata, *J. Phys. A* **37**, 44, pp. L559–L565.

[61] Juneja, O. P., Kapoor, G. P. and Bajpai, S. K. (1976). On the (p,q)-order and lower (p,q)-order of an entire function, *J. Reine Angew. Math.* **282**, 53–67.

[62] Juneja, O. P., Kapoor, G. P. and Bajpai, S. K. (1977). On the (p,q)-type and lower (p,q)-type of an entire function, *J. Reine Angew. Math.* **290**, pp. 180–190.

[63] Kajiwara, K., Masuda, T., Noumi, M., Ohta, Y. and Yamada, Y. (2004). Hypergeometric solutions to the q-Painlevé equations, *Int. Math. Res. Not. IMRN* **2004**, 47, pp. 2497–2521.

[64] Kajiwara, K., Masuda, T., Noumi, M., Ohta, Y. and Yamada, Y. (2005). Construction of hypergeometric solutions to the q-Painlevé equations, *Int. Math. Res. Not. IMRN* **2005**, 24, pp. 1441–1463.

[65] Kelley, W. - Peterson, A. (2001). *Difference Equations* (Academic Press, San Diego).

[66] Kobayashi, S. (1998). *Hyperbolic complex spaces, Grundlehren der Mathematischen Wissenschaften*, Vol. 318 (Springer-Verlag, Berlin).

[67] Korhonen, R. and Tohge, K. (2014). Second main theorem in the tropical projective space, *arXiv:1401.5584v1*.

[68] Laine, I. (1993). *Nevanlinna Theory and Complex Differential Equations* (Walter de Gruyter, Berlin–New York).

[69] Laine, I. and Tohge, K. (2011). Tropical Nevanlinna theory and second main theorem, *Proc. Lond. Math. Soc.* **102**, pp. 883–922.

[70] Laine, I. and Yang, C.-C. (2007). Clunie theorems for difference and q-difference polynomials, *J. Lond. Math. Soc.* **76**, pp. 556–566.

[71] Laine, I. and Yang, C.-C. (2010). Tropical versions of Clunie and Mohon'ko

lemmas, *Complex Variables Elliptic Equ.* **55**, pp. 237–248.

[72] Lang, S. (1987). *Introduction to complex hyperbolic spaces* (Springer-Verlag, New York).

[73] Lelong, P. and Gruman, L. (1986). *Entire functions of several complex variables, Grundlehren der Mathematischen Wissenschaften*, Vol. 282 (Springer-Verlag, Berlin).

[74] Maclagan, D. and Sturmfels, B. *Introduction to tropical geometry, preprint.*

[75] Malgrange, B. (1983). Sur les déformations isomonodromiques. I. Singularités régulières, in *Mathematics and physics (Paris, 1979/1982)* (Birkhäuser Boston, Boston, MA), *Progr. Math.* **37**, pp. 401–426.

[76] Malmquist, J. (1913). Sur les fonctions à un nombre fini des branches définies par les équations différentielles du premier ordre, *Acta Math.* **36**, pp. 297–343.

[77] Miwa, T. (1981). Painlevé property of monodromy preserving deformation equations and the analyticity of τ functions, *Publ. Res. Inst. Math. Sci.* **17**, pp. 703–721.

[78] Mohon'ko, A. (1971). The Nevanlinna characteristics of certain meromorphic functions (Russian), *Teor. Funktsiĭ Funktional. Anal. i Prilozhen* **14**, pp. 83–87.

[79] Murata, M. (2011). Exact solutions with two parameters for an ultradiscrete Painlevé equation of type $A_6^{(1)}$, *SIGMA Symmetry Integrability Geom. Methods Appl.* **7**, Paper 059, 15.

[80] Nijhoff, F. W. and Papageorgiou, V. G. (1991). Similarity reductions of integrable lattices and discrete analogues of the Painlevé II equation, *Phys. Lett. A* **153**, pp. 337–344.

[81] Nobe, A. (2006). Ultradiscretization of elliptic functions and its applications to integrable systems. *J. Phys. A* **39**, 20, pp. L335–L342.

[82] Nochka, E. I. (1983). On the theory of meromorphic curves (Russian), *Dokl. Akad. Nauk SSSR* **269**, 3, pp. 547–552.

[83] Noguchi, J., and Winkelmann, J. (2014). *Nevanlinna theory in several complex variables and Diophantine approximation, Grundlehren der Mathematischen Wissenschaften*, Vol. 350 (Springer, Tokyo).

[84] Nörlund, N. (1924). *Vorlesungen über Differenzenrechnung* (Springer-Verlag, Berlin).

[85] Noumi, M. (2007). Special functions arising from discrete Painlevé equations: a survey, *J. Comput. Appl. Math.* **202**, 1, pp. 48–55.

[86] Ormerod, C. M. (2010). Hypergeometric solutions to an ultradiscrete Painlevé equation, *J. Nonlinear Math. Phys.*, **17**, 1, pp. 87–102.

[87] Ormerod, C. and Yamada, Y. (2014). From polygons to ultradiscrete Painlevé equations, *arXiv:1408.5643*.

[88] Painlevé, P. (1900). Mémoire sur les équations différentielles dont l'intégrale générale est uniforme, *Bull. Soc. Math. France* **28**, pp. 201–261.

[89] Painlevé, P. (1902). Sur les équations différentielles du second ordre et d'ordre supérieur dont l'intégrale générale est uniforme, *Acta Math.* **25**, pp. 1–85.

[90] Periwal, V. and Shevitz, D. (1990). Unitary matrix models as exactly solvable string theories, *Phys. Rev. Lett.* **64**, 1326–1329.

[91] Quispel, G. R. W., Roberts, J. A. G. and Thompson, C. J. (1988). Integrable mappings and soliton equations, *Phys. Lett. A* **126**, pp. 419–421.

[92] Ramani, A. and Grammaticos, B. (1996). Discrete Painlevé equations: coalescences, limits and degeneracies. *Phys. A* **228**, 1–4, pp. 160–171.

[93] Ramani, A., Grammaticos, B. and Hietarinta, J. (1991). Discrete versions of the Painlevé equations, *Phys. Rev. Lett.* **67**, pp. 1829–1832.

[94] Ramis, J.-P. (1992). About the growth of entire functions solutions of linear algebraic q-difference equations, *Ann. Fac. Sci. Toulouse Math. (6)* **1**, 1, pp. 53–94.

[95] Rockafellar, R.T. (1970). *Complex Analysis* (Princeton University Press, Princeton).

[96] Ru, M. (2001). *Nevanlinna theory and its relation to Diophantine approximation* (World Scientific Publishing Co., Inc., River Edge, NJ).

[97] Shimomura, S. (2003). Growth of the first, the second and the fourth Painlevé transcendents, *Math. Proc. Cambridge Philos. Soc.* **134**, pp. 259–269.

[98] Shimomura, S. (2004). Growth of modified Painlevé transcendents of the fifth and the third kind, *Forum Math.* **16**, 2, pp. 231–247.

[99] Shimomura, S. (2004). Lower estimates for the growth of the fourth and the second Painlevé transcendents, *Proc. Edinb. Math. Soc. (2)* **47**, pp. 231–249.

[100] Shimomura, S. (2009). Meromorphic solutions of difference Painlevé equations, *J. Phys. A* **42**, 31, 315213, 19 pp.

[101] Shohat, J. A. (1939). A differential equation for orthogonal polynomials, *Duke Math. J.* **5**, pp. 401–417.

[102] Steinmetz, N. (2002). Value distribution of the Painlevé transcendents, *Israel J. Math.* **128**, pp. 29–52.

[103] Takahashi, D., Tokihiro, T., Grammaticos, B., Ohta, Y. and Ramani, A. (1997). Constructing solutions to the ultra-discrete Painlevé equations, *J. Phys. A: Math. Gen.* **30**, 22, pp. 7953–7966.

[104] Tohge, K. (2014). The order and type formulas for tropical entire functions - another flexibility of complex analysis, in *Proceedings of the Workshop on Complex Analysis and its Applications to Differential and Functional Equations*, pp. 113–164, http://urn.fi/URN:ISBN:978-952-61-1354-8.

[105] Tokihiro, T., Takahashi, D., Matsukidaira, J. and Satsuma, J. (1996). From soliton equations to integrable cellular automata through a limiting procedure, *Phys. Rev. Lett.* **76**, pp. 3247–3250.

[106] Tsai, Y.-L. (2012). Working with tropical meromorphic functions of one variable, *Taiwanese J. Math.* **16**, pp. 691–712.

[107] Veselov, A. P. (1992). Growth and integrability in the dynamics of mappings, *Comm. Math. Phys.* **145**, pp. 181–193.

[108] Wen, Z. T. (2014). Finite logarithmic order solutions of linear q-difference equations, *Bull. Korean Math. Soc.* **51**, pp. 83–98.

[109] Whittaker, J. A. (1933). The 'sum' of a meromorphic function, *J. London*

Math. Soc. **8**, pp. 62–69.

[110] Whittaker, J. A. (1964). *Interpolatory function theory, Cambridge Tracts in Mathematics and Mathematical Physics*, Vol. 33 (Stechert-Hafner, Inc., New York).

[111] Wong, P.-M., Law, H.-F., and Wong, P. P. W. (2009). A second main theorem on $\mathbb{P}^n$ for difference operator, *Sci. China Ser. A* **52**, 12, pp. 2751–2758.

[112] Yanagihara, N. (1980). Meromorphic solutions of some difference equations, *Funkcialaj Ekvacioj* **23**, pp. 309–326.

[113] Yang, C.-C. and Ye, Z. (2007). Estimates of the proximate function of differential polynomials, *Proc. Japan Acad. Ser. A. Math. Sci.* **83**, pp. 50–55.

[114] Yoeli, M. (1961). A note on a generalization of Boolean matrix theory, *Amer. Math. Monthly* **68**, pp. 552–557.

[115] Yosida, K. (1933). A generalization of Malmquist's theorem, *J. Math.* **9**, pp. 253–256.

Index

c-separated points, 198
q-order, 56
q-type, 56

admissible meromorphic solution, 229

Borel lemma, 75, 193

Cartan identity, 192
Casorati determinant, 211
characteristic function, 14, 67, 122, 191, 206
  Ahlfors, 210
  Cartan, 131
compatibility condition, 221
counting function, 14, 78, 129, 189, 198, 200
  for hyperplanes, 207
  truncated, 204

discrete Painlevé equation
  multiplicative type, 173
discrete Painlevé equation, 223

forward invariant preimage, 216

Gaussian, 58
general position, 205
genus, 156

height, 207
hyper-order, 17, 192

integrability, 173, 219, 221
iso-monodromy, 221

Lax Pair, 234
logarithmic measure, 75

max-plus differential operator, 181
max-plus semiring, 1, 121
  completed, 121

nearly everywhere, 204

order, 17, 191

Painlevé equation, 219
Painlevé property, 220
Poisson-Jensen formula, 190
pole, 5
  multiplicity, 14
proximity function, 14, 67, 104, 189, 200
  for hyperplanes, 207

ramification, 204
reduced representation, 122, 206
root, 5
  multiplicity, 14

singularity confinement, 226

transcendental, 210
tropical, 1

addition, 1
Cartan characteristic, 122
Cartan identity, 71
Casoratian, 112, 113
compact form of a polynomial, 7, 11
complete linear combination, 118
conjugate entire function, 47
degenerate linear combination, 120
degree of a polynomial, 16
degree of degeneracy of a linear combination, 120
determinant, 112
difference polynomial, 98
dimension of a linear span, 117
entire function, 30, 45
equivalent polynomials, 4
exponential function, 19
first main theorem, 69
Gondran-Minoux linear independence, 115
holomorphic curve, 122
holomorphic map, 122
indentity elements, 1
Jensen formula, 15
linear combination, 115
linear span, 117
matrix, 111
maximally represented entire function, 50
maximally represented polynomial, 2, 7, 11, 127
meromorphic function, 1, 14, 125
multiplication, 1
non-degenerate linear combination, 120
normalized entire function, 35
periodic function, 145
Poisson-Jensen formula, 12, 65
polynomial, 2, 14, 29, 49, 50
projective space, 121
proximity small function, 98
ramification, 136

rational function, 5, 13, 125, 175
regular matrix, 112
second main theorem, 78, 83, 92
small function, 98
spanning basis, 117
unit, 32
weak linear independence, 115
tropical entire function
    compact form, 51
    essential term, 51
    inessential term, 51
    standard form, 50
truncated counting function, 137, 208
    for hyperplanes, 207
truncation, 204

ultra-discrete
    q-Airy equation, 245
    cellular automata, 237
    Cole-Hopf transformation, 155, 170, 178
    equation, 157
    hypergeometric function, 29, 57, 177, 178, 184, 246
    hypergeometric series, 179, 185
    Lax pair, 235
    limit, 233, 244
    operator, 181
    Painlevé equation, 29, 172, 175, 177, 219, 232
    Painlevé property, 240
    Riccati equation, 169, 178
    singularity confinement, 237
    Taylor series, 39, 183
    theta function, 29, 57, 153
ultra-discrete limit, 173
ultra-discretization, 173, 178, 233
    formal, 29, 38, 53

Weil function, 207
Wronskian, 112, 131, 137, 201, 202

zero, 5

Printed in the United States
By Bookmasters